Build Customized Apps with Amazon Honeycode

Quickly create interactive web and mobile apps for your teams without programming

Aniruddha Loya

BIRMINGHAM—MUMBAI

Build Customized Apps with Amazon Honeycode

Group Product Manager: Alok Dhuri
Publishing Product Manager: Richa Tripathi
Senior Editor: Ruvika Rao
Content Development Editor: Urvi Shah
Technical Editor: Maran Fernandes
Copy Editor: Safis Editing
Language Support Editor: Safis Editing
Project Coordinator: Deeksha Thakkar
Proofreader: Safis Editing
Indexer: Hemangini Bari
Production Designer: Shankar Kalbhor
Marketing Coordinator: Deepak Kumar and Rayyan Khan
Business Development Executive: Uzma Sheerin

First published: June 2022

Production reference: 1080622

Published by Packt Publishing Ltd.
Livery Place
35 Livery Street
Birmingham
B3 2PB, UK.

978-1-80056-369-8

`www.packt.com`

This book is dedicated to my family, friends, teachers, and everyone else who, at any point, came into my life and enriched it with their presence. And then there are those who deserve special mentions:

First, my mother, Hemangi Loya, and the memory of my father, Arun Loya, for their unconditional love, their values, and their strength and support at every step of my life.

Next to the loving memory of my grandparents: To my grandmother, Shanti Loya, for her love and nurturing disposition, and for sowing the seeds of curiosity in my childhood. "Maa," I love you and miss you! And to my grandfather, Ramgopal Loya, for his ever-so-simple pearls of wisdom, including "to keep laughing," and "to hear everyone's views but choose what you want to do," among many others.

And finally, to my partner, Tejashvini, who has stood by me through the ups and downs, supported, and, on occasion, tolerated me, but never stopped loving me. And to my sons, Avyukt and Aarit, who continue to fill my life with joy, happiness, love, and laughter. You three are my pillars, my energy source (and sink), and my daily inspiration (and desperation).

– Aniruddha Loya

Contributors

About the author

Aniruddha Loya is an engineering leader and entrepreneur with over 10 years of experience in building teams, scalable systems, services, and large-scale enterprise and consumer products, including the first version of Amazon Honeycode.

He is passionate about applying technology to solve everyday problems, and even more so when it can be applied to improving the environment. His most recent start-up attempt was also in the ESG space.

Presently, he is working as the engineering leader at CasaHealth Tech, building and establishing the engineering site in Canada, and leading multiple teams in bringing pharmacy and healthcare offerings to every Canadian's doorstep.

I want to thank my loving wife, Tejashvini, for her support and the borrowed time to make this book a reality, and to the Packt team for helping me in this wonderful and fulfilling journey of completing my first book.

About the reviewer

Alejandro Escobar Garces has almost 10 years of professional experience working as a software engineer. He began work developing applications for Point of Services, but later on worked for an outsourcing company in a variety of roles, from QA automation to frontend engineer. In the last 7 years, Alejandro has dedicated most of his time to working as a backend engineer, with interests in operational excellence and chaos engineering. He has received a BSc. and an MSc. in system and informatic engineering, with research experience in AI and simulations. Alejandro has also worked as a university professor on a number of different courses related to computer science at several universities in Colombia.

Table of Contents

Preface

Part 1: Introduction to Honeycode

1
Amazon Honeycode – Day One

Technical requirements 4
Creating the first application 4
Creating an account on Amazon Honeycode 4
Creating a To-Do application 6
Running the application on the web and a mobile 8
Running our To-Do application on a web browser 8
Running our To-Do application on a mobile device 10
Summary 12

2
Introduction to Amazon Honeycode

Technical requirements 14
Exploring Honeycode Dashboard 14
Dashboard 16
Teams 16
Learning & Resources 17
Exploring a Honeycode Workbook 17
Left Navigation Bar 17
Tables 19
Builder 24
Automations 31
Understanding Honeycode Teams 34
Adding team members 34
Managing a team 40
Honeycode's pricing tiers 41
Summary 42

3

Building Your First Honeycode Application

Technical requirements 44

Defining the app requirements 44

Creating the app data model 46

Creating a workbook 47

Creating tables 48

Data formats and relations 54

Building the app interface 57

Creating an app from scratch 61

Creating an app using App Wizard 76

Summary 83

4

Advanced Builder Tools

Technical requirements 86

Defining the app requirements 86

Applying styles to app components 87

Style controls in Honeycode 87

Conditional-styling app components 88

Styling the To-Do App 90

Controlling the component visibility with conditions 95

Restricting data access per user using personalized views 98

Searching, filtering, and sorting data views on the fly 101

Summary 104

5

Powering Apps with Automations

Technical requirements 106

Defining the app requirements 106

Understanding variables in Honeycode 107

User-defined variables 107

System variables 108

Making apps more powerful with automations 109

Using actions for data input through forms 109

Adding functionality to delete a task 120

Processing data with automation based on triggers 123

Sending reminders based on set preference 124

Sending a task completion notification to the task creator 128

Debugging automations 131

Summary 132

Part 2: Deep-Dive into Honeycode Templates

6
Introduction to Honeycode Templates

Technical requirements 136
Getting to know Honeycode templates 136
Applicant Tracker 138
Collaborative brainstorming 140
Connect manager 142
Content tracker 146
Customer tracker 146
Employee onboarding 147
Event management 150
Field Service Agent 152
Instant Polls 153
Inventory Management 154
Launch Manager 156
Meeting Runner 157
PO Approvals 158
Simple Survey 160
Simple To-Do 161
Team Task Tracker 161
Timeoff Reporting 162
Timesheet Manager 163
Weekly Demo Schedule 165
Summary 166

7
A Simple Survey Template

Technical requirements 168
Defining the app requirements 168
Creating the Survey app 169
Reviewing the data model 170
Results 171
Scale 172
Survey 173
_README 175
Reviewing the app 175
Home 176
Thanks 178
Survey 178
Summary 181

8

Instant Polls Template

Technical requirements 184
Defining app requirements 184
Creating an Instant Polls app 185
Reviewing the data model 186
Last_Visits 187
Options 189
Questions 190
Votes 191
z_Readme 192
Reviewing the app 192
Polls 193
New Poll 196
Cast a Vote 200
Summary 202

9

Event Management Template

Technical requirements 204
Defining the app requirements 204
Creating an Event Management app 206
Reviewing the data model 208
A_Sessions 209
B_FAQ 211
D_Registrations 212
M_Category 212
M_Dates 213
M_Ratings 214
M_Speakers 214
R_Dashboard 215
_Readme 215
Reviewing the app 216
Sessions 217
Detail 223
My Agenda 224
Speakers 225
By Category 226
FAQ 226
Summary 228

10

Inventory Management Template

Technical requirements 230
Defining the app requirements 230
Creating an Inventory Management app 231
Reviewing the data model 234
Categories 235
Devices 235
Inventory 236
Manufacturers 237
Status 237
_Readme 238
Reviewing the apps 238

My Devices app 238
My Devices - Manager app 243
Summary 248

Part 3: Let's Build Some Apps

11
Building a Shopping List App in Honeycode

Technical requirements 252
Defining the app requirements 252
Translating requirements to app interactions 253
Defining the data model 254
Building the app 254
Creating a new workbook 254
Creating tables 254
Creating the app 260
Clearing the bought items list 273
Building an app using the Wizard 276
Summary 278

12
Building a Nominate and Vote App in Honeycode

Technical requirements 280
Defining the app requirements 280
Translating requirements to app interactions 281
Defining the data model 282
Building the app 282
Creating the template app 283
Setting up the tables 284
Editing the template app 288
Functionalities just for organizers 304
Restricting voting to a panel of judges 307
Summary 307

13
Conducting Periodic Business Reviews Using Honeycode

Technical requirements 310
Defining the app requirements 310
Translating requirements to app interactions 311
Defining the data model 311
Building the app 312
Creating a new workbook 312
Creating tables 312
Creating the app 322
Sending review updates to all and archiving the updates 336
Sending reminders for providing updates 341
Summary 341

14
Solving Problems through Multiple Apps within a Workbook

Technical requirements 344

Defining the app requirements 344

Translating requirements to app interactions 346

The Realtor app 346

The Client app 346

Defining the data model 348

Building the app 348

Creating a new workbook 348

Creating tables 349

Creating the Realtor app 353

Creating the Buyer app 362

Creating the Seller app 372

Discussion 377

Summary 377

Assessments

Chapter 4 – Advanced Builder Tools in Honeycode 379

Chapter 5 – Powering the Honeycode Apps with Automations 386

Chapter 7 – Simple Survey Template 389

Chapter 8 – Instant Polls Template 389

Chapter 9 – Event Management Template 391

Chapter 10 – Inventory Management Template 394

Chapter 11 – Building a Shopping List App in Honeycode 397

Chapter 12 – Building a Nominate and Vote App in Honeycode 399

Chapter 13 – Conducting Periodic Business Reviews Using Honeycode 401

Chapter 14 – Solving Complex Problems through Multiple Apps Within a Workbook 402

Index

Other Books You May Enjoy

Preface

Amazon Honeycode enables users to build fully managed, customizable, and scalable mobile and web applications for personal or professional use with little to no code.

With this practical guide to Amazon Honeycode, you'll be able to bring your app ideas to life, improving your and your team/organization's productivity. You'll begin by creating your very first app from the get-go and use it as a means to explore the Honeycode development environment and concepts. After that, you'll learn how to set up and organize the data to build and bind an app on Honeycode, as well as deconstructing different templates to understand the common structures and patterns that can be used. Finally, you'll build a few apps from scratch and discover how to apply the concepts you've learned.

By the end of this app development book, you'll have gained the knowledge you need to be able to build and deploy your own mobile and web applications and share them with people who you want to collaborate with.

Who this book is for

Like the platform, this book is meant for anyone—professional and citizen developers alike—who wants to build and deploy apps for their personal or professional use as an individual or as a team. However, for professional developers, it is important to note that the book will not focus on advanced use cases with features such as using public APIs and 3P integrations.

What this book covers

Chapter 1, Amazon Honeycode – Day 1, starts the book by enabling you to discover the power and value of Honeycode straight away with a to-do app that runs on both mobile and web browsers, built and deployed within 2 minutes by a few simple clicks on the UI. Being able to build and deploy your own app is the key value proposition of Honeycode.

Chapter 2, Introduction to Honeycode, provides the typical introduction to any new platform. It walks through Honeycode's layout and different sections and explains the various components and terminology used.

Chapter 3, Building Your First Honeycode Application, is a step-by-step guide to building the same to-do app as the first chapter, but from scratch and not using the template. While doing so, it applies the topics covered in *Chapter 2, Introduction to Honeycode*, and introduces the next set of details needed to build the app.

Chapter 4, Advanced Builder Tools in Honeycode, builds on the previous chapter, in which the to-do app we built was functional but, in all fairness, it was very basic and left much to be desired. But Honeycode offers a lot more tools for customization. In this chapter, we'll explore some of the advanced functionality that enables builders to improve the presentation with conditional styling, control the visibility of different components, add controls for filtering and sorting views, and even personalize the view for each user of the app.

Chapter 5, Powering the Honeycode Apps with Automations, focuses on the processing of user inputs from the app to add or change data, as well as indirect triggers such as time reached and changing certain values using automation.

Chapter 6, Introduction to Honeycode Templates, introduces you to templates, how they work, and the use cases that are currently enabled out of the box through these templates. Honeycode provides a diverse set of templates aimed at helping users to jump-start their journey. The chapter also provides a showcase of capabilities and use cases that can be built using Honeycode.

Chapter 7, Simple Survey Template, gets you started with setting up a simple survey out of the box.

Chapter 8, Instant Polls Template, gets you started with setting up polls out of the box.

Chapter 9, Event Management Template, gets you started with an out-of-the-box app for event management.

Chapter 10, Inventory Management Template, gets you started with an inventory management app out of the box.

Chapter 11, Building a Shopping List App in Honeycode, explains how to build an app for an everyday use case of maintaining an updated shopping list. The app we build will have the option to build lists for the different stores you may be purchasing from, move items from one list to another, and clear out items you've already bought. And finally, we have some exercises to further enhance the functionality.

Chapter 12, Building a Nominate and Vote App in Honeycode, asks you the question *"have you ever found yourself searching for a tool/application that is not email or some word document or spreadsheet that lets you:*

Run a contest where everyone can submit nominations and later vote for selecting a winner?

Decide by majority on the next movie to play on movie night or game(s) to play on games night?

Collect questions from an organization before an all-hands meeting, vote, and rank them to help pick the most-asked question?"

In this chapter, we'll build an app, using an existing template, that will enable you to answer these questions by adding functionality for running a contest.

Chapter 13, Conducting Periodic Business Reviews Using Honeycode, looks at the use case that most businesses have, where they hold periodic review meetings requiring collaboratively generated data and reports that are often shared and updated over email, requiring someone to collect all the threads and compile them for the final report.

In this chapter, we'll build an app that lets teams collaboratively work on a single source of truth and, therefore, always be able to see the latest information. We'll also make use of automation for sending reminders and generating a post-review email containing all the updates.

Chapter 14, Solving Complex Problems through Multiple Apps within a Workbook, explains that it is not always possible to solve a use case with a single app. Often, there are permissions, separate views, or even Honeycode limitations that require us to make use of multiple apps within the same workbook. In this chapter, we take the example of a realtor managing their buying and selling clients and servicing them to showcase the power of Honeycode in allowing multiple apps to provide different customized data views based on the user type.

To get the most out of this book

You'll need to have access to Amazon Honeycode, which requires a laptop with a web browser, preferably Google Chrome, and optionally a mobile device running either a Honeycode-supported version of Android (currently requires Android 8.0 and up) or iOS (currently requires iOS 11 or later). No programming knowledge or experience is expected to start creating basic apps. However, working knowledge of Microsoft Excel or similar spreadsheet tools and a general understanding of logical statements are helpful to get the most out of this book.

Software/hardware covered in the book	Operating system requirements
Amazon Honeycode	Windows, macOS, or Linux

If you have questions, new ideas to discuss, or require help or consultation on your Honeycode projects, come and join this LinkedIn group: **Honeycode builders | Groups | LinkedIn**.

Download the color images

We also provide a PDF file that has color images of the screenshots and diagrams used in this book. You can download it here: `https://static.packt-cdn.com/downloads/9781800563698_ColorImages.pdf`.

Conventions used

There are a number of text conventions used throughout this book.

`Code in text`: Indicates code words in text, database table names, folder names, filenames, file extensions, pathnames, dummy URLs, user input, and Twitter handles. Here is an example: " `=Tasks[[#Headers], [Assignee]]` and `=Tasks[[#Headers],[Notes]]`."

Bold: Indicates a new term, an important word, or words that you see onscreen. For instance, words in menus or dialog boxes appear in **bold**. Here is an example: "In the previous subsection, we added the two new fields on the **Add New Task** screen. However, as we had noted earlier, if you try to add a new task with these fields filled, the values will not be retained, and the fields would show up empty in the **My Tasks** list view. So, now let's update the automation defined on the **Done** button and learn how we can persist the information from a given form to our tables."

> **Tips or Important Notes**
> Appear like this.

Get in touch

Feedback from our readers is always welcome.

General feedback: If you have questions about any aspect of this book, email us at `customercare@packtpub.com` and mention the book title in the subject of your message.

Errata: Although we have taken every care to ensure the accuracy of our content, mistakes do happen. If you have found a mistake in this book, we would be grateful if you would report this to us. Please visit `www.packtpub.com/support/errata` and fill in the form.

Piracy: If you come across any illegal copies of our works in any form on the internet, we would be grateful if you would provide us with the location address or website name. Please contact us at `copyright@packt.com` with a link to the material.

If you are interested in becoming an author: If there is a topic that you have expertise in and you are interested in either writing or contributing to a book, please visit `authors.packtpub.com`.

Share Your Thoughts

Once you've read *Build Customized Apps with Amazon Honeycode*, we'd love to hear your thoughts! Scan the QR code below to go straight to the Amazon review page for this book and share your feedback.

`https://packt.link/r/1-800-56369-8`

Your review is important to us and the tech community and will help us make sure we're delivering excellent quality content.

Part 1: Introduction to Honeycode

In this part, you will understand the basics of Honeycode. You will start with an out-of-the-box "Simple TO-DO" app with just a few simple clicks and then use it to explore and learn about the key components of Honeycode. You will then put all the pieces together and build the same TO-DO app from scratch. Finally, you'll cover advanced topics to improve the app's presentation layer and user experience as well as apply the power of automation to complex data processing.

This section comprises the following chapters:

- *Chapter 1, Amazon Honeycode – Day 1*
- *Chapter 2, Introduction to Honeycode*
- *Chapter 3, Building Your First Honeycode Application*
- *Chapter 4, Advanced Builder Tools in Honeycode*
- *Chapter 5, Powering the Honeycode Apps with Automations*

1
Amazon Honeycode – Day One

Amazon Honeycode's key value proposition is to enable users to build and deploy mobile and web applications with minimal to zero need of programming. Therefore, we'll start this book by enabling you to discover this power and value right at the first step.

You will first create your Honeycode account, and then in less than 2 minutes, you will be able to create and use a simple **To-Do** app that runs on both mobile devices and on laptops/desktops following some simple steps and a few clicks.

Are you wondering what Amazon Honeycode is? What are the key features, major components, and where are the other details that are typically covered at the start of a book? Do not worry – we have you covered in *Chapter 2, Introduction to Honeycode.* However, before all those details, let's zoom through the core product and learn what it has to offer.

In this chapter, we're going to cover the following main topics:

- Creating the first application
- Running the application on the web and mobile devices

Technical requirements

In order to follow this and all the subsequent chapters in this book, you'll need to have access to Amazon Honeycode, which requires a laptop with a web browser, preferably Google Chrome, and optionally a mobile device running either a Honeycode-supported version of Android (it currently requires Android 8.0 or upward) or iOS (it currently requires iOS 11 or later).

Creating the first application

In this section, we'll create a Free Tier account on Amazon Honeycode and then quickly create a to-do application.

Creating an account on Amazon Honeycode

To create an account on Amazon Honeycode, let's follow these steps:

1. We will start by using our browser to navigate to `www.honeycode.aws` and click on the **Try for free** button at the top-right corner of the page. Provide your details in the form, verify the email, and complete the signup.
2. On the first login, you will be greeted with a **Welcome to Honeycode!** popup. It is up to you whether you want to provide the requested data review the content of the popup or skip it. In our next chapter, we'll cover in detail the information provided on the pop-up screens:

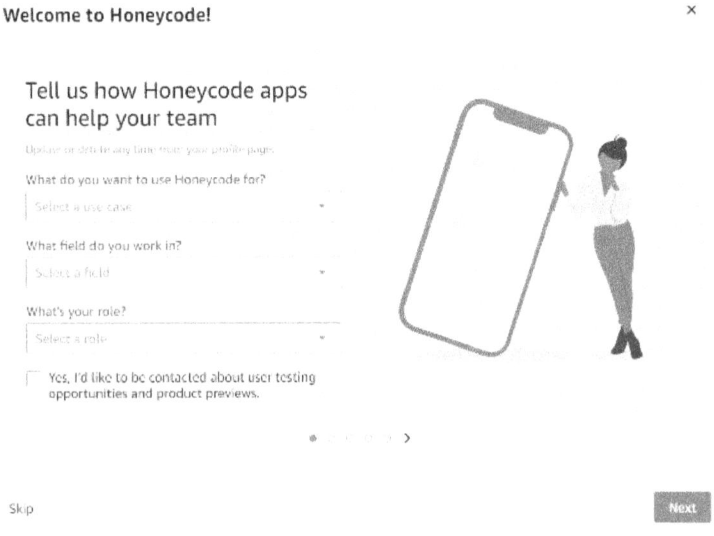

Figure 1.1 – The first Welcome to Honeycode! pop-up screen

3. After logging in and skipping through the popups and nudges, we land on a screen that is called **DASHBOARD** (there'll be more on the dashboard in *Chapter 2, Introduction to Honeycode*), which is the place where we'll start creating our **To-Do** application:

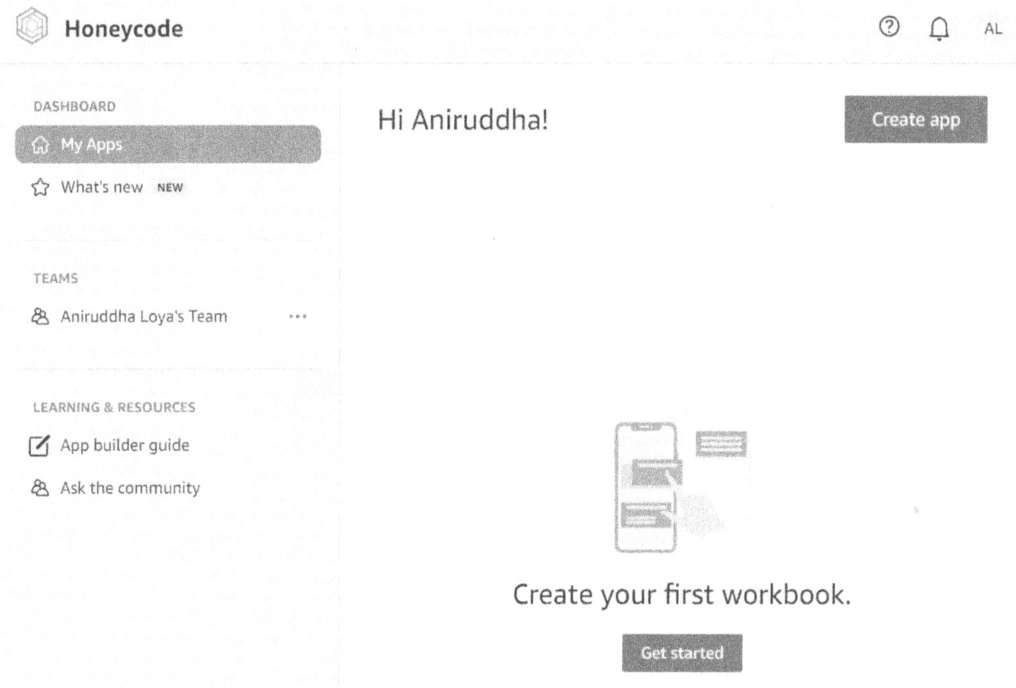

Figure 1.2 – An empty dashboard on the first login

Creating a To-Do application

Let's understand how to create a **To-Do** application with the following steps:

1. On the **DASHBOARD** screen, locate the **Create app** button in the top-right corner and click it.
2. This loads a popup with quite a number of options for us. Under the **USE A TEMPLATE** header, locate a tile with the **Simple To-Do** label and click it:

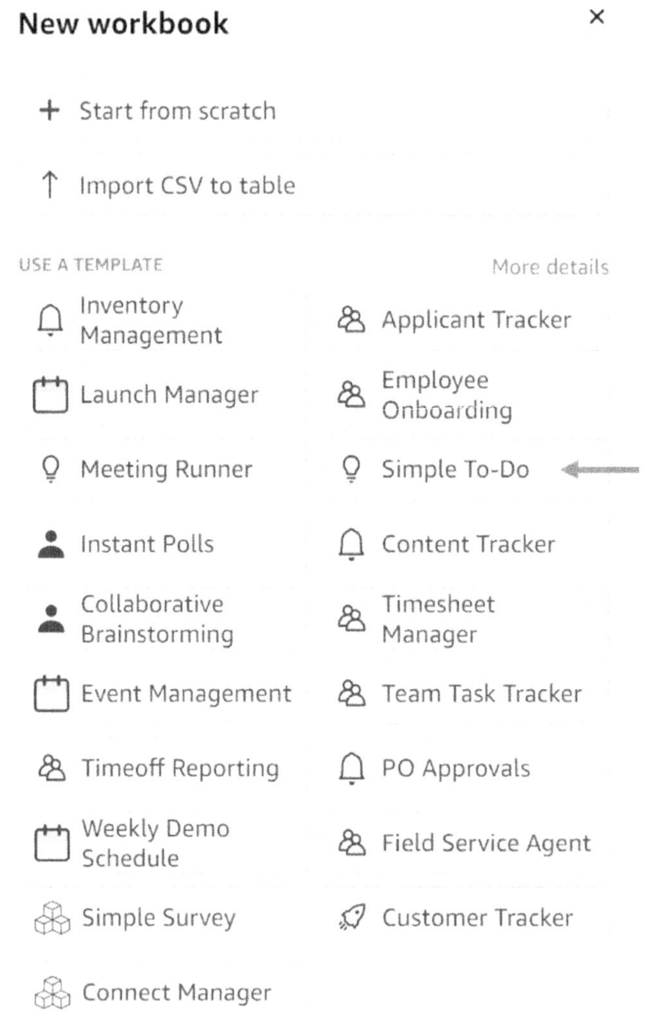

Figure 1.3 – Selecting the Simple To-Do application template

3. Next comes a popup to name the workbook and choose a team. For now, let's leave the default values and click on the **Create** button:

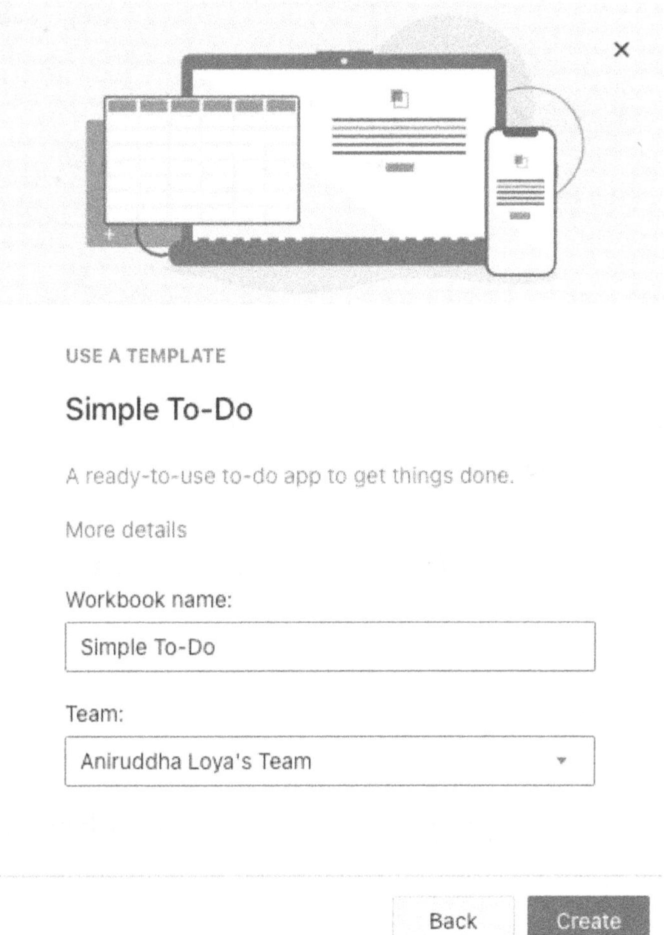

Figure 1.4 – Providing a workbook name and team details to create the workbook

4. This creates and loads our **Simple To-Do** workbook and, given it is our first time using it, another set of popups with links and information about the components of the workbook.

   It is up to you whether you want to review the links and follow the popups or skip them.

5. After reviewing or skipping the popups, click on the **Honeycode** icon in the top left, which navigates back to the dashboard.

6. Your **Honeycode** dashboard now has the newly created **Simple To-Do** app:

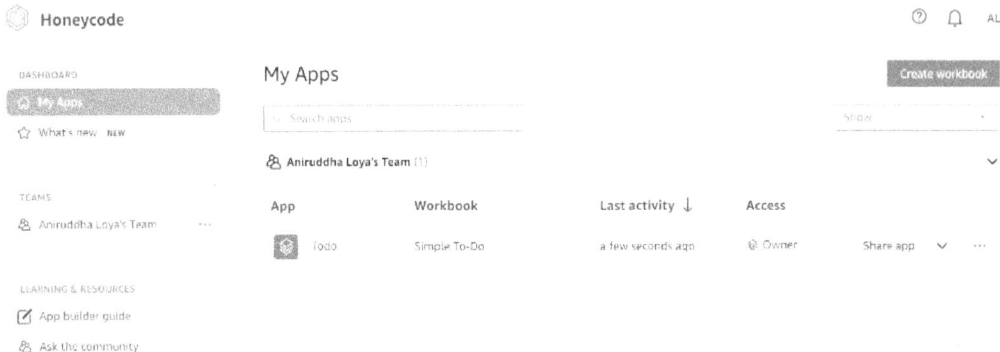

Figure 1.5 – The dashboard showing the newly created application

And that's it. In six simple steps, you have created your first Honeycode application that is ready to be used. In the next section, we'll learn how to start using this application through browsers and mobile devices.

Running the application on the web and a mobile

In the previous section, we created the Amazon Honeycode account and used it to log in and create our first application – **To-Do**. In this section, we'll learn how to run this application on the web and mobile devices.

Running our To-Do application on a web browser

Let's see how to run the To-Do app:

1. If not already logged in, log in to your Honeycode account.
2. Locate the **To do** application in **DASHBOARD** and click on it:

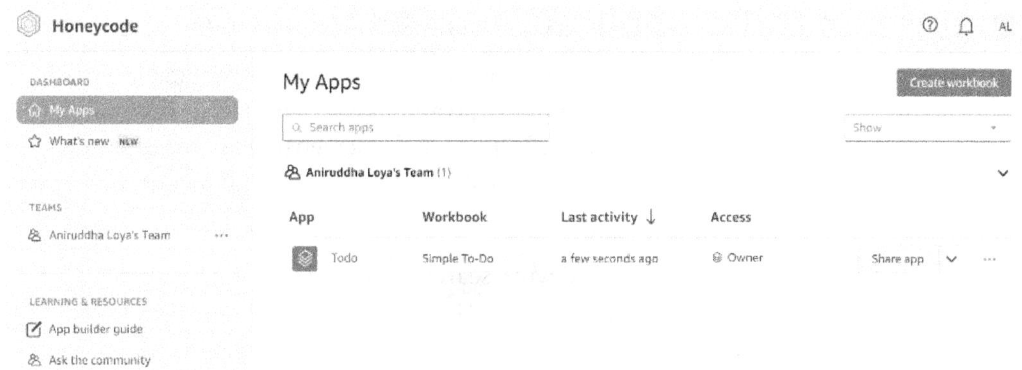

Figure 1.6 – Launching the To-Do web application from the dashboard

And there we have our **To-Do** application ready to use. Try it out by adding a new task, marking tasks complete, and reviewing the screens:

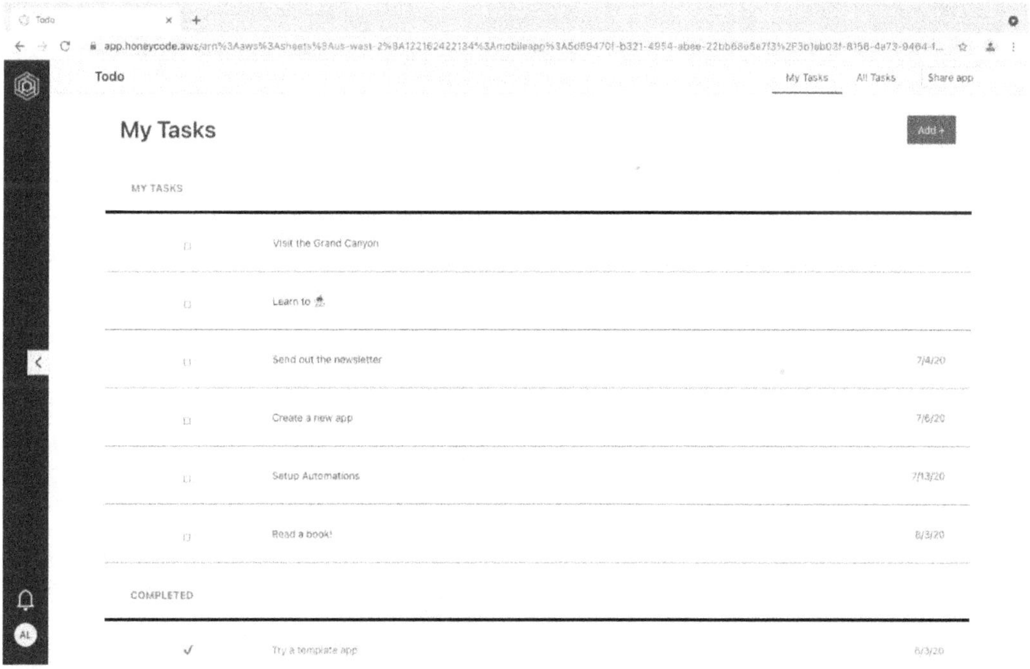

Figure 1.7 – The To-Do application running in a web browser

Running our To-Do application on a mobile device

The next set of steps will require a mobile device running either a Honeycode-supported version of Android (it currently requires Android 8.0 and upward) or iOS (it currently requires iOS 11 or later). For the purpose of illustration, the screenshots in this section are taken from an iPhone, so there might be some differences in look and feel on Android devices:

1. Go to the app store on your mobile device, search for `Amazon Honeycode`, and download it.
2. Log in using the credentials of the Honeycode account we created in the first section.
3. The **Apps** screen loads with the **To-Do** app we created. Tap on it to launch the app:

Figure 1.8 – The To-Do application on a mobile

And there we have our **To-Do** application ready to use. Try it out by adding a new task, marking tasks complete, and reviewing the screens:

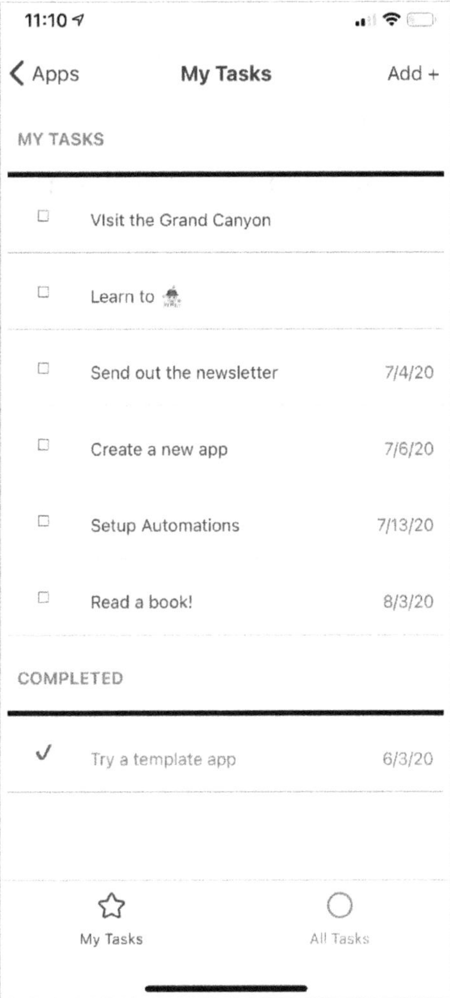

Figure 1.9 – The To-Do application running on a mobile device

> **Bonus**
>
> Open both the web and mobile applications. Make a change in one and note that the change is reflected in the other app in near real time. In a later chapter, we will learn more about how this sync hronization enables real-time collaboration among the team members using Honeycode apps and across different platforms and devices too.

Summary

In this chapter, in less than 10 steps, we created and tested a new mobile and web application without writing a single line of code.

Now that we know the power of Honeycode and, more importantly, how to navigate and run apps on both mobile devices and the web, it's time to dive into the details of different aspects and terminology associated with Honeycode.

In the next chapter, we'll use our **Simple To-Do** workbook and the **To-Do** app to get ourselves acquainted with Honeycode.

2
Introduction to Amazon Honeycode

Amazon Honeycode is a fully managed low-code/no-code application development tool offered by AWS. We learned about this key value proposition in the previous chapter when we deployed our own to-do list application in less than 10 steps and tested it on both mobile and the web. With the power of AWS behind it, a simple to-do application barely scratches the surface, as the platform offers scalability, performance, security, and compliance, which we are accustomed to expecting from AWS.

This enables citizen developers to focus on their core ideas and convert them into real-world applications with ease. They can then share these ideas with their friends, family, or teams to test them out and iterate quickly without having to worry about the required technology infrastructure had they built and deployed a similar application natively. Alternatively, they might just want to get control of their data and, thereby, want to have a small application simply for their personal use or an application to improve the team's productivity. The use cases are many, and we'll explore a range of them in the following sections and chapters of this book.

In this chapter, we will look at the basic concepts of Honeycode. This will equip you with the knowledge to identify the different components of the platform and understand how they relate and interact with each other. Additionally, it will enable us to share a common vocabulary to use in the remainder of the book.

We will also look at another important aspect of Honeycode – *collaboration*. We'll learn how we can use Honeycode Teams to collaborate with others in the process of building our application or just invite them as an app user only.

Following this, we will look at the pricing model for using Honeycode to enable you to understand the key differences between the different pricing tiers and to empower you to choose the appropriate tier based on the functional needs of the application that you want to build.

In this chapter, we're going to cover the following main topics:

- Exploring the landing screen
- Exploring a Honeycode workbook
- Understanding Honeycode Teams
- Honeycode's pricing tiers

Technical requirements

In order to follow this and all the subsequent chapters in this book, you'll need to have access to Amazon Honeycode, which requires a laptop with a web browser, preferably Google Chrome, and, optionally, a mobile device running either a Honeycode supported version of Android (currently, this requires Android 8.0 or higher) or iOS (currently this requires iOS 11 or later).

In this chapter, we'll continue to use the To-Do workbook and application from the previous chapter. While the previous chapter is not a prerequisite for following this chapter, it is highly recommended that you do, as we will use the workbook to illustrate and describe some of the concepts.

Exploring Honeycode Dashboard

Honeycode's **Dashboard** is the default landing page of the product and is designed to provide quick and easy access to all the key components of Honeycode. The default view for the Dashboard is the list of apps that you can start using with a single click. The Dashboard is composed of a header panel at the top, and a navigation panel on the left that governs the data displayed in the center.

The header panel has the Honeycode icon on the left, and three icons on the right for Help, Notifications, and Profile respectively. The Help icon, *Figure 2.1 (a)*, provides you with a quick search for various Honeycode resources available at your disposal. It also provides links to get help from the vibrant Honeycode community, as well as to report issues with the product, or provide a suggestion to the product team. The bell icon, *Figure 2.1 (b)*, is for notifications. You might receive notifications from one of your applications, or a request to join the team, or for an app or workbook that was shared with you, and so on. The Profile icon, *Figure 2.1 (c)*, opens the menu required to access your account settings, and to sign-out.

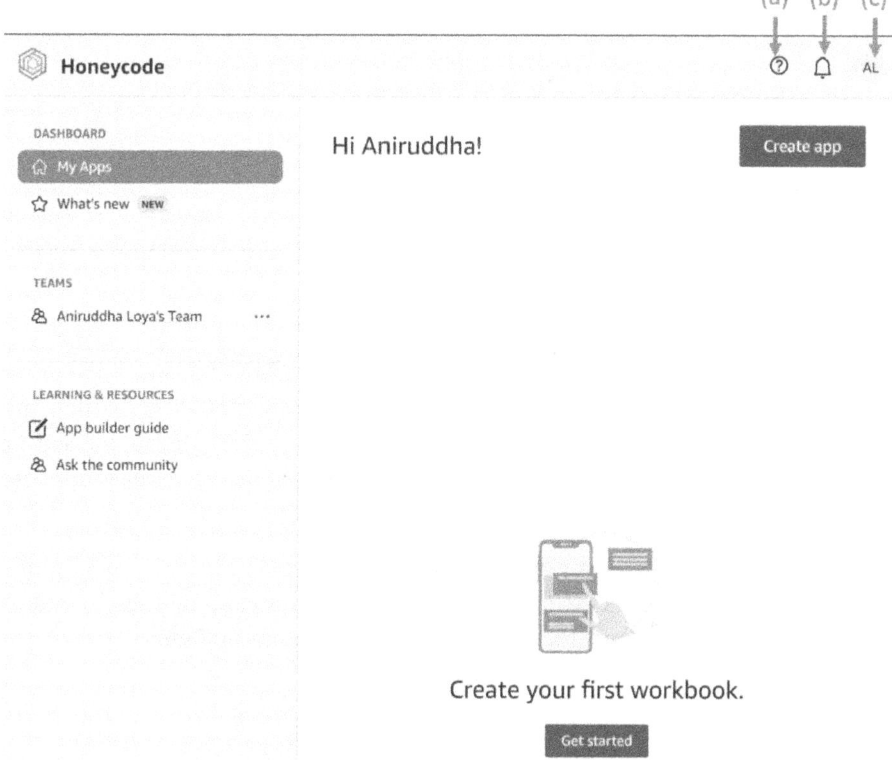

Figure 2.1 - Honeycode Dashboard

The navigation panel consists of three sections:

1. Dashboard
2. Teams
3. Learning & Resources

Let's check out each of these sections in detail.

Dashboard

The **Dashboard** sub-section of the navigation panel typically consists of two navigation links: **My Apps** and **What's new**, as shown in *Figure 2.1*. However, there is a hidden navigation link for **Workbooks** that is only visible when there is a workbook that does not contain any apps (see *Figure 2.2*). This hidden link, and the default view it takes you to will not come as a surprise to users, as the key use of Honeycode is not storing data, but being able to build and use apps on top of it with ease.

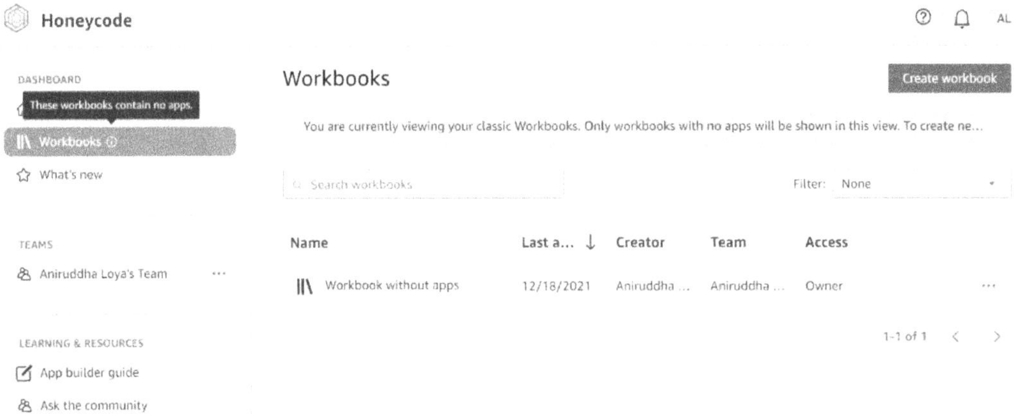

Figure 2.2 Dashboard sub-section with Workbook link

Both the **My Apps** and **Workbooks** list views provide the capability to search, filter, and sort the list of apps and workbooks displayed, allowing you to quickly locate the app or workbook you are interested in. The view also provides collapsible grouping of apps by the teams they belong to.

The **What's new** section is where you can find the highlights of the latest updates and releases by the Honeycode product team, with links to learn more about them. Look for the **New** icon that shows up against the link whenever there is an update.

Teams

This sub-section lists all the teams you are a part of. You can click on each team link and view the members of that team, along with their roles. Based on your assigned role within the team, you can perform other actions to manage the team and its members. We will cover **Honeycode Teams** in detail later in this chapter, in the section *Understanding Honeycode Teams*.

Learning & Resources

This section is essentially quick access to a collection of getting started guides provided in the **Knowledge Center**, and to getting help from the **Honeycode community**.

Now we know about Honeycode's Dashboard and the different resources we can access from it, let's jump into the core of the product and start exploring a Honeycode Workbook.

Exploring a Honeycode Workbook

In this section, we'll explore a Honeycode workbook and dive into the three key components that it is made up of, namely:

- Tables
- Builder
- Automations

For this section, we'll use the Simple To-Do workbook created in *Chapter 1*, Amazon *Honeycode - Day 1*, to illustrate and explain the components. So, let's load our workbook by clicking on the pencil icon next to the **ToDo** app from the **Dashboard** (see *Figure 2.2*). Our workbook will load into the Builder view, displaying the ToDo app. Also, note the black bar on the left. This is the **Left Nav Bar**, which contains different icons to help you navigate to the different components of the workbook, as well as shortcuts for other key components in Honeycode. Given that we will need to use this bar extensively for moving around the different components of our workbook, let's review it before diving into the workbook components.

Left Navigation Bar

The **Left Navigation Bar**, or **Left Nav Bar** for short, is the dynamic bar on the left side of the window. The bar is context-aware and changes depending on whether you are currently accessing the workbook or using the webapp. The bar is essential for moving around in the workbook as we go about building our apps, so let's take a quick look at all the navigable icons present on this bar.

- The Honeycode icon (*Figure 2.3 (a)*) allows you to navigate back to the **Dashboard** from anywhere in the product.
- The **Tables** icon (*Figure 2.3 (b)*) is for listing all the tables present in the workbook.
- The **Builder** icon (*Figure 2.3 (c)*) is for listing the apps in the workbook.

- The **Automations** icon (*Figure 2.3 (d)*) is for listing all the workbook automations.

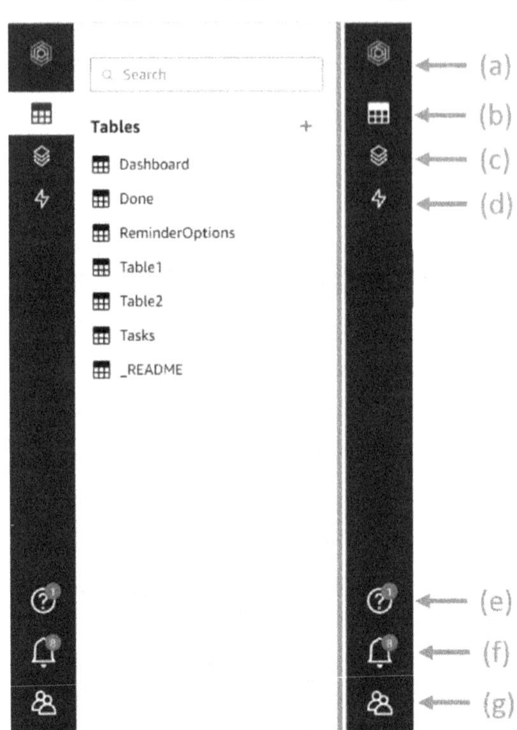

Figure 2.3 - Left Navigation Bar – pinned with Tables icon selected of left; unpinned to only display the icons for pulling out a slider to navigate

- The help icon (*Figure 2.3 (e)*) directs you to the various helpful resources available at your disposal.
- The bell icon (*Figure 2.3 (f)*) is for notifications that could be from one of your applications, or perhaps a request to join a team.
- The teams icon (*Figure 2.3 (g)*) opens a menu with links to pages for any teams that you are part of.

The Profile icon (*Figure 2.3 (h)*) opens a menu with options to access your account settings, or to sign-out. You can also report issues with the product, or provide a suggestion from this icon.

The Left Nav Bar can be pinned to always list the menus for the associated icons, or be left as an on-click slide-in. In the case of web apps, the entire bar can be hidden by clicking on the left point arrow in the middle of the bar.

Tables

Tables are the data store for Honeycode applications. A workbook can have multiple tables, up to a current maximum of 100, and each table can be used by multiple applications, enabling you to build several different applications using the same underlying data.

While a developer could make a direct association to tables in the context of databases, the interface used to present and model data is that of a spreadsheet, with support for most of the Excel functions and formats, making it intuitive for users to get started. *Figure 2.4* lists all of the components of the **Tables** interface in Honeycode:

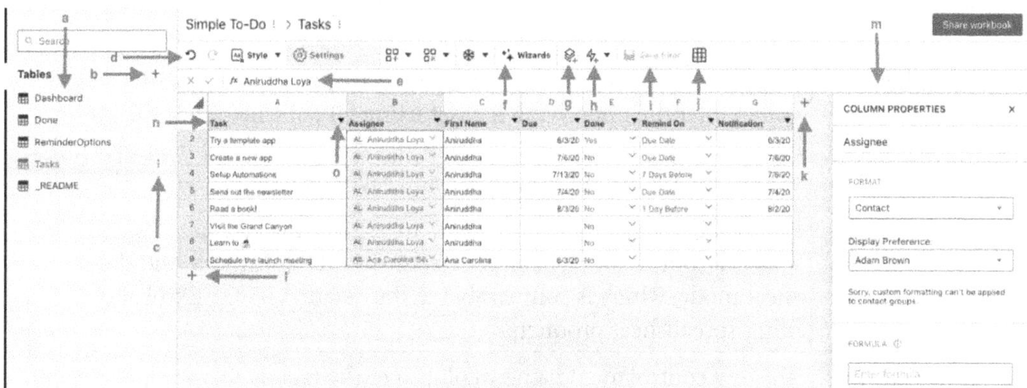

Figure 2.4 – The Tables interface in Honeycode

Components of the Tables interface

Here are the components of the **Tables** interface:

	Component	Description
a	Table list	This is the list of all the tables created in this workbook.
b	Add table	This is a control to create a new table.
c	Table actions	This is a control consisting of operations to rename, delete, open in a new tab, and import data into the table. This can also be revealed using the right mouse click.
d	Toolbar	This is a collection of various tools and controls, to create, format, and manage data, and other shortcuts.
e	Formula bar	This displays the content of the selected cell and can be used to edit or add content to the selected cell.

	Component	Description
f	Wizards	Wizards in Honeycode guide you in the creation of some commonly needed items. Currently, Honeycode provides four wizards. Two are for data modeling and visualization, namely Dashboard and Create Picklists, and two for automation creation, namely Set Alert and Set Reminder.
g	Build app	This is a shortcut to kick off the app creation process; it has no direct relation to the table being viewed while taking this action.
h	Create automation	This is a shortcut for creating automations. This shortcut also provides the option for the two pre-built automation templates that are also available under the wizards, namely Set Alert and Set Reminder.
i	Save filter	Using the table column actions, users can open the control that allows you to filter the underlying dataset using UI controls. Once the filter has been created, this control can be used to persist that filter by converting it into a filter formula and saving it to any cell on a sheet in the workbook.
j	See sheets view	This is a control to toggle between the default table mode and the sheet mode, which is comparable to the default view in Excel and other spreadsheet products.
k	Add column	This is a control to add a new column to the table.
l	Add row	This is a control to add a new row to the table. This action is also achieved by pressing enter inside any cell in the last row of the table as long as the row is within the limit of table rows for that tier.
m	Column properties	This is the right-hand side panel for displaying the properties set on a table column level, which then propagates to every cell in that column.
n	Table header row	This is the table row containing the column headers that are typically an indicator of what data is contained in those table columns.
o	Table column actions	This is a control to open a list of action items to filter and sort data in a specific table column.

Importing data to tables

Being able to easily build and deploy an app is the key value proposition of Honeycode. However, not every app is meant for data collection, and not every app needs to be built on an empty dataset. There are use cases for easier representation and access of existing data via apps that can be built with Honeycode. To facilitate this, we need to be able to import this existing data into Honeycode.

In Honeycode, you can easily import your existing data by uploading a CSV file. The data can be uploaded to a new table or an existing table following the steps listed in the following sub-sections.

Importing data to a new table

1. Click on the control to create a new table, and then select **Import CSV file** from the list, as shown in *Figure 2.5*:

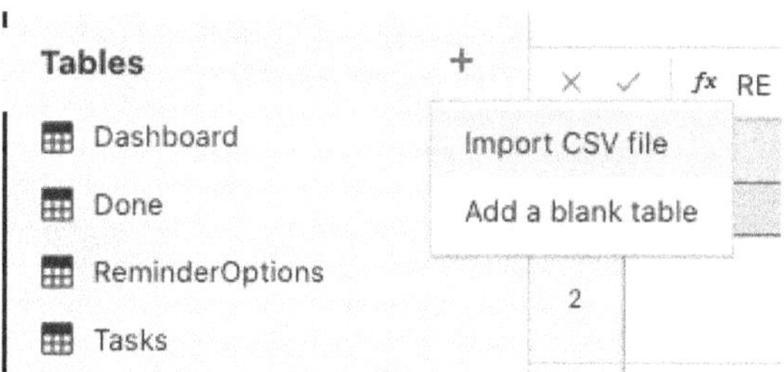

Figure 2.5 – Creating a new table by importing data from the CSV file

2. Next, a popup opens, as shown in *Figure 2.6*, that lets you choose whether the data imported through the CSV file has a header row or not. If you do not select this option, Honeycode will provide the default column names for each of the data columns created, namely Column1, Column2, and so on:

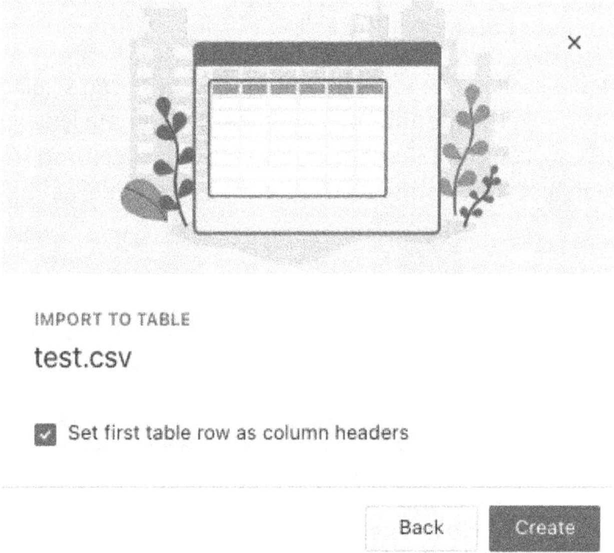

Figure 2.6 – The follow-up dialog to declare whether there is a header row for the data or not; if unselected, default column headers will be applied

Importing data to an existing table

1. Right-click on the table to which you'd like to import the data. Alternatively, click on the 3-dot icon on the right-hand side of the table name and then select **Import to this table** from the list of possible actions, as shown in *Figure 2.7*:

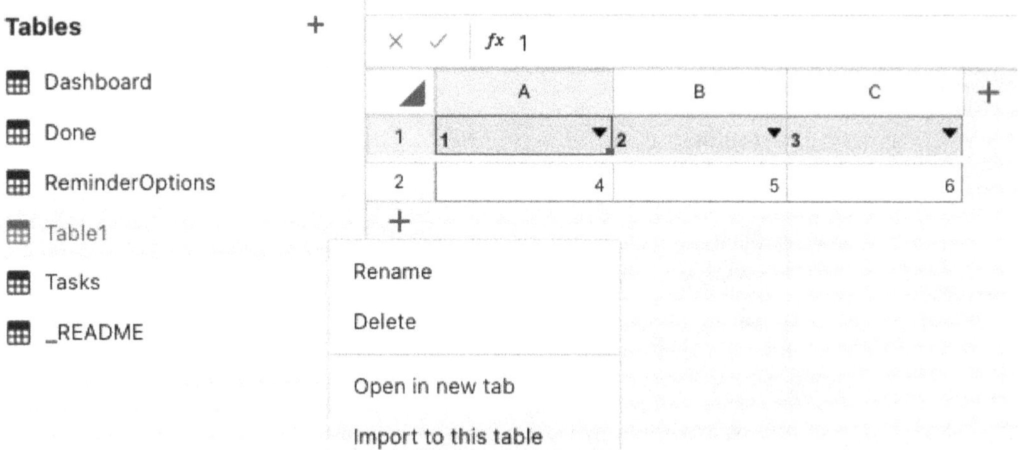

Figure 2.7 – Importing data from the CSV file to an existing table

2. Next, a popup opens that allows you to choose whether the data imported through the CSV file has a header row or not and also provide the mapping from imported data columns to those in the existing table:

    I. If the imported data has a header row option that has been selected, the popup reads the first row of the file and populates the field on the left-hand side with those values. Now you can select which imported data columns should be mapped to which of the existing data columns in the table, as shown in *Figure 2.8*:

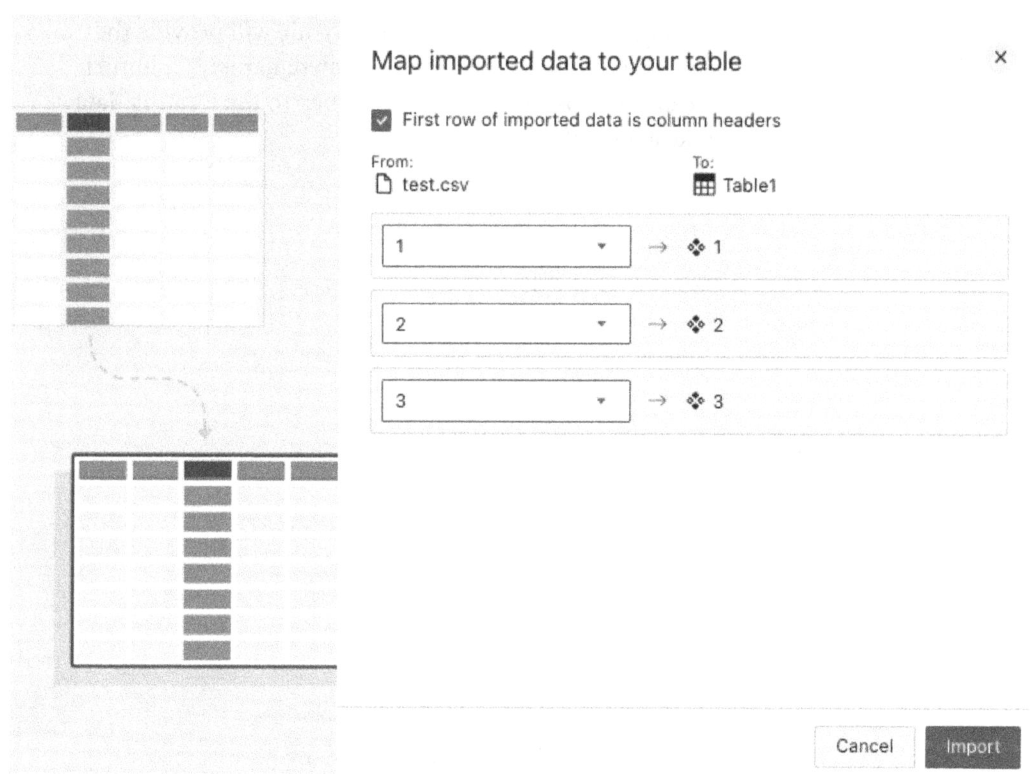

Figure 2.8 – Mapping imported data columns to your existing table when the imported data also has column headers

II. If the imported data does not have a header row, Honeycode will provide the default column names for each of the data columns created, namely Column1 and Column2, and you can use those as a reference to map to the existing data columns in the table. Take a look at *Figure 2.9*:

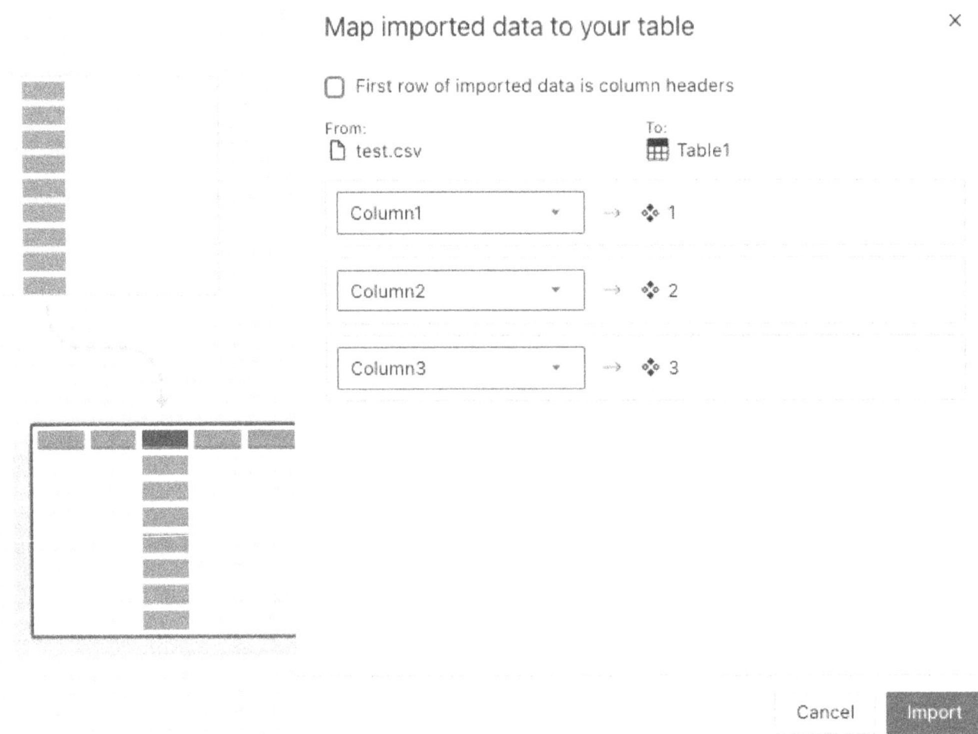

Figure 2.9 – Mapping imported data columns to an existing table when the imported data does not have column headers

Builder

Builder is where the magic happens in Honeycode. This is where we'll create our apps for both mobile and the web. In this section, first, we'll take a look at the components of the Builder interface, as shown in *Figure 2.10*. Following that, we will dive in and explore the app screen and its parts. Finally, we will list and describe all the objects or controls that you can add to the screen:

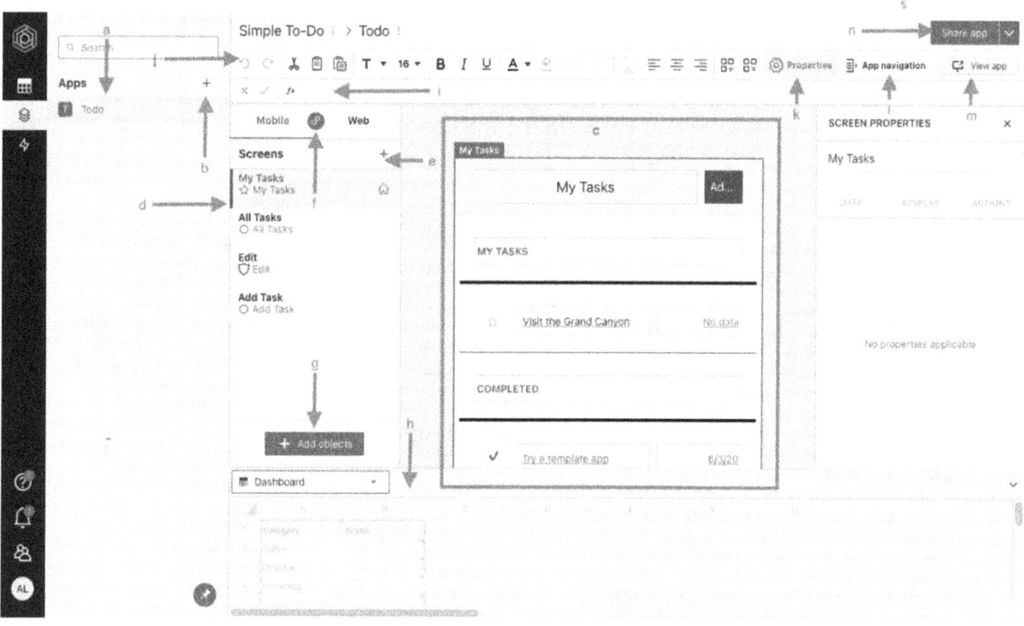

Figure 2.10 – The Builder interface

Components of the Builder interface

Here are the components of the **Builder** interface:

	Component	Description
a	App list	This is a list of all the apps created in this workbook.
b	Build app	This is a control to create a new app.
c	Screen editor	This is the primary area where you'll add components and design layouts for your app screens.
d	Screen list	This is a list of all the screens in the selected app.
e	Create new screen	This is a control to add a new screen in the selected app.
f	Unlink Mobile and Web layout	In Honeycode, by default, you only design the layout once and it automatically adjusts to both the mobile and web layouts. However, on occasion, we might want to have a separate treatment for mobile versus the web, especially since the web offers a much larger screen space and can, therefore, potentially have more controls and objects in its layout. This control enables us to break the link between the two apps and enable independent future development for the two apps.
g	Add objects	This control lists all the objects supported in Honeycode.

	Component	Description
H	Peeking sheet	This is a collapsible peeking sheet to enable a quick view of different tables in the workbook. It also supports point and click functionality for binding different data columns to the controls on different screens.
i	Formula bar	This displays the content of the controls that supports source data-binding, allowing you to edit the formula.
j	Toolbar	This is a collection of tools and controls to create, format, and configure the app.
k	Properties	This is a shortcut to toggle the display of properties of the selected control within the right-hand side panel.
l	App navigation	This is a shortcut to toggle the right-hand side panel to configure which screens should be visible to a user as part of the app navigation bar and which should be not.
m	View web app	This is a control to quickly launch the app in web view,
n	Share app	This is a shortcut to share the app as you build.

App screens

In general, an app will be made of multiple screens, and the builder lists all of the screens in the column to the left-hand side of the screen editor. Each screen in the app has two separate identifiers for each screen, as marked by **a** and **b** in *Figure 2.11*. Sometimes, it can be confusing to know which one is for what, and in general, the best option is to keep them the same:

- **Local name**: This is the identifier marked **a** in the following screenshot and is used for local references when building the app, for example, identifying the screen when building navigations through automation. This name is not exposed outside of the builder context and is, therefore, not visible to the end user.
- **Screen name**: This is the identifier marked **b** in the following screenshot and is the value exposed to the end users of the app. The value is what you'll see in the screen header and the global navigation menu of the app.

Every app screen consists of three main sections:

- **Header**: This contains two parts:

  A. A content box with the name of the screen

  B. An optional button to define an action at the screen level

- **Body**: This is where we build our application layout and define our business logic.
- **Global Navigation**: This section is located at the bottom of the screen for mobile apps and at the top of the screen for web apps. It is not editable from the screen editor and is configured through the right-hand side panel that opens up when you click on the section in the screen editor or the app navigation control from the toolbar.

Finally, earlier in this section, we noted that in Honeycode, by default, we only design the layout once as it automatically adjusts to both the mobile layout and the web layout. We can view the screen layouts for mobile and web apps using the control labelled **Mobile** and **Web** on either side of the **Unlink Mobile and Web Layout** control, as shown in *Figure 2.10 (f)*.

The sections for **Screens** in the mobile app view and the two separate app identifiers are displayed in *Figure 2.11*:

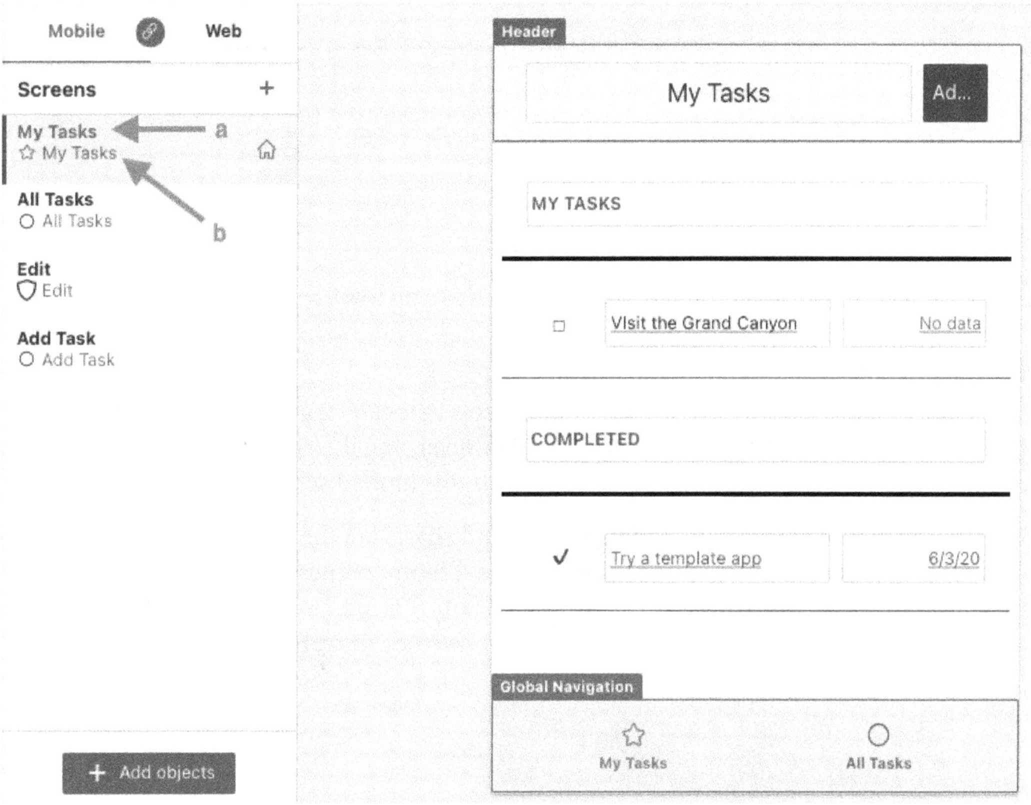

Figure 2.11 – The sections in a screen for a mobile app

Similarly, *Figure 2.12* shows the sections for the screens in the web app view. Note that the **Global Navigation** menu is located at the top in comparison to where it is located in the mobile app:

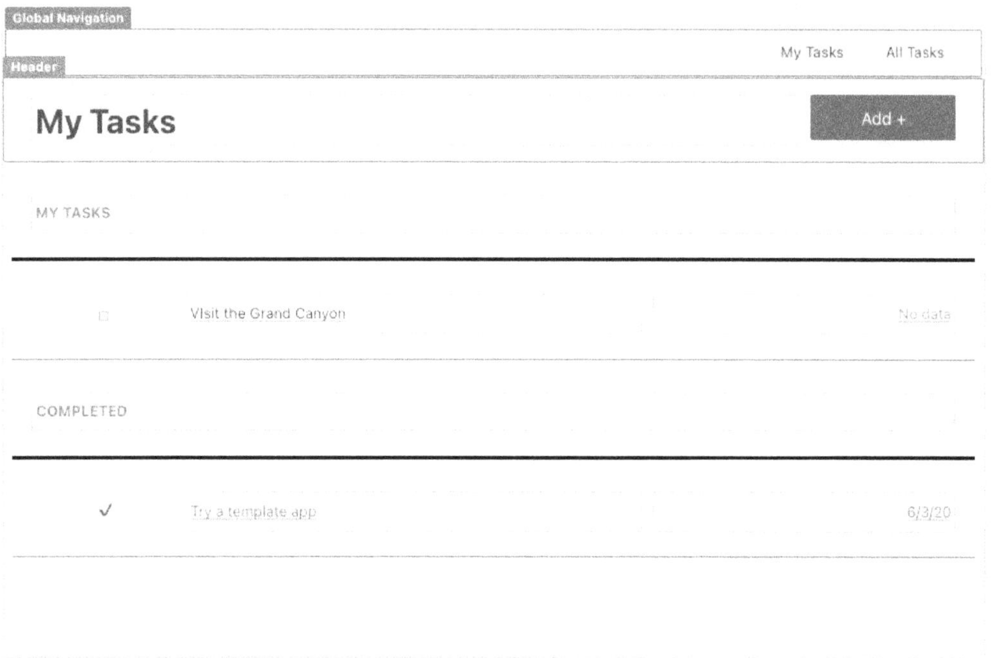

Figure 2.12 – The sections in a screen for a web app

App objects and controls

Honeycode offers a collection of objects and controls that can be added to your app, which, at the outset, might seem quite limited. However, you'll be pleasantly surprised at how complex applications can be built even with this limited set.

However, the lack of controls for adding images and charts are the two most noticeable misses to make even richer apps and presentations. Feature requests to support images (`https://honeycodecommunity.aws/t/picture-picture-images/16453`) and charts (`https://honeycodecommunity.aws/t/graphs/14047`)ranks among the top asks on Honeycode community:

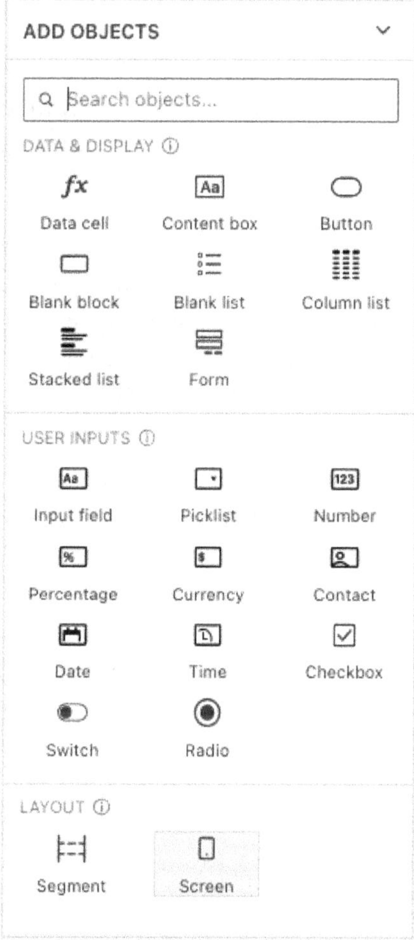

Figure 2.13 – App objects and controls

Let's take a look at some of the objects displayed in *Figure 2.13*.

The DATA & DISPLAY objects

Let's examine the data and display objects:

- **Data cell**: As its name suggests, this object is used to show data from the workbook through a mapping to a table or by using a formula to derive values using data from the workbook.
- **Content box**: This object allows you to add free-form data to your app.
- **Button**: This control allows you to define actions on the app such as submit data, navigate to another screen, and more.

- **Block**: This object can be used to display a single row from a table or used as a container for other controls such as data cells, buttons, and more.
- **List**: This is a list object type that allows you to display multiple rows from the source. These can be a table or a formula that returns one or more rows such as a `Filter` function:
    - **Blank list**: This is a free-form configurable list that users can configure as needed.
    - **Column list**: This is an adapted form of a list that lets you choose which columns to choose from a table and displays them in a columnar fashion on the screen. Additionally, it provides the option to create a detailed screen that can be navigated to upon clicking and used to display additional information for the selected data row.
    - **Stacked list**: This is an adapted form of a list that lets you choose which columns to choose from a table and displays them in a stacked fashion on the screen. Additionally, it provides the option to create a detailed screen that can be navigated to upon clicking and used to display additional information for the selected data row.
- **Form**: This is a pre-built control that allows you to quickly create a new screen or add a set of controls to the existing screen that enables data entry into the configured source table.

The USER INPUTS objects

Most of the controls provided in this section are essentially preconfigured **Data cell** objects to specific data types that are easy to relate to different data types supported in other spreadsheet products. However, there are a few controls that are specific to Honeycode, as follows:

- **Picklist**: This is probably the most powerful of the custom controls that are unique to Honeycode. While the rendering might look like a simple drop-down menu, conceptually, this allows you to build relations among the different tables in the workbook.
- **Contact**: This is another custom control that allows you to display user information. In *Chapter 4*, Advanced builder tools in Honeycode, we'll see how this control and the corresponding data type can be used to control and customize the data displayed to each app user.
- **Checkbox/Switch**: These controls allow you to provide toggle-like behavior on values in the app, for example, on/off, yes/no, and more.
- **Radio**: This is another representation of **Picklist** where the data values are displayed in the form of a radio control instead of a drop-down selection of a picklist.

LAYOUT

- **Segment**: Similar to **Block**, this object also acts as a container for other objects. However, unlike a block object, this is purely a container and cannot have data binding added to it. The object is useful for the custom control grouping of objects and customizing the layout in an app screen.
- **Screen**: This is a duplicate control to add a new screen to the application. The same outcome can be achieved using the + button to add a screen, as listed in *Figure 2.9*.

Automations

Automations in Honeycode power both the application interactions and the automated data processing that could be triggered by user actions, such as a specific date/time, which can even run periodically, and more. Honeycode categorizes automations into two types: workbook automations and app automations. In the following sub-sections, we'll familiarize ourselves with the interfaces of both these types. However, further details of different actions that we can configure for automations will be covered in *Chapter 5, Powering the Honeycode apps with Automations.*

Workbook automations

Workbook automations are useful for automated data processing and are either triggered by a change in data in a table or when the configured date and time is reached. *Figure 2.13* lists all the components of the workbook automations interface:

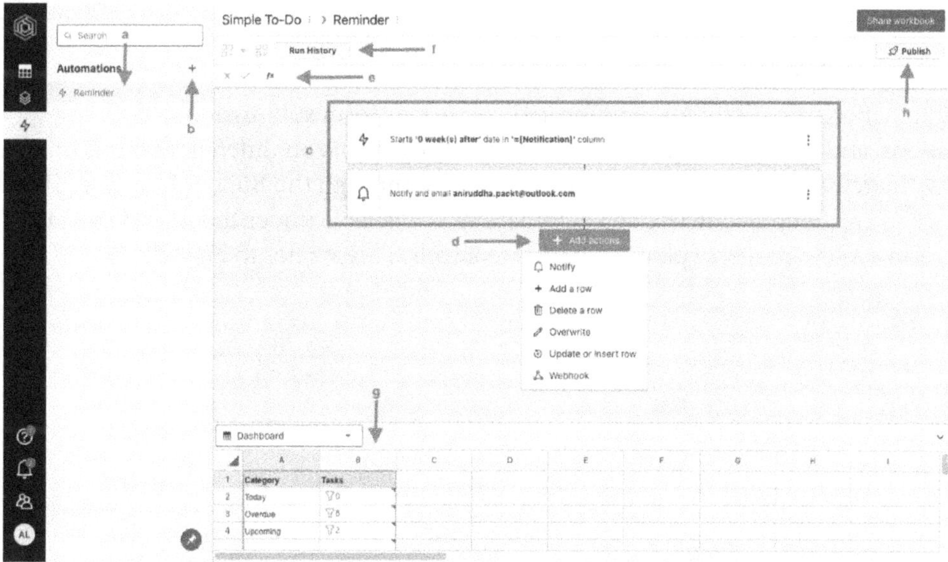

Figure 2.14 – The workbook Automations Builder interface

Workbook Automations Builder components

Here are the components of the workbook Automations Builder interface:

	Component	Description
a	Automation list	A list of all workbook automations created in this workbook.
b	Build automation	A control to create a new workbook automation.
c	Automation editor	The primary area where you'll add actions and configure logic for the workbook automation.
d	Add actions	A control to add a new action to the automation.
e	Formula bar	This displays the content of the controls that supports source data-binding, allowing you to edit the formula.
f	Run history	A control to be able to audit historical runs of this automation.
g	Peeking sheet	A collapsible peeking sheet that allows a quick view of different tables in the workbook. It also supports point and click functionality for binding the different data columns to controls on the different screens.
h	Publish automation	A control to publish the automation. When created or edited, workbook automations are in a draft state to prevent the accidental processing or updating of data as we build it up. An explicit publish action is required to indicate that the automation is now live.

App automations

App automations are useful for defining actions such as form submission, filter, navigations, and more to be performed on touch/click events on different controls on the screens. Therefore, these automations are configured through the Builder interface. The interface to create/update/delete app automations is available under the **ACTIONS** tab, which is inside the property panel of the app controls. Please refer to *Figure 2.15*:

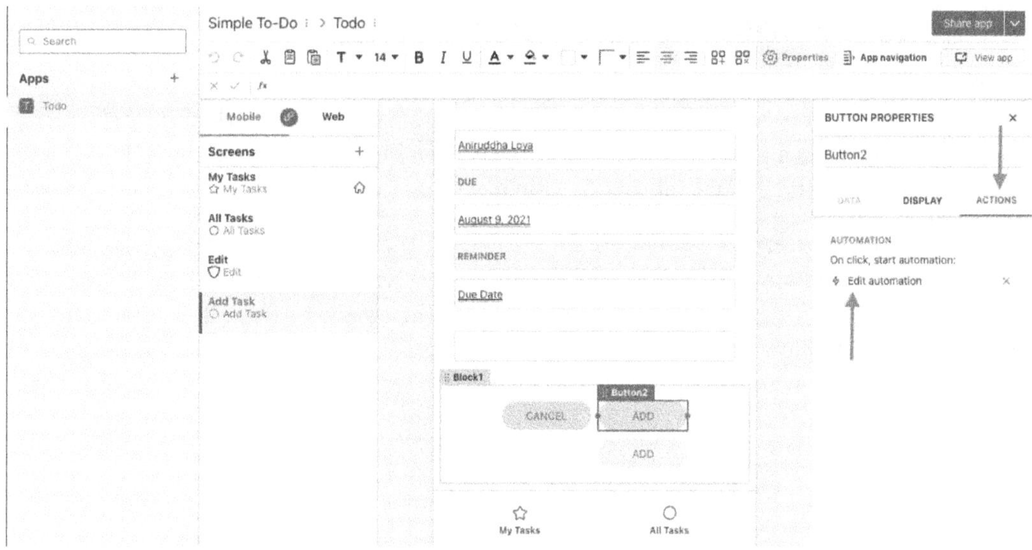

Figure 2.15 – Accessing app automations

In terms of the automation editor itself, it will be the same as that for workbook automations and consists of the same actions and controls, including **Run History** (please refer to *Figure 2.16*). The only difference from workbook automations is that app automations are not required to be published, that is, they are available the moment they are added to the control:

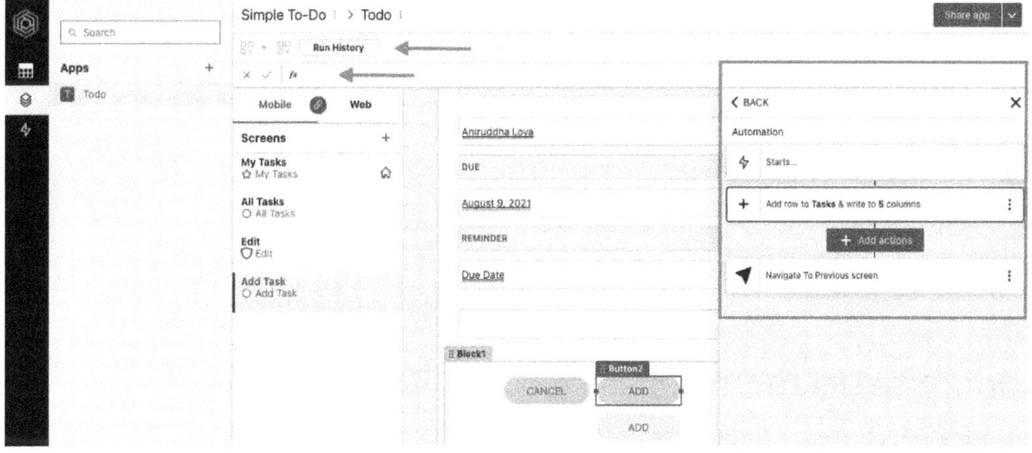

Figure 2.16 – The app automations editor displaying automation on the ADD button in the to-do app

Understanding Honeycode Teams

Honeycode Teams is the essential piece to know once you decide to graduate from building apps for yourself to being able to collaborate and share your work and/or start to use your application as a team. In this section, we'll learn how to use Honeycode Teams to enable these sharing and collaborating functionalities in Honeycode.

In Honeycode, every user has a default team created for them at the time of sign-up, and all the workbooks and apps you create will default to this team. It is important to note that, at the time of writing, Honeycode does not allow the creation of an additional team as a user, but it does allow you to be part of multiple teams. *Figure 2.17* shows the selected **Teams** icon in the left navigation bar along with the two teams that I'm a member of listed in the open slider:

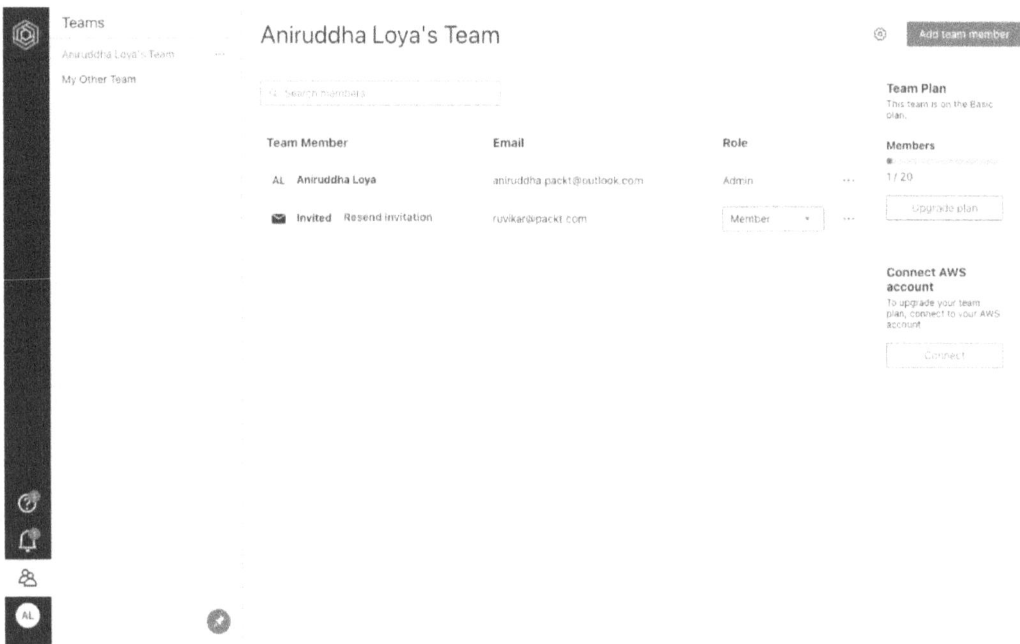

Figure 2.17 – Honeycode Teams

Adding team members

In Honeycode, there are three ways to add someone to your team:

- Invite members.
- Share a workbook or an app with a person that is not part of your team.
- Request access to the workbook or app not shared with you.

Therefore, the invites that are sent show up in the invitee's email inbox. If the person invited already has a Honeycode account associated with that email ID, they'll also receive a notification in Honeycode stating that they have been invited to join your team, and they can accept or dismiss this invite from the notification. Similarly, there is both an email and an in-app notification, where anything that is shared with team members or a share and invite is sent to a new member.

Inviting team members

This is the most commonly understood paradigm where an invitation is sent to join the team, and the receiver chooses to accept or reject the invite. To add someone to your team using this method, all you need to know is their email ID. Then, follow these three easy steps:

1. Go to the team page where you'd like to add the person.
2. Click on the **Add Team Member** button in the upper-right corner.
3. In the popup that appears, provide the email of the person, choose the right role for them, and press the **Invite** button.Note that you can provide more than one email, each separated by a comma, to invite multiple members in a single attempt. Do note that there is a limit of 20 free team members beyond which you'll have to start to pay for each seat. Details about the pricing are discussed in the next section:

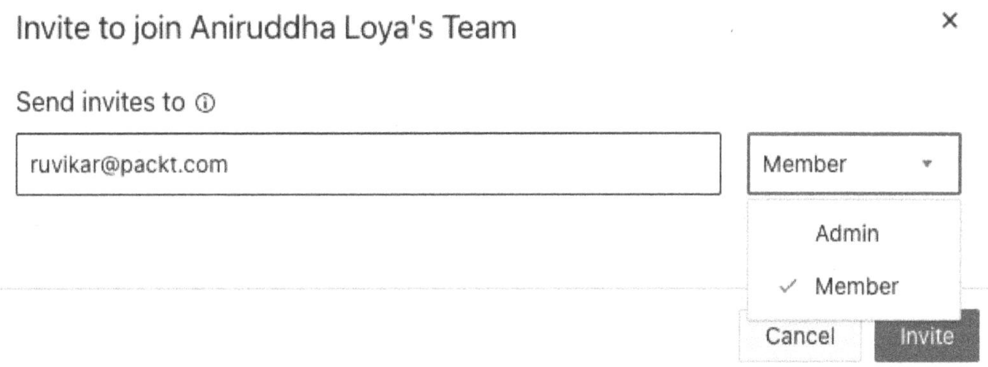

Figure 2.18 – Inviting someone to join the team with Add Team Member

Inviting team members by sharing the app or workbook

This is a simplification of the invite-first model of adding team members that the Honeycode team has provided to its users by including the invitation along with the sharing flow. This takes away the overhead of sending out an explicit invite or checking whether the person the invite is being shared with is part of the team or not.

To add someone to your team using this method, simply share the app or workbook that you'd like them to have access to (details on how to share can be found in the next section). First, they'll get an invite to the team, and upon accepting, they'll be able to access the shared resource. On clicking the share action on the workbook from the **My Drive** screen, the following popup is displayed, as shown in *Figure 2.19*, listing all the members with current access to the workbook and an option to quickly copy the link to the workbook, which can be sent separately if needed or used for bookmarking:

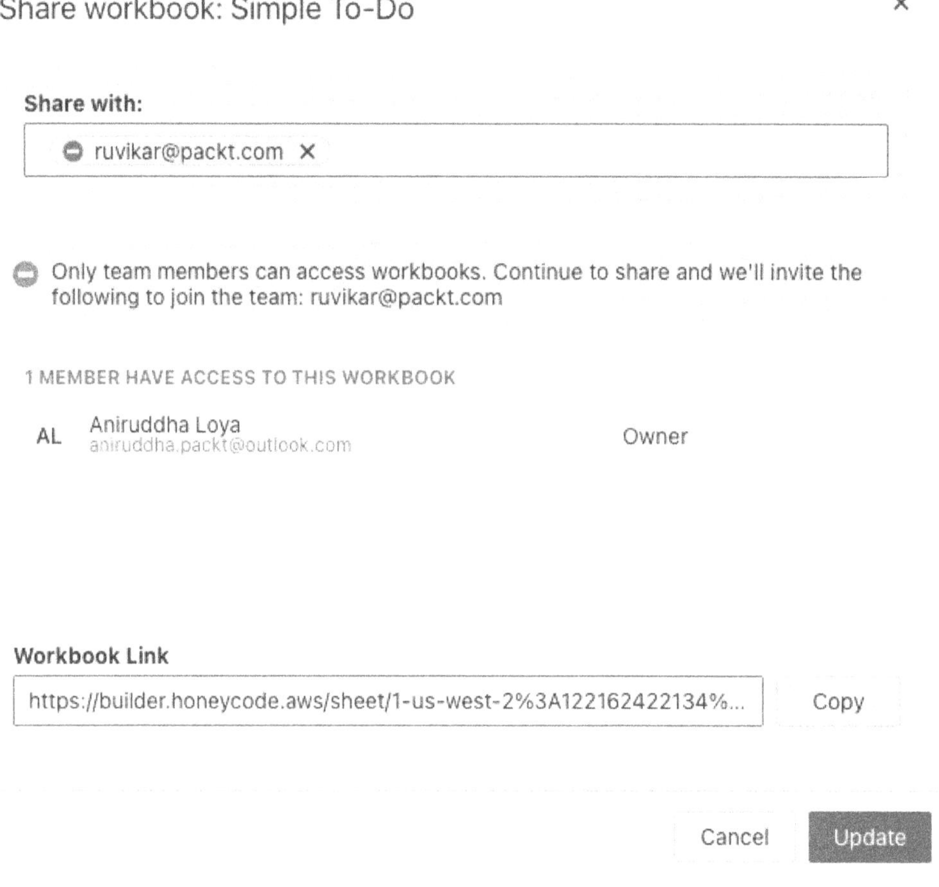

Figure 2.19 – Inviting someone with Share workbook

Similarly, the share action on the app from **My Drive** displays the following popup, as shown in *Figure 2.20*, listing all the members with current access to the app and an option to quickly copy the link to the app, which can be sent separately if needed or be used for bookmarking. Note the presence of an additional section of **Advanced Settings** that is not present in the popup for sharing the workbook. We'll cover this setting in the next section:

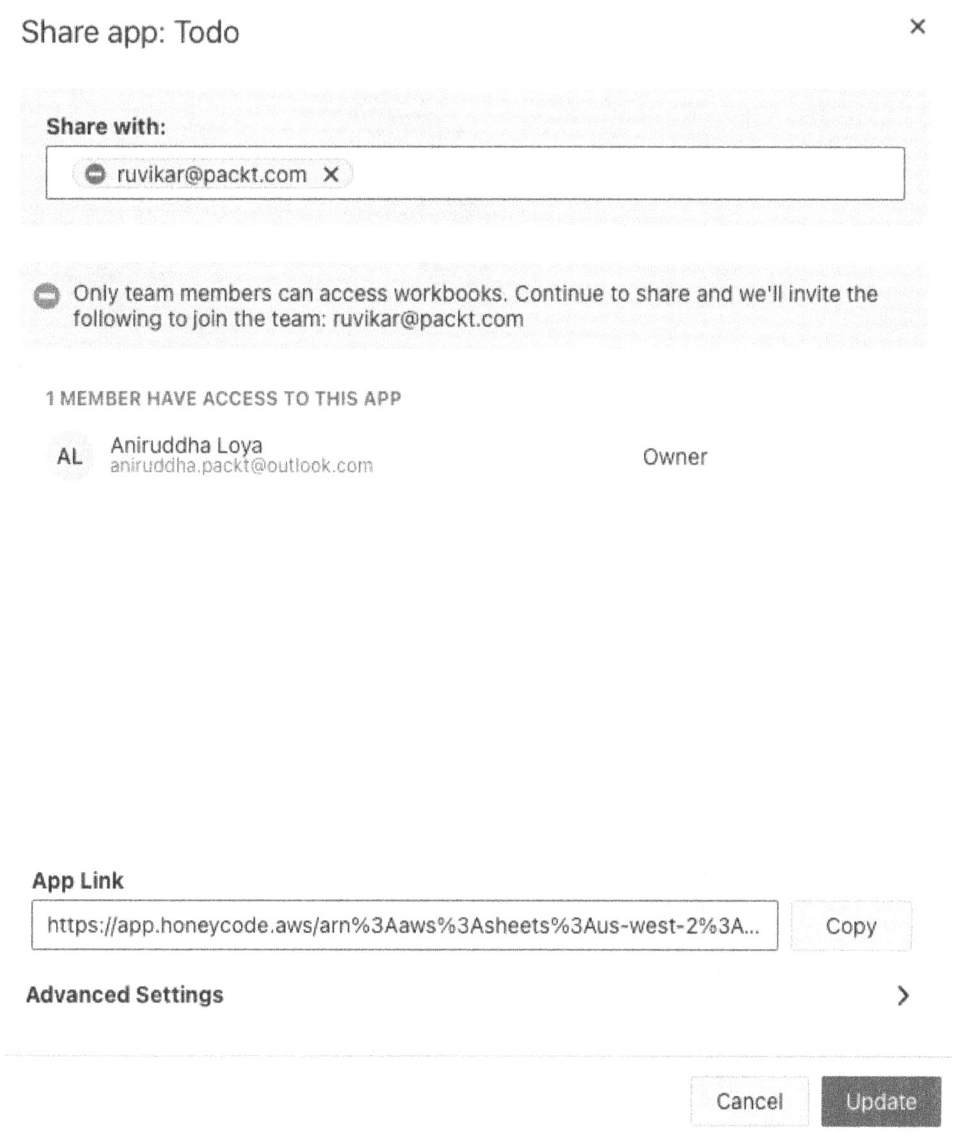

Share app: Todo

Share with:

ruvikar@packt.com

Only team members can access workbooks. Continue to share and we'll invite the following to join the team: ruvikar@packt.com

1 MEMBER HAVE ACCESS TO THIS APP

AL Aniruddha Loya
aniruddha.packt@outlook.com
Owner

App Link

https://app.honeycode.aws/arn%3Aaws%3Asheets%3Aus-west-2%3A...
Copy

Advanced Settings

Cancel
Update

Figure 2.20 – Inviting someone with Share app

Adding team members by approving access requests

In contrast to the previous two methods that are invite-accept models, in this request-approve flow, anyone with the link to your app can request access to the same. *Figure 2.21* shows such an example with a button to **Request access** to the app you wish to access:

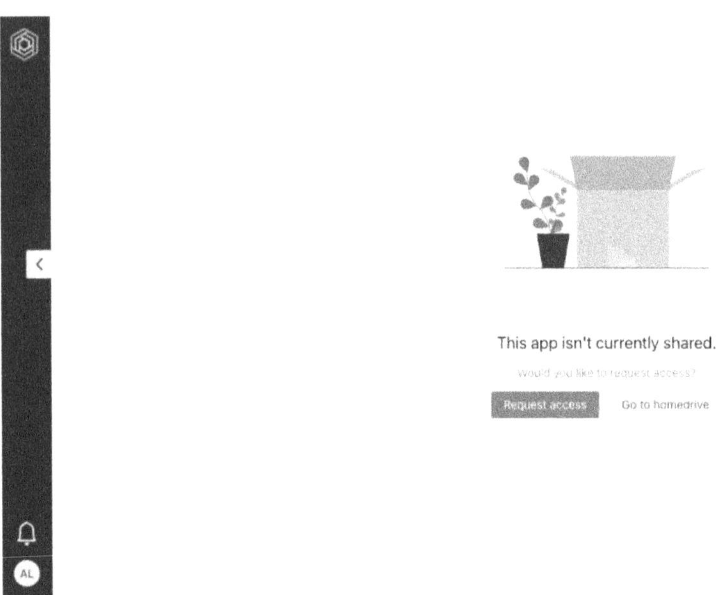

Figure 2.21 – The Request access button when trying to load an app that hasn't been shared

The requests will show up in your inbox under the **Requests** tab, as shown in *Figure 2.22*

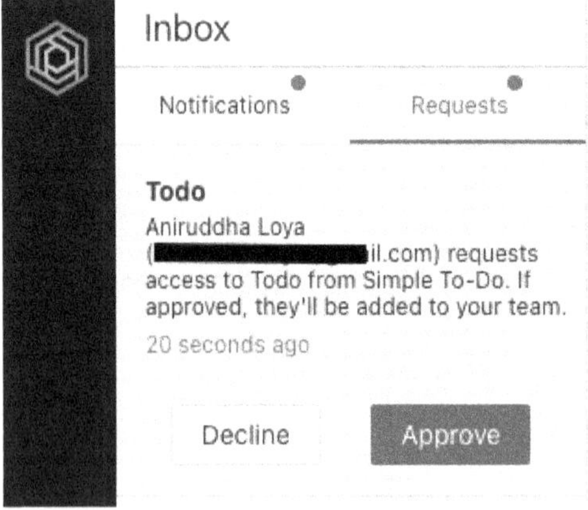

Figure 2.22 – An access request approval notification

> **Note**
>
> There is one caveat to this model – for apps, you can control whether people can request access or not, as shown in *Figure 2.23*

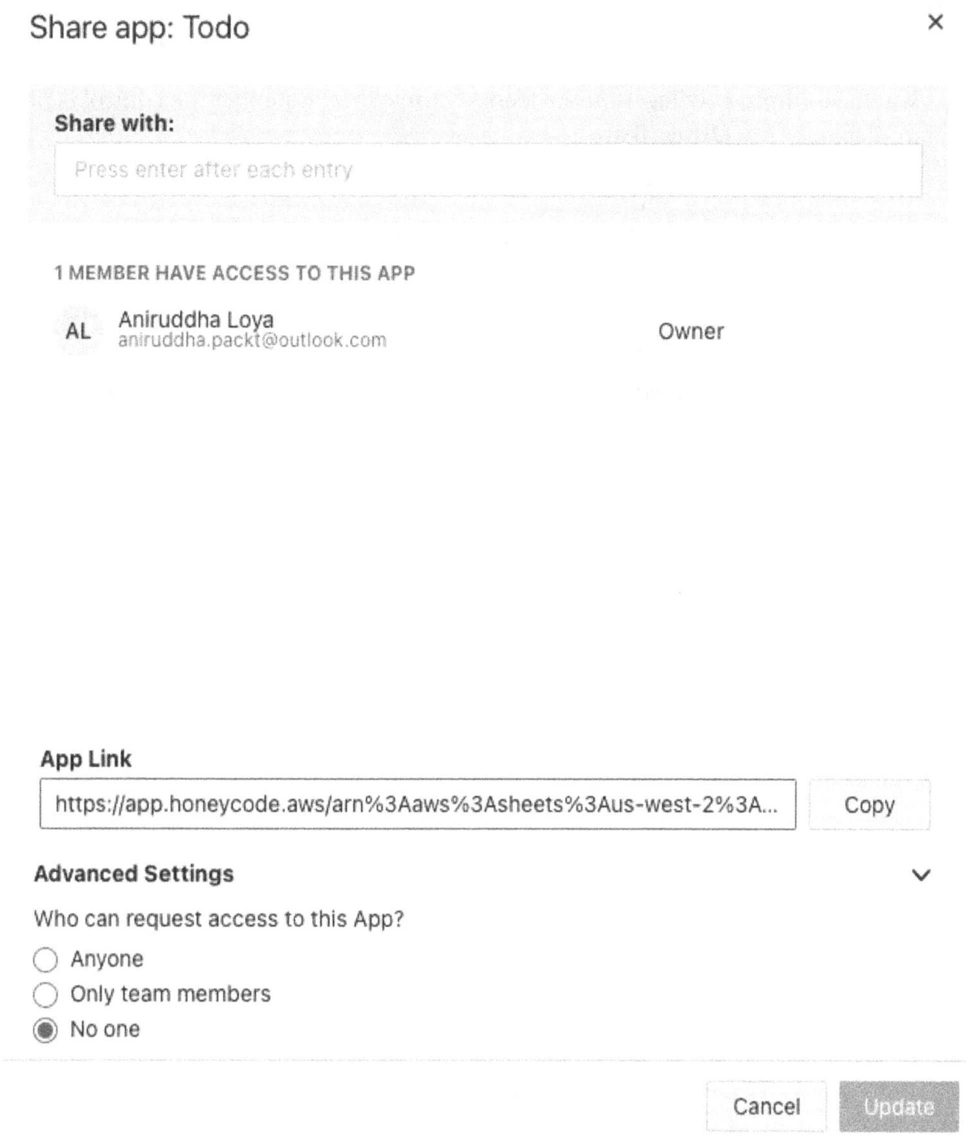

Share app: Todo

Share with:

Press enter after each entry

1 MEMBER HAVE ACCESS TO THIS APP

AL Aniruddha Loya aniruddha.packt@outlook.com Owner

App Link

https://app.honeycode.aws/arn%3Aaws%3Asheets%3Aus-west-2%3A... Copy

Advanced Settings

Who can request access to this App?

- Anyone
- Only team members
- No one

Cancel Update

Figure 2.23 – The settings to restrict requesting access to an app

Managing a team

In Honeycode, controls for team management are spread across the screen, which might take some time to get used to. *Figure 2.24* highlights all the places where you can find different controls for configuring and managing your team or teams in which you have admin privilege:

A. This provides options to rename and delete teams. This will only be available if you have admin privileges on the team. For instance, note that the control is not available for **My Other Team**.

B. This setting provides invite and integration controls for team members.

C. This control is context-aware, and its options include **Cancel Invite**, **Leave**, and **Remove**.

D. This is a simple drop-down menu with two options to define the role of each team member – **Admin** or **Member**.

E. These two controls together enable the upgrade of your Honeycode account from a free tier to a paid tier. For more details on pricing tiers, please refer to the next section:

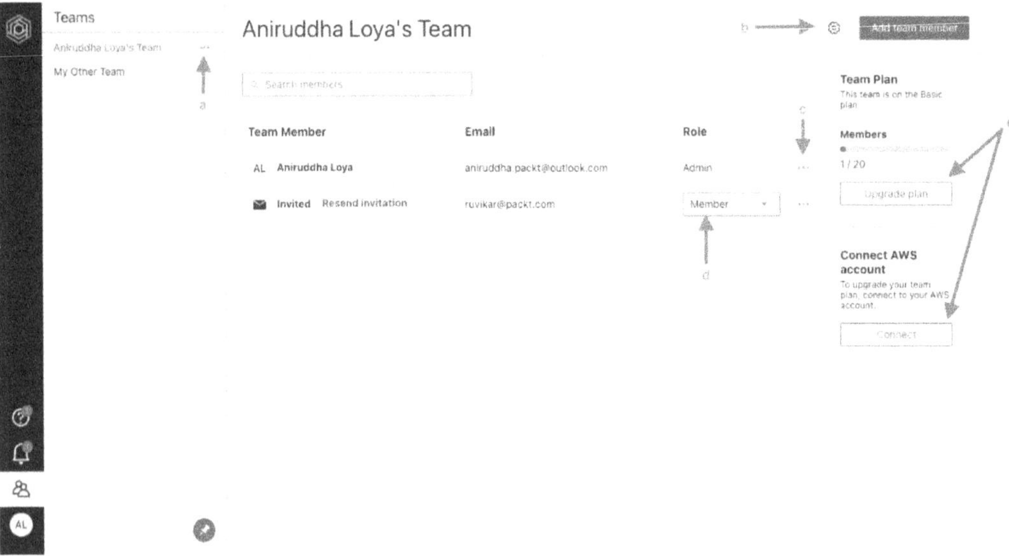

Figure 2.24 – Various controls for managing teams

Honeycode's pricing tiers

In this section, we'll learn about different pricing tiers that are available for you to use so that you can understand when to choose what.

Currently, Honeycode offers three pricing tiers – Basic, Plus, and Pro. The key differentiation between the free and paid tiers is the number of rows you can have in your workbooks, the ability to add more team members, and the capability of single sign-on. All three of these options are primarily for enterprise usage. What is important to note is that the number of team members included for all the tiers is 20, so with each additional member that you add at one of the paid tiers, your monthly cost increases significantly. Each additional member in the Plus tier is 50% of your monthly subscription, and in the Pro tier, the cost is 66% of the monthly plan. This might feel exorbitant as it does not add to the other important limit, that is, the number of rows in the workbook.

At the moment, Honeycode does not offer any data export functionality, as that would limit the size and type of application you can build with this tool, especially for mid-to-large enterprise usage and for applications that require you to maintain the data history, which implies that you cannot periodically clean the data. However, if you are building a hobby project or something to share with friends or your community, you should look at the Basic tier to do your job. This doesn't require an AWS account, so removes the additional necessity for sign-ups and account management, which is a major plus.

Another limitation of Honeycode is that the apps cannot be shared publicly, that is, the apps that are created can only be shared among team members, which you or another team admin/member has to invite and share. In any case, with the pricing based on the number of users, you would definitely not want to be sharing these apps publicly. So, if you, as a citizen developer, are looking to launch your next big idea, then this is not the platform for you. You might be able to prototype here and test among friends, but you will have to find developers to build the final publicly available version of your idea. However, this is a good tool when you want to automate some of the processes in your team or smaller organizations.

Summary

In this chapter, we learned about the terminology used in Amazon Honeycode, the different components of the platform, and gained an understanding of how they relate and interact with each other. Additionally, we learned how teams are created, managed, and used. We also learned about the pricing model of Honeycode to help us make better decisions regarding when to use this platform and when not to.

Now that we are equipped with a common vocabulary of terms and components, we are ready to dive into Honeycode and start building.

In the next chapter, we'll rebuild the To-Do app, but this time, we won't use a template. Instead, we'll start from scratch to learn how to structure an app and translate an idea into an app, both visually as well as with the underlying data model.

3
Building Your First Honeycode Application

In the first chapter, we created an out-of-the-box To-Do application using an existing template. Honeycode offers many more templates to provide a starting point for a myriad of use cases; however, the true power of the platform lies in the flexibility and ease it provides to adapt existing templates or build something from the ground up.

There are two key components in building an app – **data modeling** and **interface creation**. And to start our learning journey for both these components, we'll reuse the example of a To-Do app and, this time, build the application from the ground up. Moreover, we'll continue to use the application we built here in *Chapter 4, Advanced Builder Tools in Honeycode,* and *Chapter 5, Powering the Honeycode Apps with Automations,* to add new capabilities to our app as we learn the advanced topics in the builder and automations.

In the previous chapter, we learned about the different components of a Honeycode workbook along with detailed descriptions of the interfaces of each of those components. In this chapter, we're going to build on that and learn how to put them into action. We'll use the app builder to build our application interface and the tables interface to create the necessary tables needed for modeling our To-Do data. We'll also learn how to use the App Wizard to create an app. However, we'll focus on the advanced builder concepts as well as the creation of automations in the later chapters.

In this chapter, we're going to cover the following main topics:

- Defining the app requirements
- Creating the app data model
- Building the app interface
- Binding the app interface with the data model

Technical requirements

In order to follow this chapter, you'll need to have access to Amazon Honeycode, which requires a laptop with a web browser, preferably Google Chrome, and optionally a mobile device running either a Honeycode-supported version of Android (it currently requires Android 8.0 and up) or iOS (it currently requires iOS 11 or later).

Furthermore, we'll use the Honeycode terminology and refer to the components that we covered in *Chapter 2, Introduction to Honeycode*, and therefore, we recommend you complete that first.

Defining the app requirements

Before building any app, it is useful to list the requirements or the use cases that the app is expected to fulfill. This helps in conceptualizing the application interface, defining the data model, and visualizing the interactions between various onscreen elements as well as the data displayed.

Before reading any further, take a few minutes to list the requirements for a To-Do app and then come back to see how your list compares to the one we have here:

- I must be able to add a new task.
- I must be able to change an existing task.
- I must be able to set a due date on the task.
- I must be able to set a reminder preference on the task from the following values – due date, 1 day before, and 1 week before.
- I must be able to mark a task as complete once it's done.
- I must be able to reopen a completed task.
- I would like to see incomplete tasks ordered by the due date.
- I would like to see completed tasks ordered by the completion date.

Here are some additional requirements that I'd like to see in my To-Do list app, and we'll cover the implementation of these in the following chapters:

1. I'd like my team to use my app, but every user must be able to see only their own tasks.
2. I'd like to see my overdue tasks highlighted in red.
3. I must receive a task-due reminder based on my preference set for each task.

Neither of the two preceding lists is exhaustive by any means, and I'm confident that your lists will have a few more use cases that you would like to see in this app. But for the purpose of learning the data and interface modeling in Honeycode, this list will suffice. And upon completion of these, you should be able to make small changes and add the missing features from your list.

Based on the requirements from the first list, we can imagine our app to have two lists:

- One for incomplete tasks
- Another for completed tasks

It is a personal choice whether we want to display them on a single screen or as two separate screens. Furthermore, we require some type of form to create new tasks, along with some means to edit existing tasks. Again, the edit option can be enabled on the same screen as the list or built with a new screen.

In terms of the data model, we can identify the need for a table for storing our tasks, and for now, the rest of the related information such as due date, status, and completion date can reside in the same table. Furthermore, looking at the requirements, we note that for each task, we have an option to set a reminder preference from one of three values:

- On the due date
- A day before the due date
- A week before the due date

These values, therefore, need to be predefined and should be available for selection for all the existing tasks as well as the new ones we add. Given that the only data store we have in Honeycode is tables, the implication is that we'll need to create another table that contains these values.

Now that we have some idea of our interface as well as our data model based on the requirements, let's move on to the next sections and start building these.

Creating the app data model

In *Chapter 2, Introduction to Honeycode* we learned how tables in Honeycode serve as the data store for the applications. Therefore, creating a data model for the app entails creating tables and defining the relationship between them through the use of picklists, filters, and rowlinks. In the previous section, we identified that based on our requirements, we would need two separate tables:

- One that contains all the tasks and their related information (also referred to as the **metadata**)
- One for defining the three available reminder preferences

So, let's start to create these tables and define our data model, but before we do that, we also need to create a new workbook.

Creating a workbook

On the Dashboard, locate the **Create Workbook** button on the top-right corner and click it:

1. In the popup that loads, click on **Start from Scratch**, as shown in *Figure 3.1*:

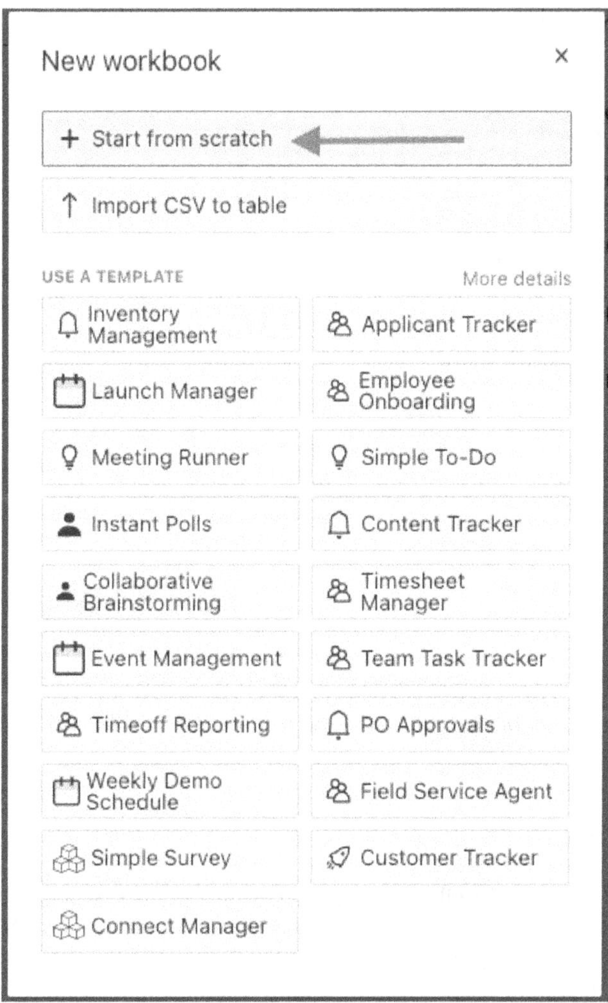

Figure 3.1 – Creating a new empty workbook

2. Next comes a popup to name the workbook and choose a team. Let's name this `My To-Do`, leave the default value for the team, and click on the **Create** button:

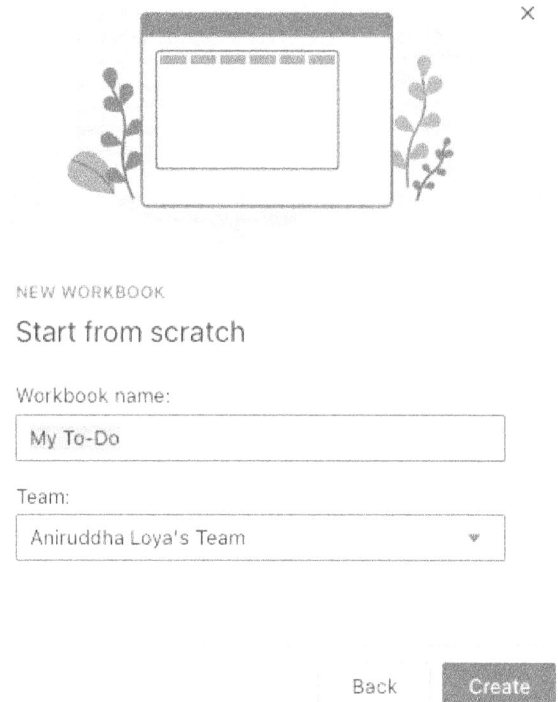

Figure 3.2 – Provide a workbook name and team details when creating the workbook

With this, we now have our new workbook that loads with a table view, displaying a default **Table 1** to get started.

Creating tables

In the earlier section, we noted that we'll require two tables for creating our data model for this app. The **Tasks** table will contain the information about our To-Do tasks and their associated metadata, such as the due date and reminder preference. The second one will be the **Reminder preferences** table, containing the three predetermined reminder options we want. But before we go about setting up these tables, let's learn some table operations that we'll be needing for this.

How to rename a table

To rename a table, you can do any of these six operations:

1. Right-click on the table name in the table list and select **Rename**.
2. Double-click on the table name in the table list, which makes the name editable.
3. Click on the table actions control and select **Rename**. Refer to *Figure 2.3* in *Chapter 2*.
4. Right-click on the table name in the navigation at the top of the **Tables** interface and select **Rename**.
5. Double-click on the table name in the navigation at the top of the **Tables** interface, which makes the name editable.
6. Click on the control with three dots next to the table name in the navigation at the top of the **Tables** interface and select **Rename**:

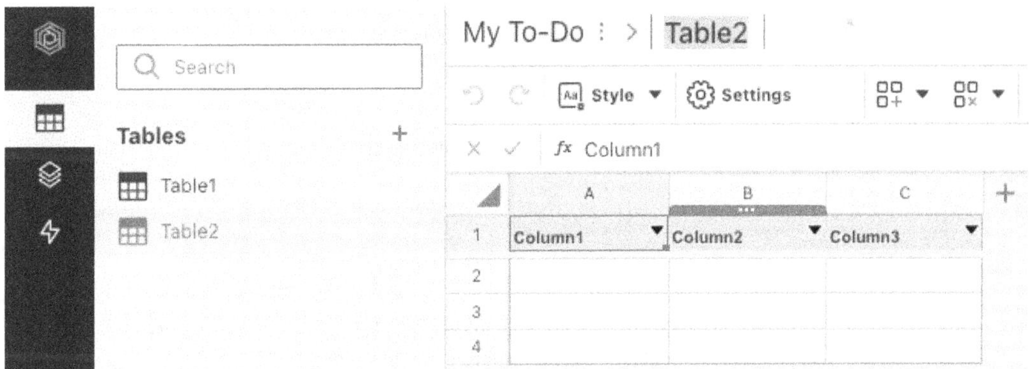

Figure 3.3 – Renaming Table 2 using the navigation information at top of the Tables interface

How to rename a table column

A table column can be renamed using one of the three ways listed here:

- Click on the cell containing the column name and start typing the new name for that column. This simply overwrites the content of that cell.
- Double-click on the cell. This will put the cell into edit mode. This can be used to edit or update the cell value.

- On the toolbar, click on the **Settings** control to open the right panel with the column properties. Click in the field with the column name and update it:

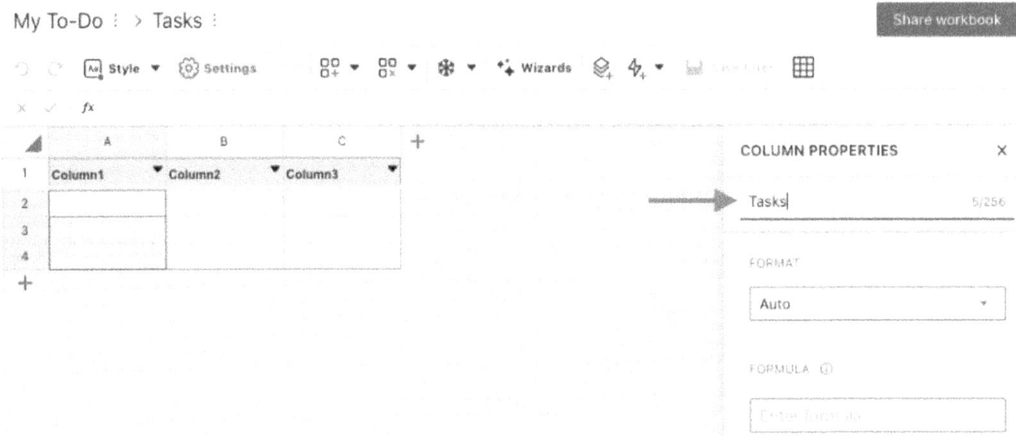

Figure 3.4 – Renaming the Column 1 table column with the COLUMN PROPERTIES panel on the right

How to add a new table column

We can add a new table column using one of the following three ways:

1. Use the **Add Table Column** control. Refer to *Figure 2.3* in *Chapter 2 , Introduction to Honeycode.*
2. Right-click on the table column, or any cell in that column, next to which you want to add a new column, and select the **Insert Column** action either to the left or the right, as per your requirement:

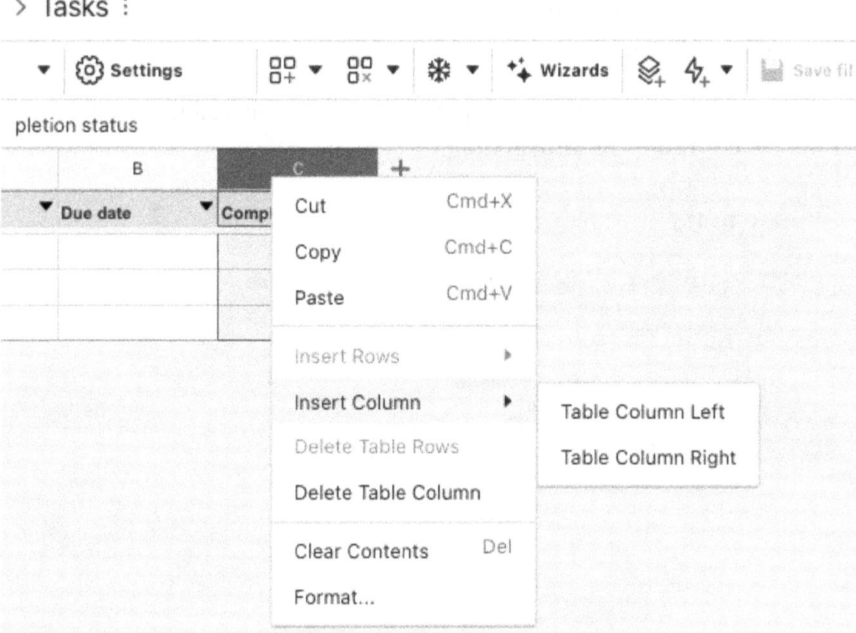

Figure 3.5 – Adding a new table column by right-clicking the menu on an existing table column

3. Select the cell or column next to which you want to add a new column and use the shortcuts in the toolbar to add the column:

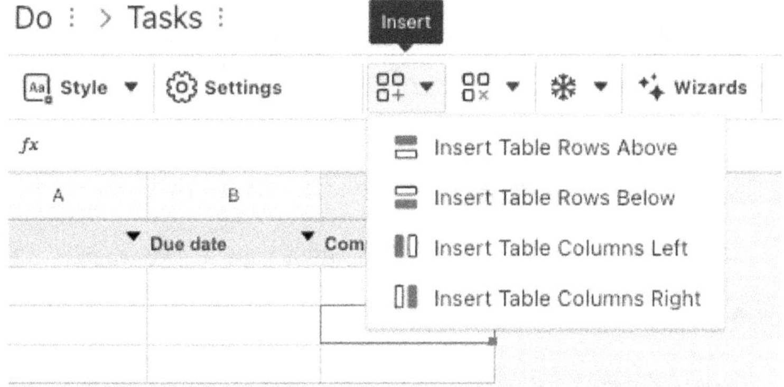

Figure 3.6 – Adding a new table column using the toolbar shortcut

How to delete a table column

We can add new table column using one of the following two ways:

1. Right-click on the table column you want to delete, or any cell in that column, and select the **Delete Table Column** action:

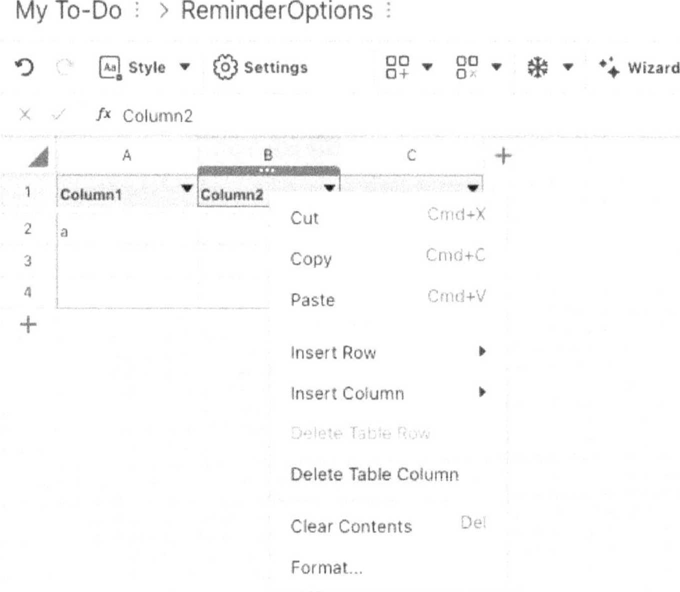

Figure 3.7 – Deleting a table column by right-clicking the menu on the table column to be deleted

2. Select the cell or column which you want to delete and use the shortcuts in the toolbar to delete the column:

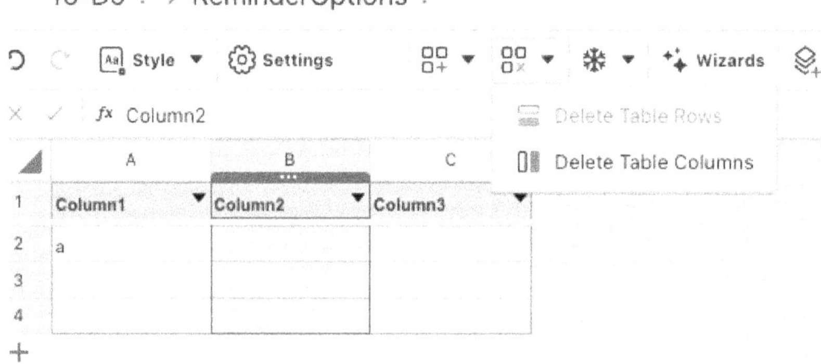

Figure 3.8 – Deleting a table column using the toolbar shortcut

Now that we have learned some of the commonly performed operations on table and table columns, we are ready to set up our two tables by following the steps in the next two subsections:

Creating a tasks table

Let's understand how to create a table:

1. Click on the **Create Table** (+) control and select **Add a blank table.**

   This adds a new table, **Table 2**, to the list of tables, with default columns similar to the existing **Table 1** table.

2. Rename the **Table 2** table `Tasks`. See the *How to rename a Table* section for details.

   With this, we now have our tasks table, but it's just a shell. We need to identify how a task and its details, namely the due date, completion status, and reminder preference, are organized within this table. We achieve this by assigning each table column to one of these data values and indicating that by renaming the columns.

3. Rename **Column1** to `Task`, **Column2** to `Due date`, and **Column3** to `Completion Status`. See the *How to rename a table column* section for details.

4. We are one column short of our required data model. Let's add a new column, (see the *How to add a new table column* section) and rename it `Reminder preference`:

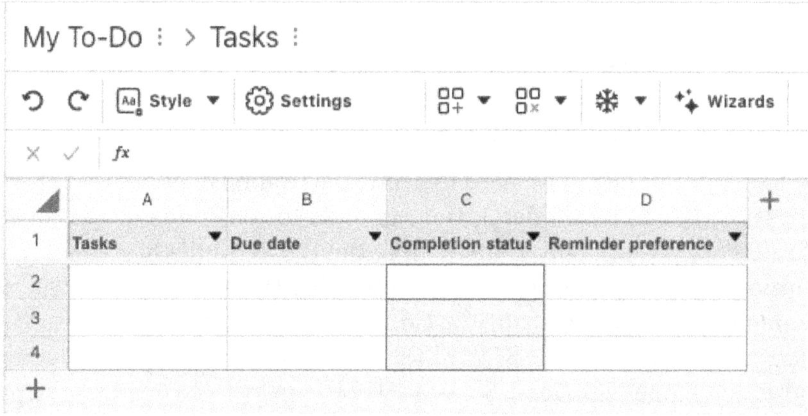

Figure 3.9 – A tasks table

Creating a reminder options table

1. For this table, let's simply reuse the existing **Table1** table and rename it `ReminderOptions`.
2. At the moment, we only need a single column containing values for the three preferences we will support, so let's rename **Column1** `Remind On`.
3. Delete **Column2** and **Column3**, as they are not needed. See the *How to delete a table column* section for details.
4. Add the following values to the three rows in the **Remind On** column – `Due Date`, `1 Day Before`, and `1 Week Before`:

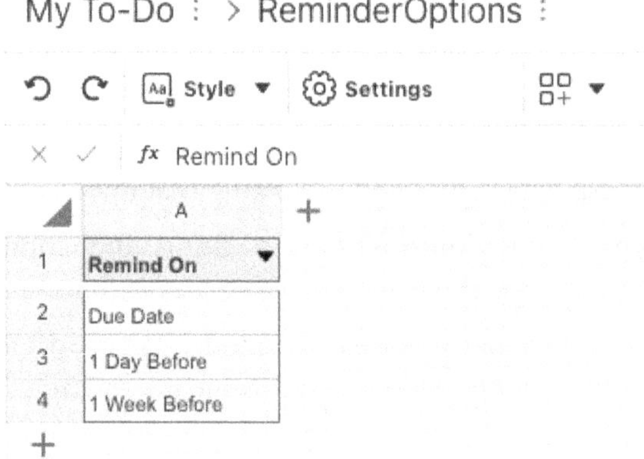

Figure 3.10 – The ReminderOptions table

Data formats and relations

Now that we have our two tables, we need to define the relationships between them by linking the tasks table to **ReminderOptions** in order to constrain the values in the **Reminder preference** column only to the values from the **Remind On** column from the **ReminderOptions** table. Furthermore, we are aware that **Due date** is a date-type value, and **Completion status** values can either be **Completed** or **Not Completed**. Such pairs are often referred to as Boolean type, with other more common examples including **Yes/No**, **True/False**, and **Done/Not Done**. So, let's set these up by following these steps:

1. Select the **Due date** column and open the **Column Properties** panel by clicking on the **Settings** shortcut in the toolbar.
2. Using the **Column Properties** panel, change **FORMAT** from **Auto** to **Date** by selecting it from the drop-down list:

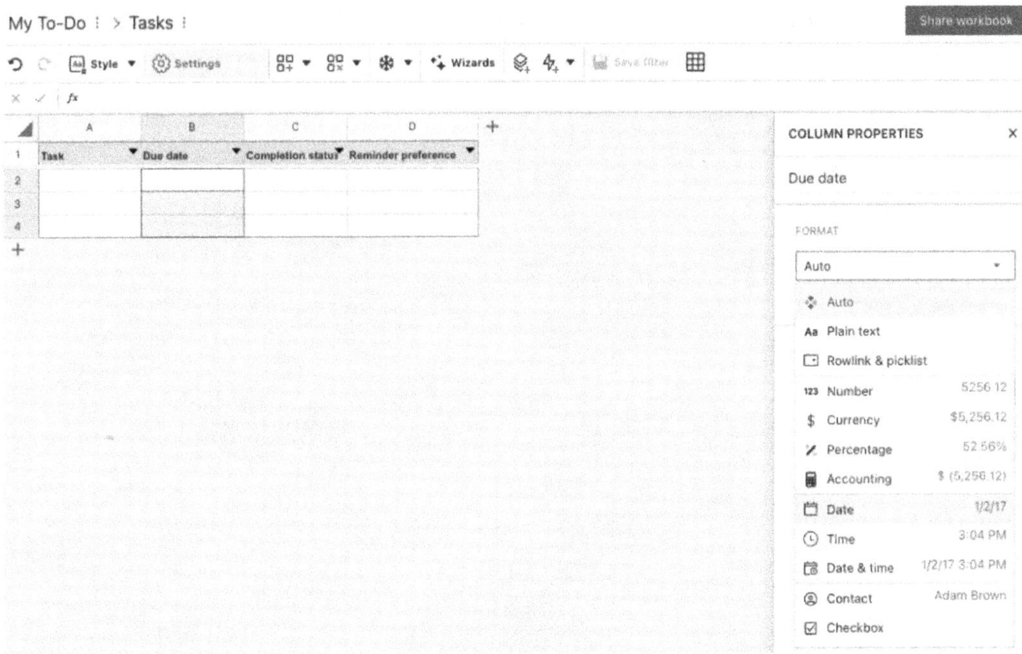

Figure 3.11 – Setting the Date format on the Due date column

You can choose to update the display preference of the date or leave it to its default style of **MM/DD/YY**.

3. Similarly, select the **Completion status** column and set its format to **Checkbox** to create a Boolean-type field.

   This updates the cells in this column to start displaying an unchecked box that, in our case, will represent an incomplete task:

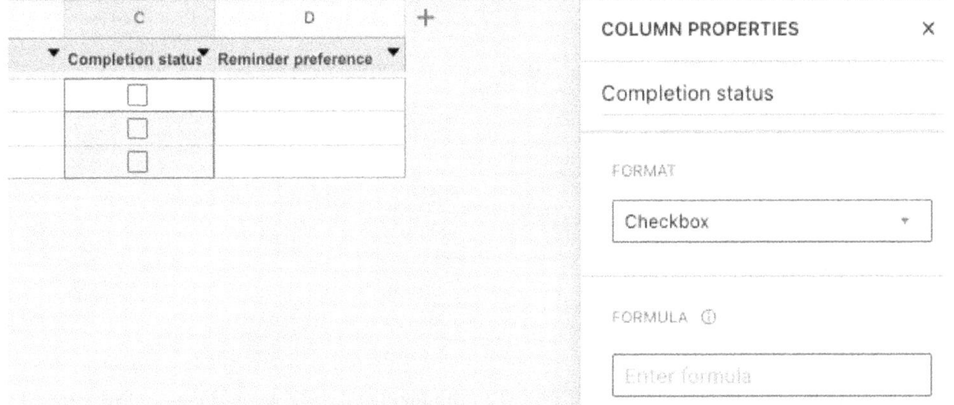

Figure 3.12 – The Checkbox format set on the Completion status column

4. Lastly, set the **Rowlink & picklist** format on the **Reminder preference** column, and set the source to the **ReminderOptions** table on the dropdown:

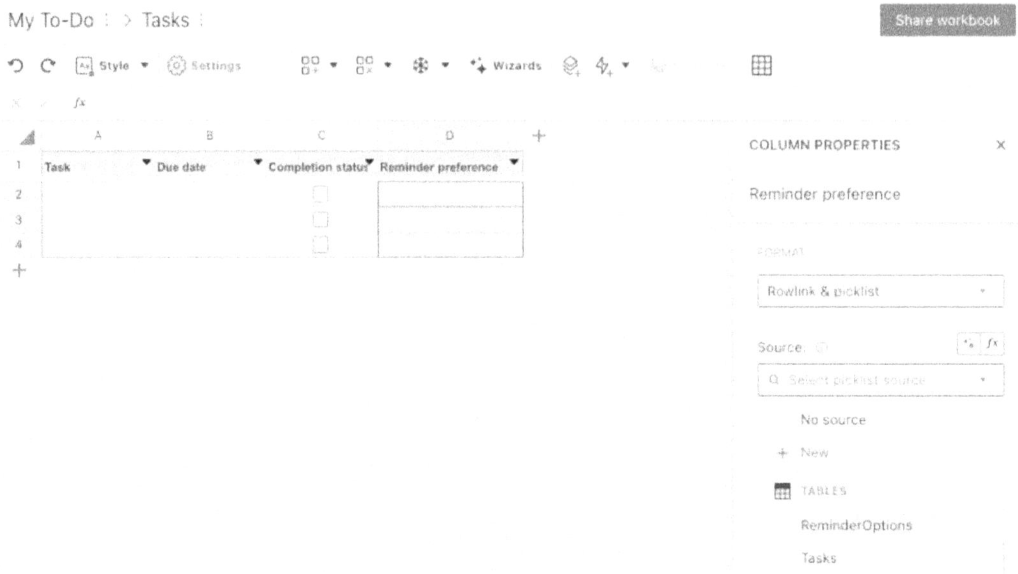

Figure 3.13 - Setting the Rowlink & picklist format on the Reminder preference column

This updates the cells in this column to display a notch at the right, indicating the presence of a drop-down list, which in Honeycode is referred to as a **picklist**:

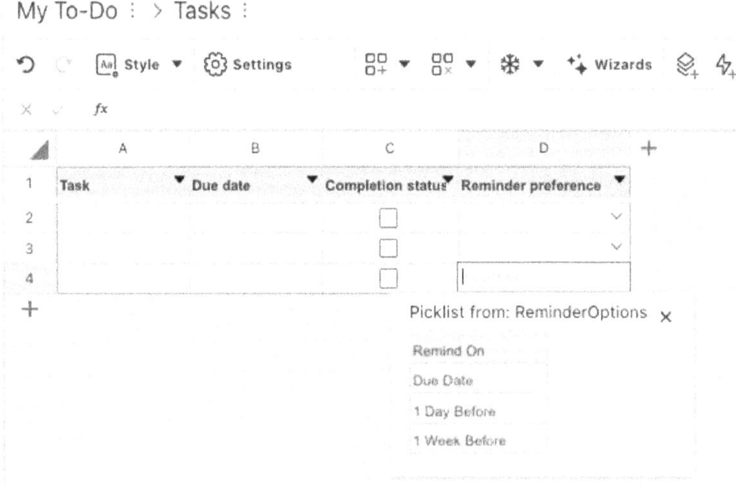

Figure 3.14 – The Reminder preference column with the Picklist format set and displaying a picklist from one of the cells in the column

And with this, we have now completed setting up our data model for the app and are ready to start building the app itself.

Building the app interface

Earlier in this chapter, in the *App requirements* section, we identified the data model requirements and defined a visualization of the app interface at a very high level. We used that as a guide for building our data model, and we'll continue to refer back to it to build our app interface as well.

In the requirements, we identified the need for a screen with a list of all incomplete tasks, and we need to make a choice of whether to display completed tasks on the same screen or create a separate one. Moreover, we require a form-like input screen to allow the creation of new tasks and some means to edit existing tasks.

In the following subsections, we'll build our application based on these requirements. But before we go about building the app, let's learn some app operations that we'll be needing for this.

How to rename an app screen

Recall that in *Chapter 2, Introduction to Honeycode,* we made a distinction between two different identifiers used for app screens, namely the **local name** and the **screen name**. Therefore, we have different ways to rename them, which are described in this section.

An app screen's local name can be changed using one of the four ways listed here:

1. Right-click on the name in the screen list and select **Edit**. This makes the screen's **Local Name** box editable.
2. Right-click on the name in the screen list and select **Properties** to open the right panel with **SCREEN PROPERTIES**. Click in the field with the screen's **Local Name** box and update it.
3. Double-click on the screen's **Local Name** box to activate edit mode on the field.

4. On the toolbar, click on the **Properties** shortcut to open the right panel with **SCREEN PROPERTIES**. Click in the field with the screen's **Local Name** box and update it:

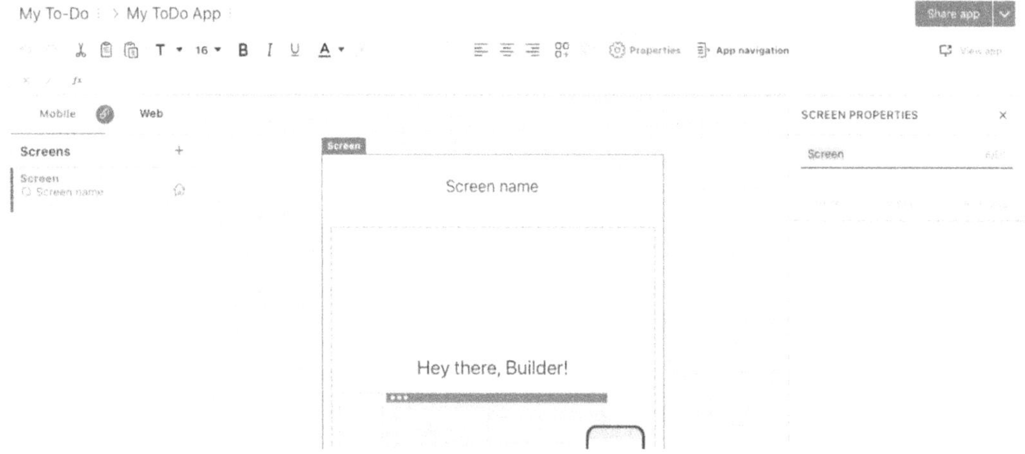

Figure 3.15 – Changing a screen's local name, Screen, from the SCREEN PROPERTIES panel on the right

An app screen's **Screen name** can be changed using one of the three ways listed here:

1. Double-clicking on the **Screen name** content box in the screen editor activates edit mode in the field to update the screen name. See the content box labeled **(a)** in *Figure 3.16*.
2. Right-click on the **Screen name** content box and select **Edit**.
3. On the toolbar, click on the **App Navigation** shortcut to open the right panel with **APP NAVIGATION PROPERTIES**. Click in the field with **Screen name** and update it:

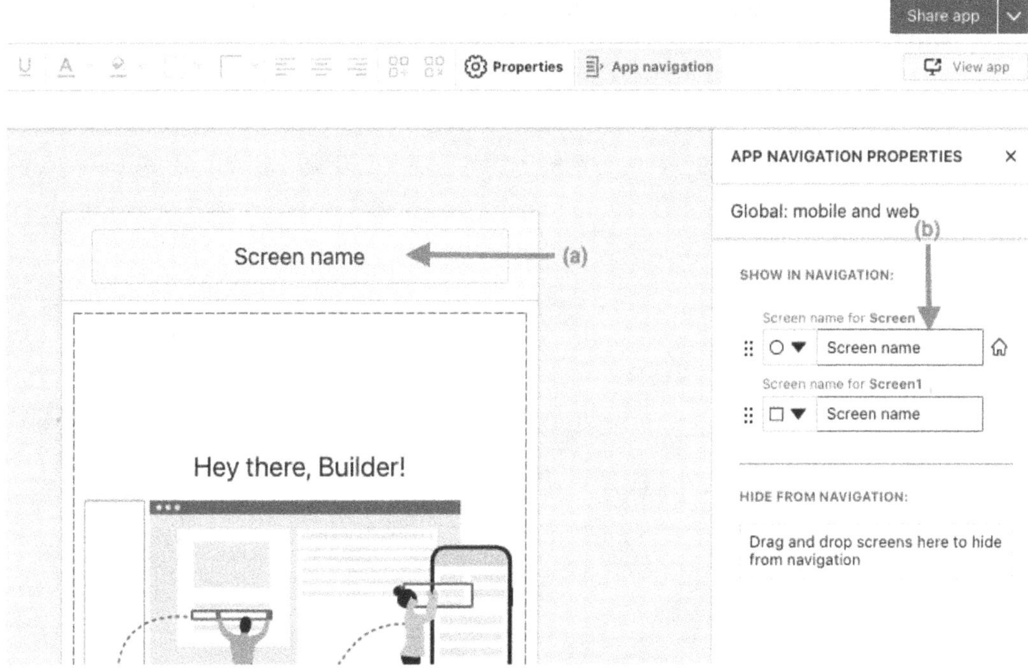

Figure 3.16 – Changing Screen name (a) using the content box and (b) from the APP NAVIGATION PROPERTIES panel on the right

How to add a new app screen

We can add a new screen to our app using one of the following two ways:

1. Using the **Create new screen** control – refer to *Figure 2.9* in *Chapter 2*
2. By selecting a **Screen** object from the list of objects that can be brought up through any of these actions:

   I. Click the **Add Objects** button at the bottom of the list of screens.

   II. Click the **Add object shortcut** from the toolbar.

III. Right-click anywhere in the screen editor or the screen list area and select **Add Objects**:

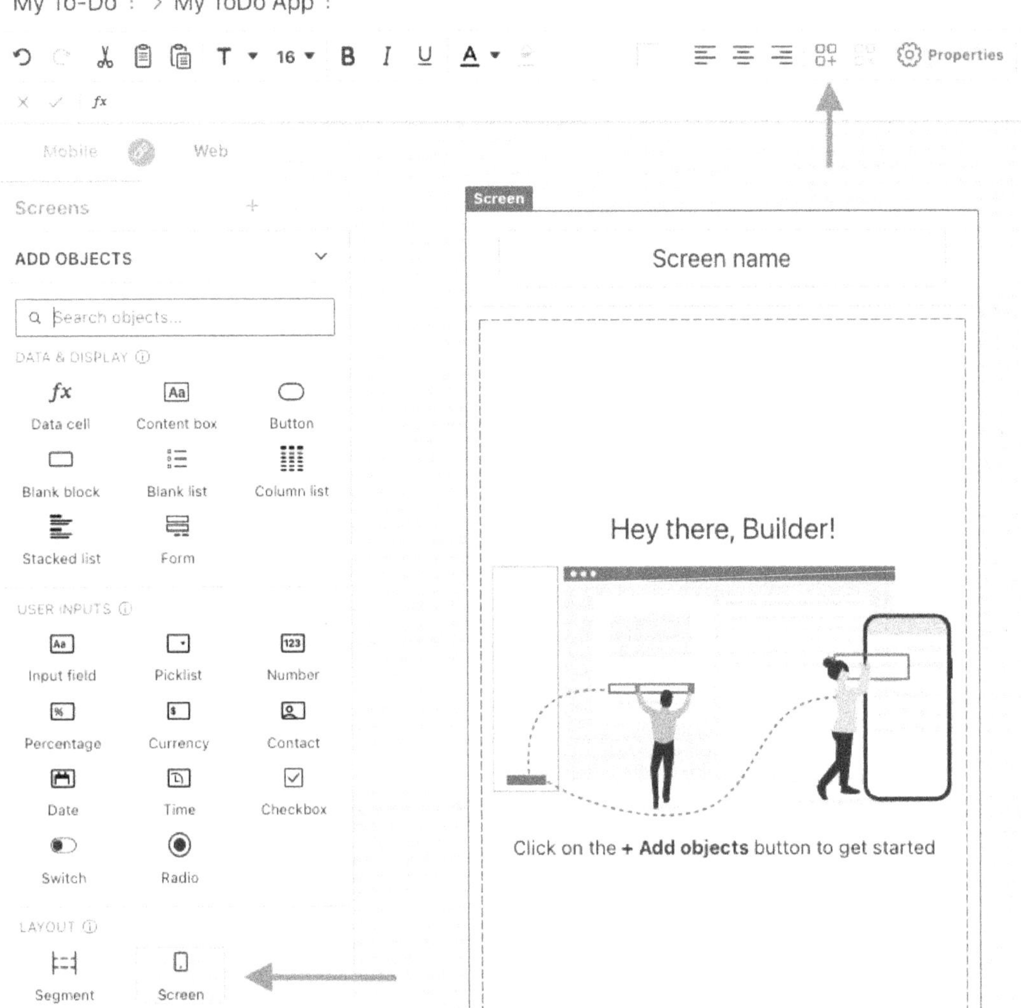

Figure 3.17 – Adding a new app screen

Now that we have learned the fundamental operations needed for creating the app, we are ready to get back to building our app.

Creating an app from scratch

To create an app, we need to navigate to the Builder interface using the left navigation bar and click on the **Build app** control (see *Chapter 2,Figure 2.9*). This brings up a popup that says **How do you want to build your app?**. In this section, we'll go with the **Build from scratch** option. We'll cover the **Use app wizard** option in a later section:

Figure 3.18 – The popup to choose the mode for creating the app

This creates an empty app, **App1**, with a default empty screen. Let's first rename our app and call it `My ToDo App`. Similar to tables, we can rename the app in the same manner as one of the six ways listed for renaming a table:

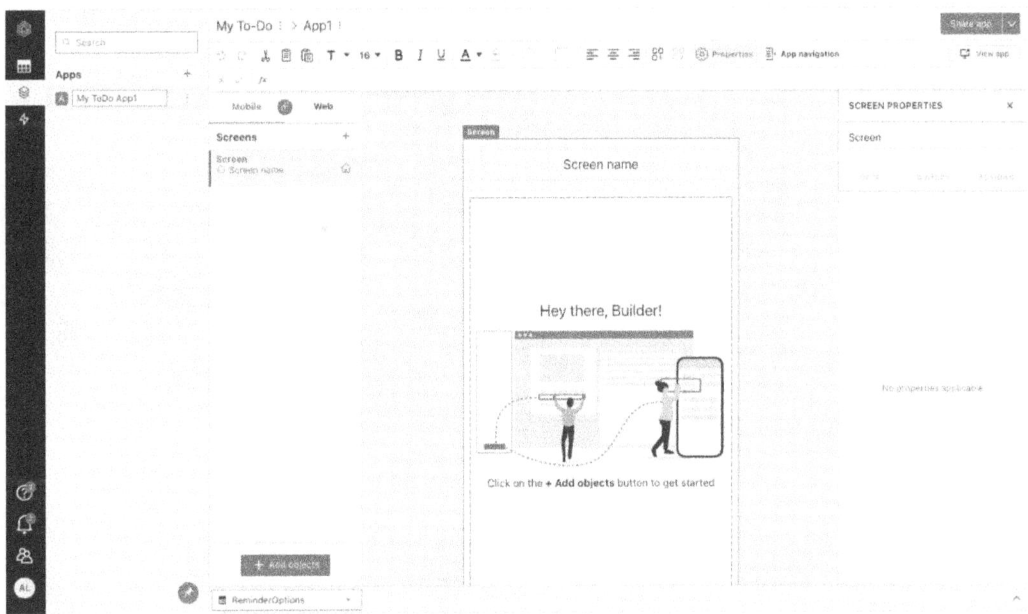

Figure 3.19 – An empty app created using the Build from scratch option and being renamed

Now, let's start filling up this empty shell app by creating the required screens and arranging the objects on those screens to present the data.

Creating the My Tasks screen

1. Rename the default screen to **My Tasks**.
2. From the **ADD OBJECTS** menu, add a **Blank list** object to this screen to display the incomplete tasks.

This adds a blank list to the screen, and the cursor is set on the **SOURCE** field in the right panel displaying **LIST PROPERTIES**, which also displays a warning, stating that it should return table rows:

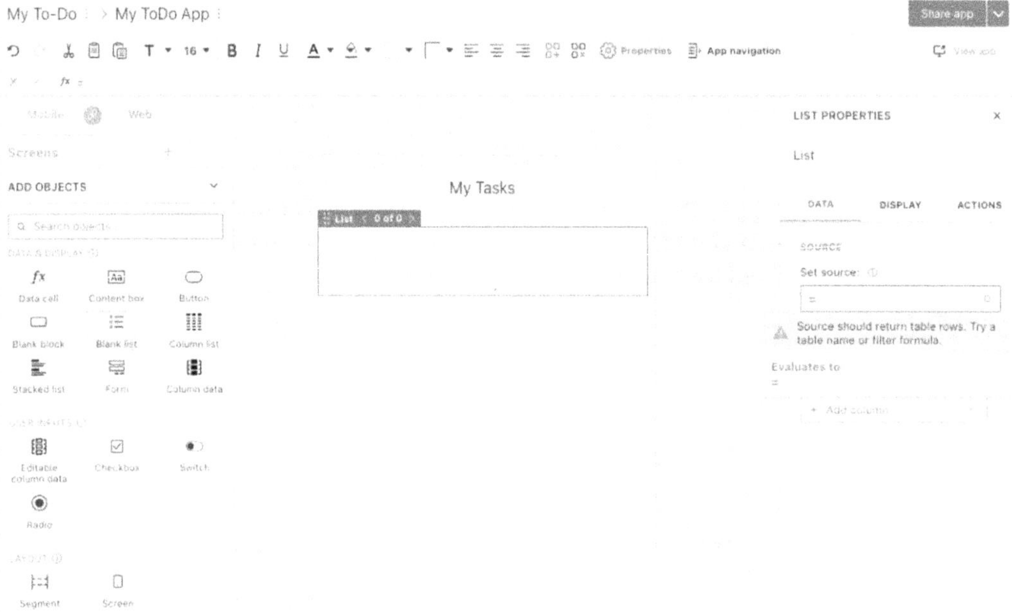

Figure 3.20 – A blank list added to the My Tasks screen

3. Next, let's fix the warning by providing a source for this list. Given that we only want to display tasks that are not completed, we need to write the following `Filter` function to only retrieve table rows that have incomplete tasks and set that as a source:

```
Filter(Tasks, "Tasks[Completion Status] = FALSE")
```

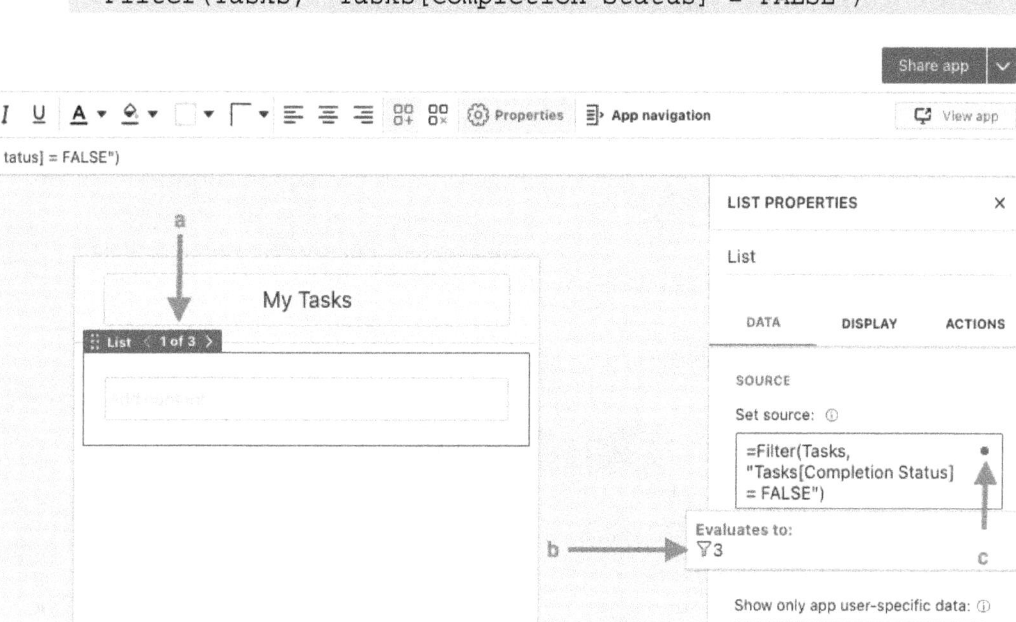

Figure 3.21 – A source added to the blank list to display the incomplete tasks

4. As we complete this step, note a few things that changed on the UI:

   I. The list control now shows how many rows are currently being returned and will be displayed in the app (see **a** in *Figure 3.21*).

   II. After pasting the formula, wait for the hollow blue icon to turn solid blue and then click inside the box again. This will bring up a tooltip-like box, also displaying the number of rows returned by the `filter` function (see **b** in *Figure 3.21*).

   III. The hollow blue circle becomes solid, indicating that the formula entered in the source field is syntactically correct (see **c** in *Figure 3.21*).

IV. Try deleting the closing bracket and clicking outside of the box; the blue circle will be replaced by a red warning symbol. Now, click on that symbol to display the error message (see *Figure 3.22*):

Figure 3.22 – An error indicator for the source formula along with information about the error itself

Now that we have our data being returned in the form of table rows, let's add controls on the list to display it:

1. I prefer to have a control to mark my task done at the far left-hand side, so the first thing I'll add is a **Checkbox** object that will reflect the data from the **Complete status** column from the table row. The mapping to the column is done through the **Set shared source** field from the **PROPERTIES** panel on the right (see *Figure 3.23*):

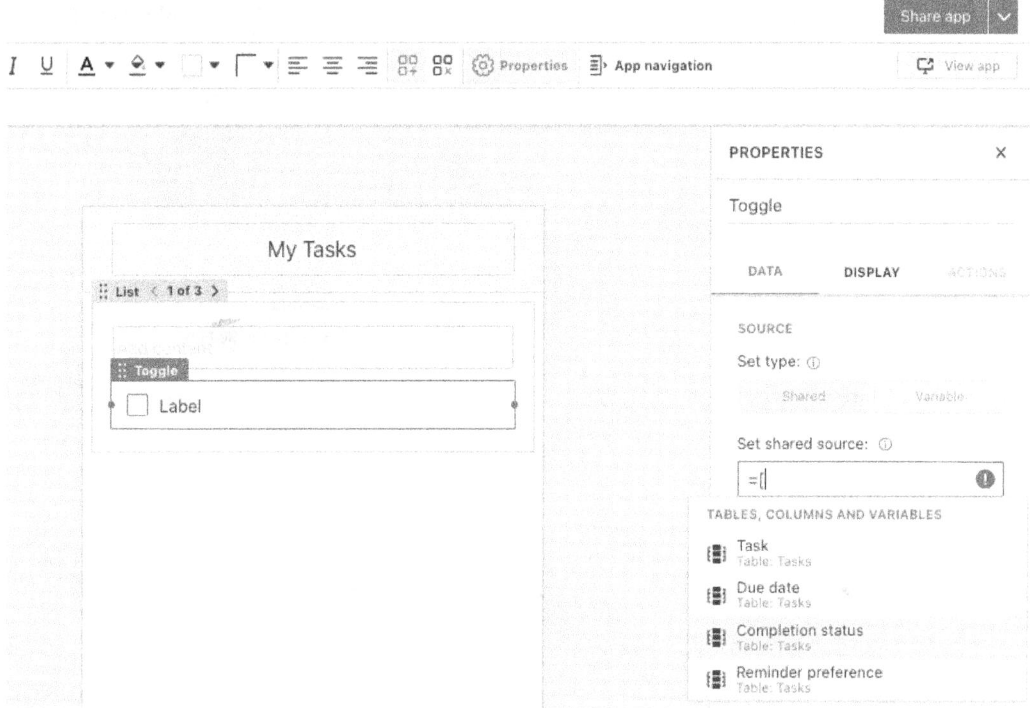

Figure 3.23 – Add a checkbox object and set its source

Furthermore, the checkbox control itself is self-sufficient, and we can, therefore, clear out the **Label** by deleting the value from the **Display data from** field under the **DISPLAY** tab in the **PROPERTIES** panel (refer to **a** in *Figure 3.24*).

And finally, we delete the default content box and then resize the **Toggle** control by clicking on the right edge of the control and dragging it to the left (indicated by **b** in *Figure 3.24*):

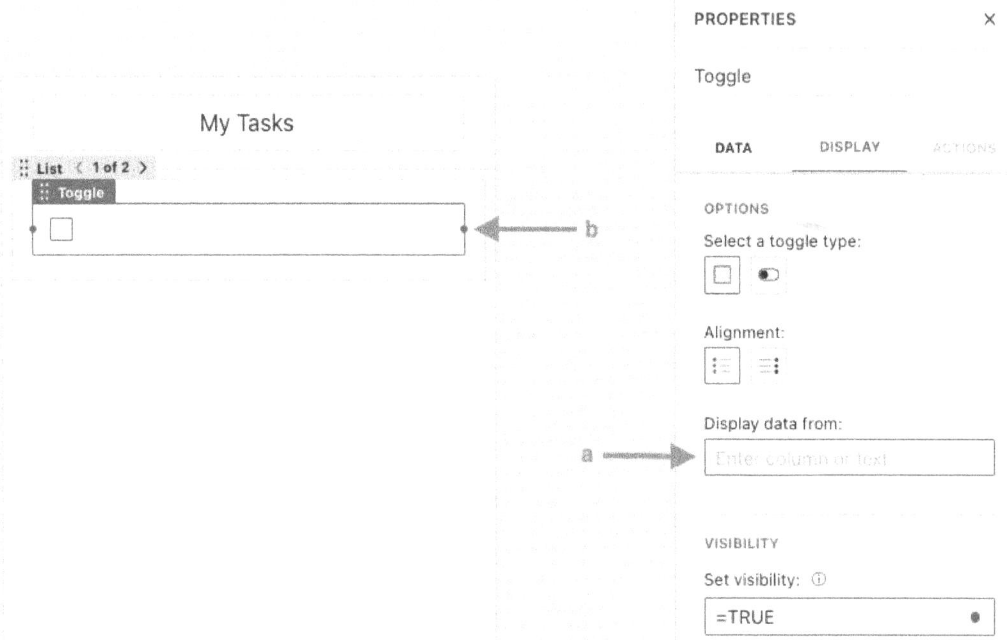

Figure 3.24 – Resize the Toggle control and clear the checkbox label

2. Next, we add a **Data cell** control from **App Objects**, set the **Task** column as its shared source, and resize it to leave space for adding the due dates of the tasks to the view:

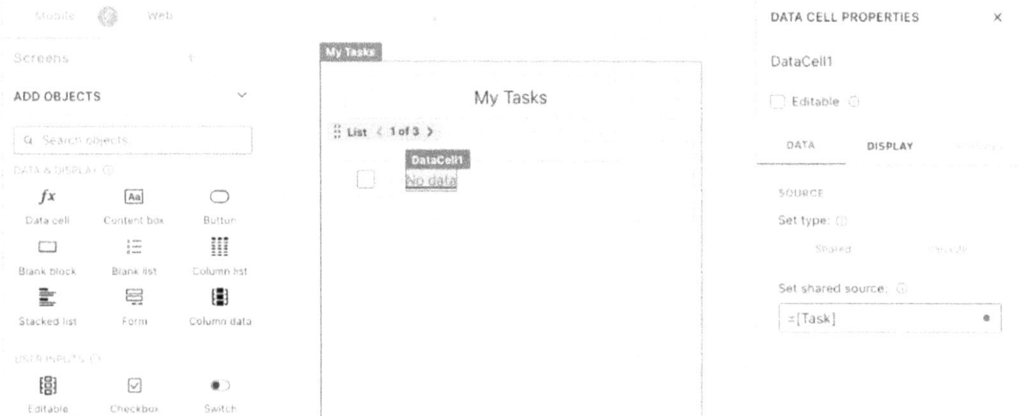

Figure 3.25 – Adding a Data cell control to display the task

3. Finally, we will add the **Due date** column to the screen using the **peeking sheet**. Keep the selection on the **List** container and select the **Due date** column from the **Tasks** table in the peeking sheet by clicking on the plus symbol against the column name:

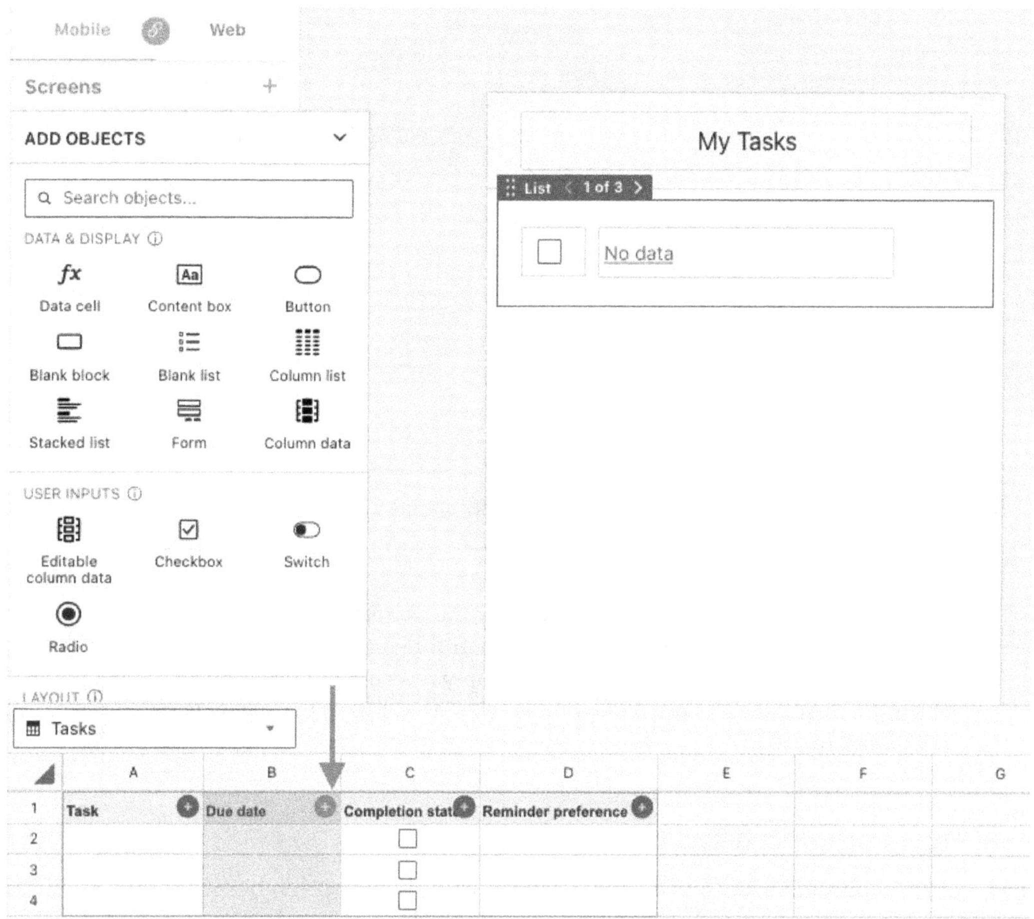

Figure 3.26 – Adding the Due date field by selecting from the peeking sheet

Note that, similar to **Task**, you can also add the **Due date** field by first adding a data cell and setting its source, but this is a simpler, easier, and faster way to achieve the same result:

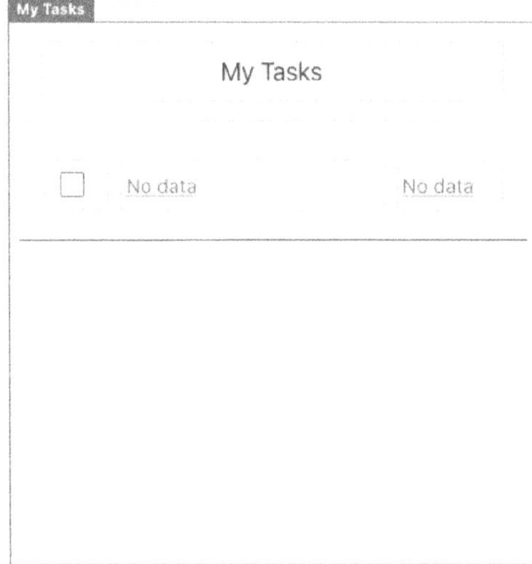

Figure 3.27 – The My Tasks screen after adding all the controls

> **Tip**
>
> Open the app in your mobile as you develop or the web app in another tab of the browser to see how the addition of each control changes the layout in real time. Furthermore, add dummy data to the empty rows in the **Tasks** table, which will then show up both in the Builder view as well as in the app on your mobile and/or web.

Creating a completed Tasks screen

This screen will be similar to our **My Tasks** screen with the exception being that it will display tasks that are completed. Moreover, since the tasks are completed, we do not require a **Due date** field to be displayed:

1. Copy and paste the **My Tasks** screen using *Ctrl* + *C* / *Ctrl* + *V* or *Cmd*+ *C* / *Cmd* + *V*.
2. Rename the screen from **My Tasks Copy** to `Completed Tasks`.
3. Delete the control for the due date.
4. Update the source on **List** to show the completed tasks by using the following `Filter` function:

```
Filter(Tasks, "Tasks[Completion Status] = FALSE")
```

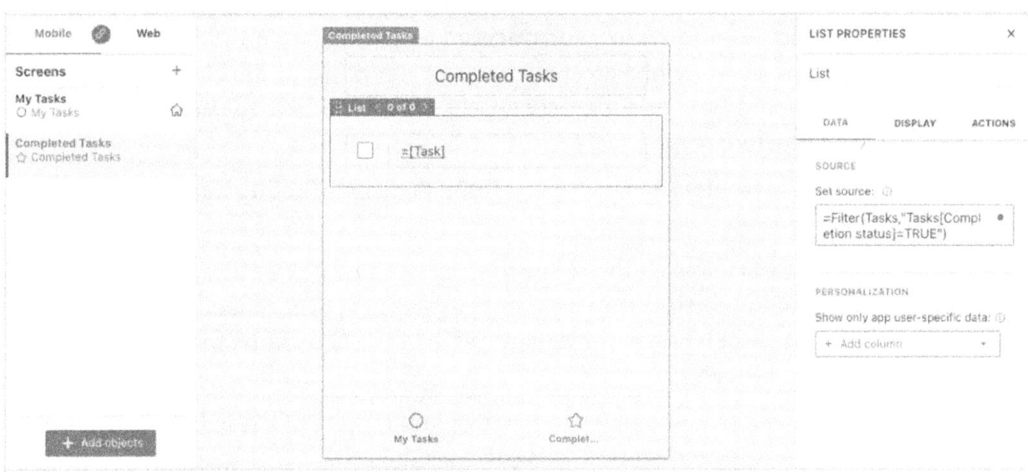

Figure 3.28 – The Completed Tasks screen

Creating an Add New Task screen

While defining our app requirements, we identified the need for a means to add new tasks to our To-Do list and also noted that it will be done using some type of form to collect the input. So, we'll now build our screen using the **Form** object by following the steps here:

1. With the **My Tasks** screen selected, click on the **Form** control from the **Add Objects** list. It will open a popup to configure the input form:

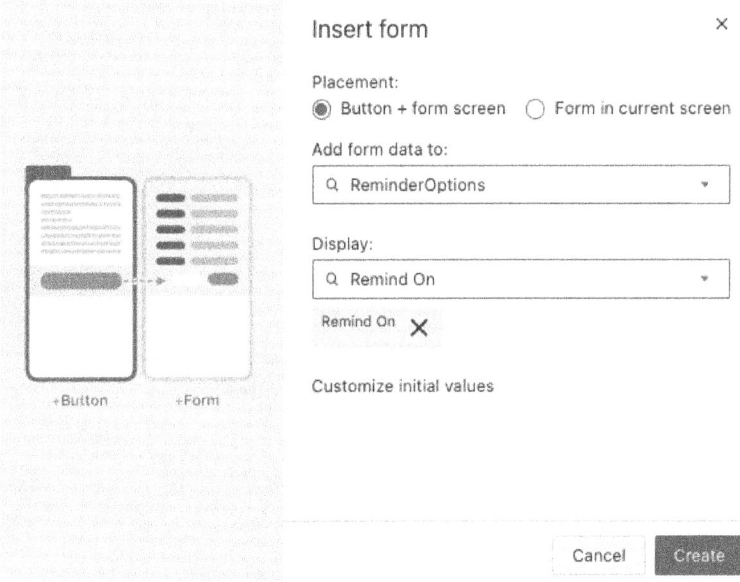

Figure 3.29 – Adding a Form object from the My Tasks screen

2. We leave the selection of **Button + form screen** as is, since we want to create this form as a new screen. If we had created an empty screen for adding a new task beforehand, we could have had that screen selected and then used the option to add the form to the current screen.
3. Change the **Add form data to** field to the **Tasks** table by selecting it from the dropdown.
4. **Display** field updates show all the columns from the **Tasks** table, from which we'll remove the **Completion status** column, as any new task created will be in incomplete status, which is also the default value of the **Checkbox** format that we have set on that table column.
5. Lastly, we don't need to customize the initial values for any of the fields in our form and will, therefore, not make any changes there either:

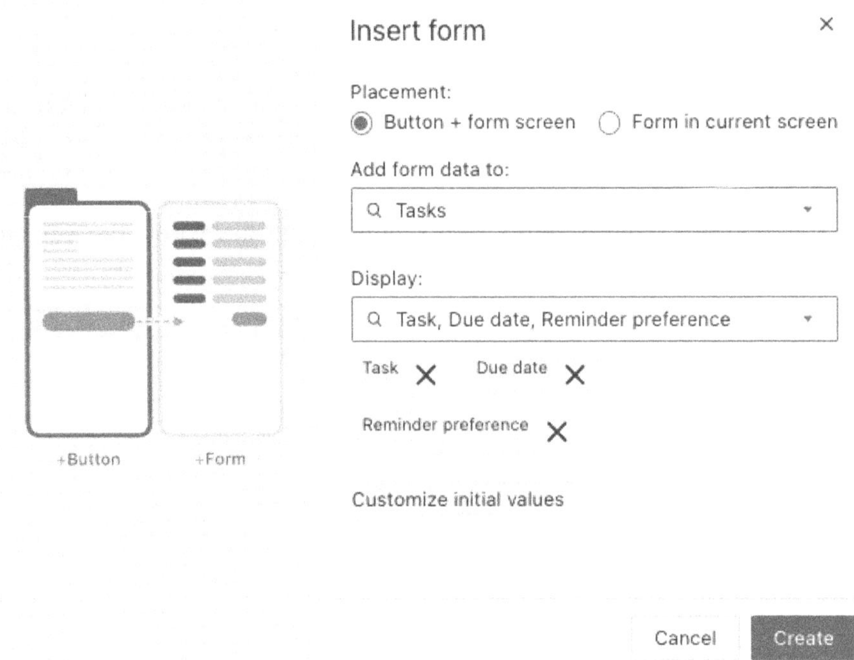

Figure 3.30 – The configured form

6. Click **Create**. This adds an **Add Task Row** button after the list control on the **My Tasks** screen and has also created a new screen, **Tasks form**, with the three fields configured in the popup.
7. Let's rename the **Tasks form** screen to `Add New Task` to make its purpose explicit:

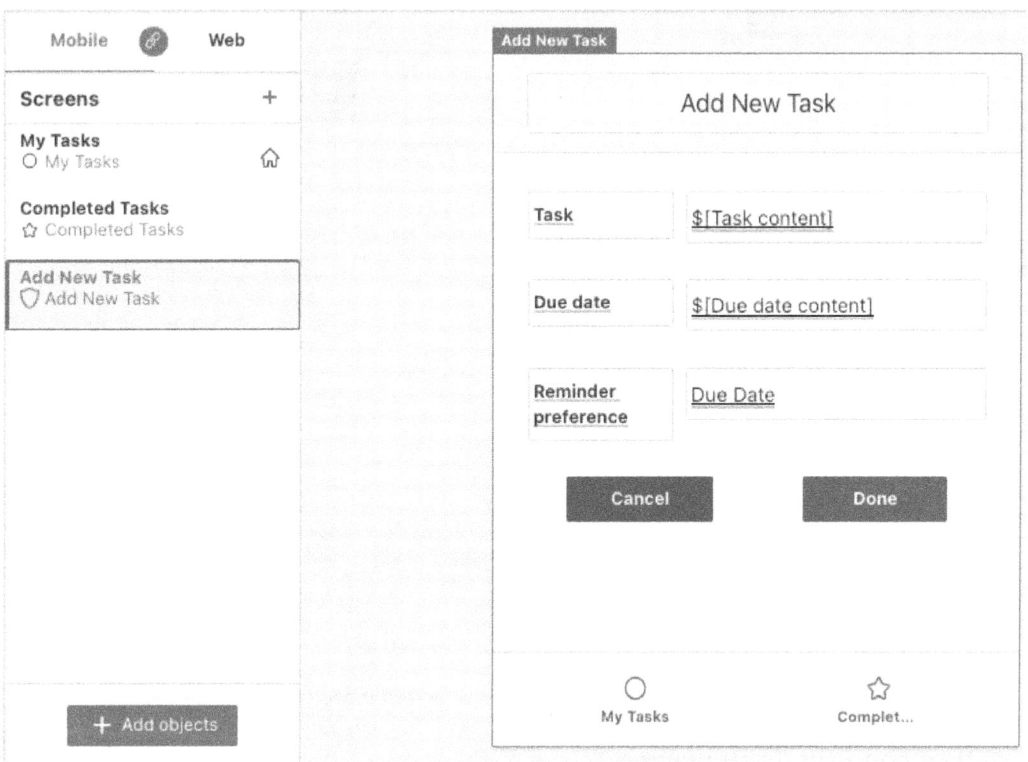

Figure 3.31 – The Add New Task screen

8. Finally, we'll rename the button added in the **My Tasks** screen `Add Task` and also move it, using the cut and paste operation, to the **Header** section so that the control is always available at the top for easy access, in the event of our To-Do list becoming too long and requiring a scroll:

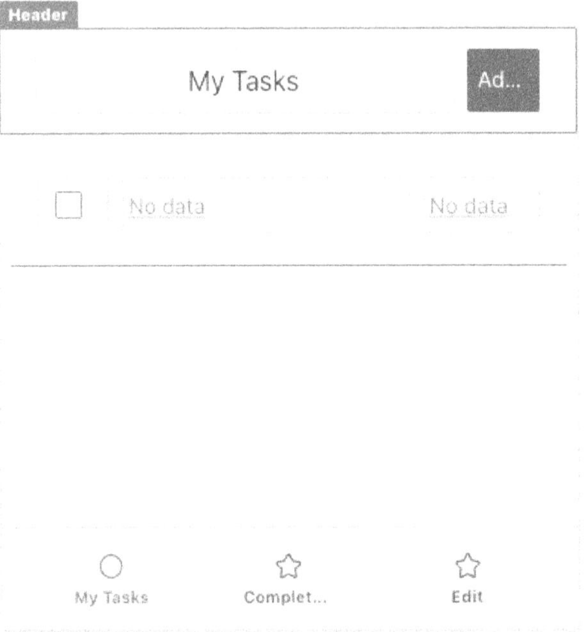

Figure 3.32 – The My Tasks screen with the Add Task button moved to the Header section

Creating the Edit Task screen

Sometimes, we may want to change something for the tasks, such as the due date, the reminder preference, or any other field we may have added for tasks. So, let's create a screen that will provide this edit functionality whenever we click on a task in our lists of incomplete tasks:

> **Tip**
>
> For the next steps, it's highly recommended to first add dummy data to the empty rows in the **Tasks** table. It will then show up in the Builder view as you configure various controls and thereby help visualize what values will be displayed for a particular field.

1. Go to the **My Tasks** screen and select the **List** control. For the **On click** property under the **ACTIONS** tab, open the dropdown and select + **Create a new screen**:

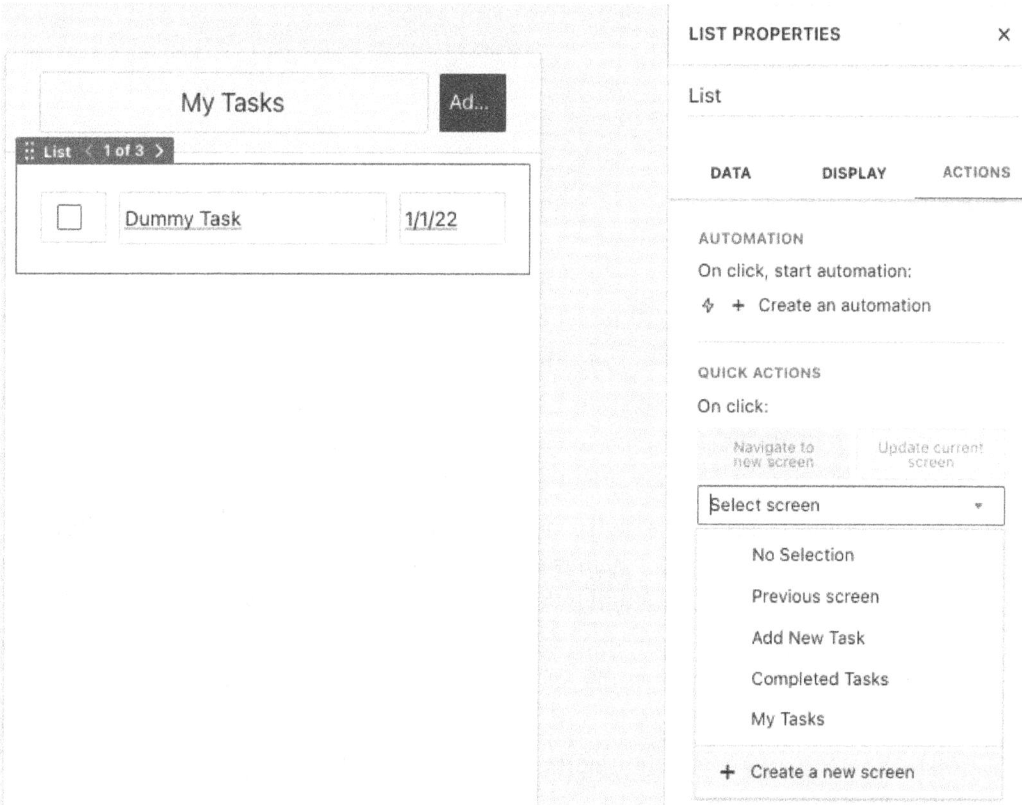

Figure 3.33 – Setting the On click property for the list and creating the new screen to be navigated to

> **Tip**
>
> Note the formula that gets populated for the **Set variables** field. This is telling the app to pass the currently selected table row as an input to the next screen as we navigate to it from here.

2. Locate the newly created screen in the screen list and rename it `Edit Task`:

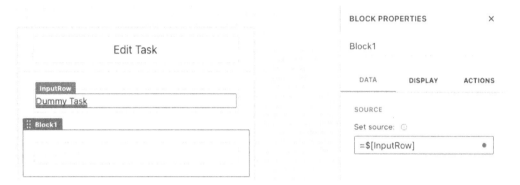

Figure 3.34 – The Edit Task screen

Note the presence of a data cell named **InputRow**, which was the variable we saw was assigned the value of the selected row in *step 1*. Also, review the fields set for this control under the **Display** tab in its **properties** panel.

Also, note **Block 1** below it with its source set to `=$[InputRow]`, which is telling our app to use the table row, which is set in our data cell named `InputRow` as the source of data for fields contained in this block.

3. Next, select the content box in the lower block and update its text to `Due date:` by double-clicking on the control and typing in it. Also, resize this control and place it on the left of the screen.
4. Next, add a data cell to the right of the content box. Make it editable and set its source to `=[Due date]`:

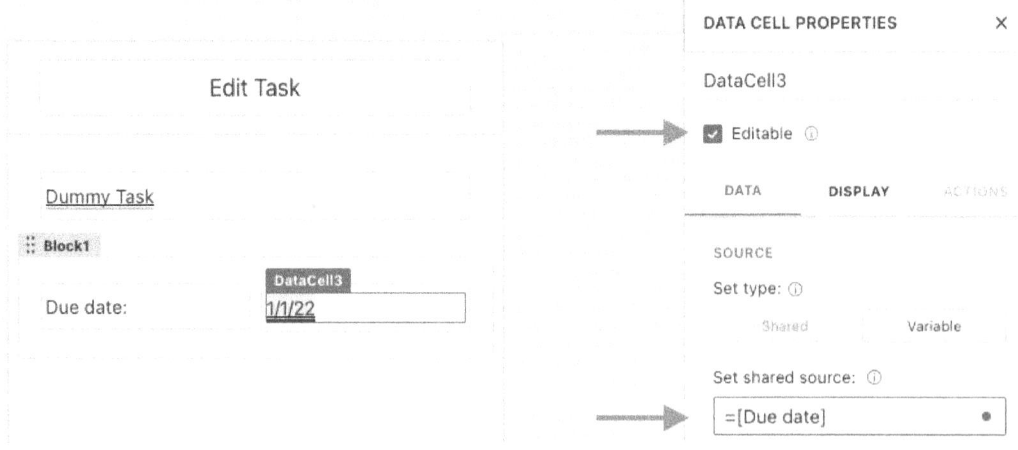

Figure 3.35 – Renaming the input field

This step is basically configuring our app to use the value from the **Due date** column of the table row that is set as the source of the containing block.

1. With the block selected, add another content box with text as `Reminder preference:`, along with the corresponding data cell to its right. Set the source of the data cell to = `[Reminder preference]`.
2. Lastly, let's add a **Button** control to go back from this edit task screen. Select the screen and add a **Button** control. Double-click on the button to edit the display text to `Done`. Finally, configure the **On click** action to navigate to the screen we came from by setting the value to **Previous screen**:

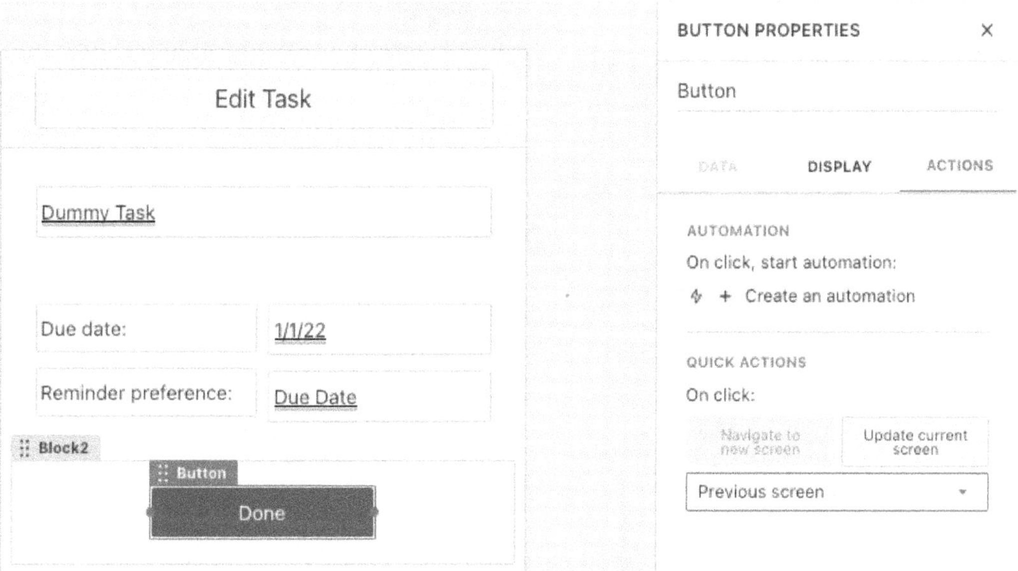

Figure 3.36 – Setting the On click action to navigate back to the source screen

3. Finally, we do not need this edit screen to be accessible from global navigation – that is, we do not need a separate icon for this screen in our navigation bar. So let's remove it by doing the following:

   I. Open the **APP NAVIGATION PROPERTIES** panel using the **App navigation** shortcut in the toolbar.

II. Drag the field displaying the **Edit Task** screen from the **SHOW IN NAVIGATION** section to below the line to the **HIDE FROM NAVIGATION** section:

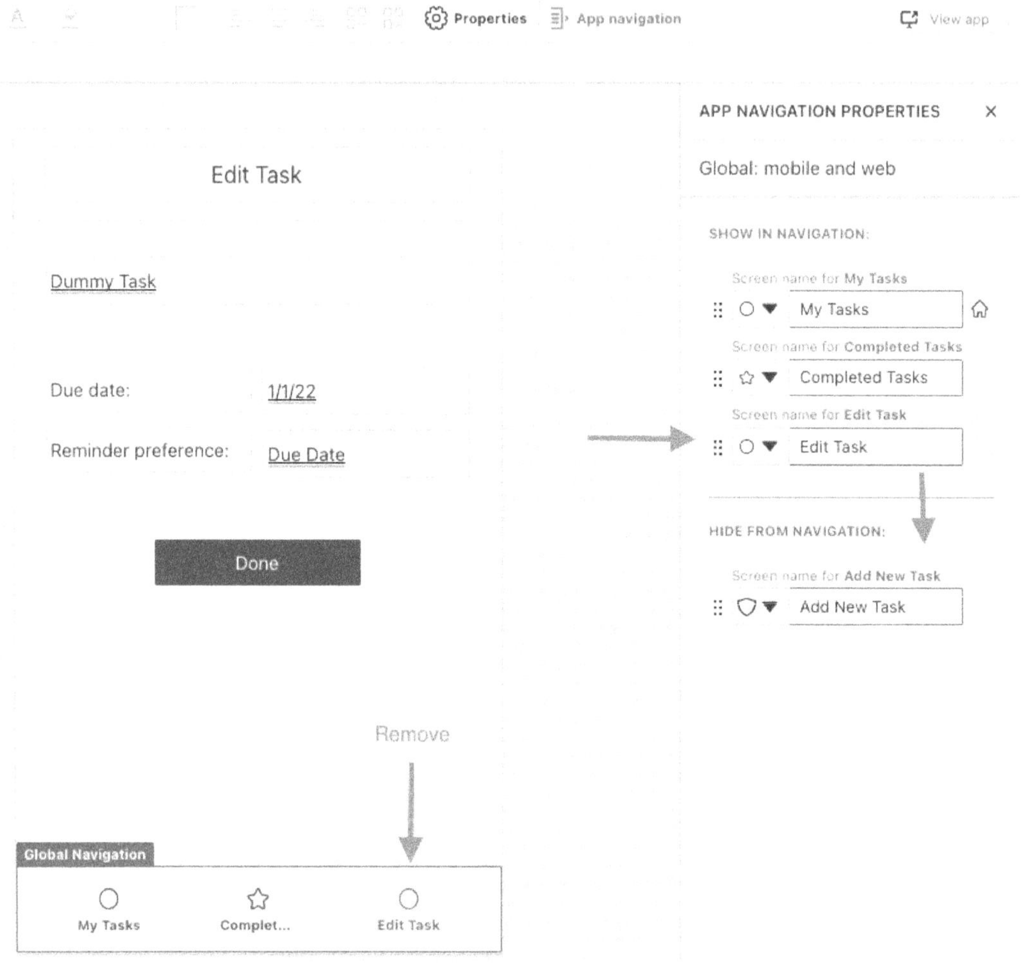

Figure 3.37 – Configuring app navigation

That's it! Our basic To-Do app is ready.

Creating an app using App Wizard

Now that we have learned the basics of creating an app from scratch, let's take a look at how App Wizard can help speed up this process:

1. In the *Creating an app from scratch* section, we chose the **Create from scratch** option. In this instance, select **Use app wizard**.

2. In the wizard screen that pops up, select the **Tasks** table as the source:

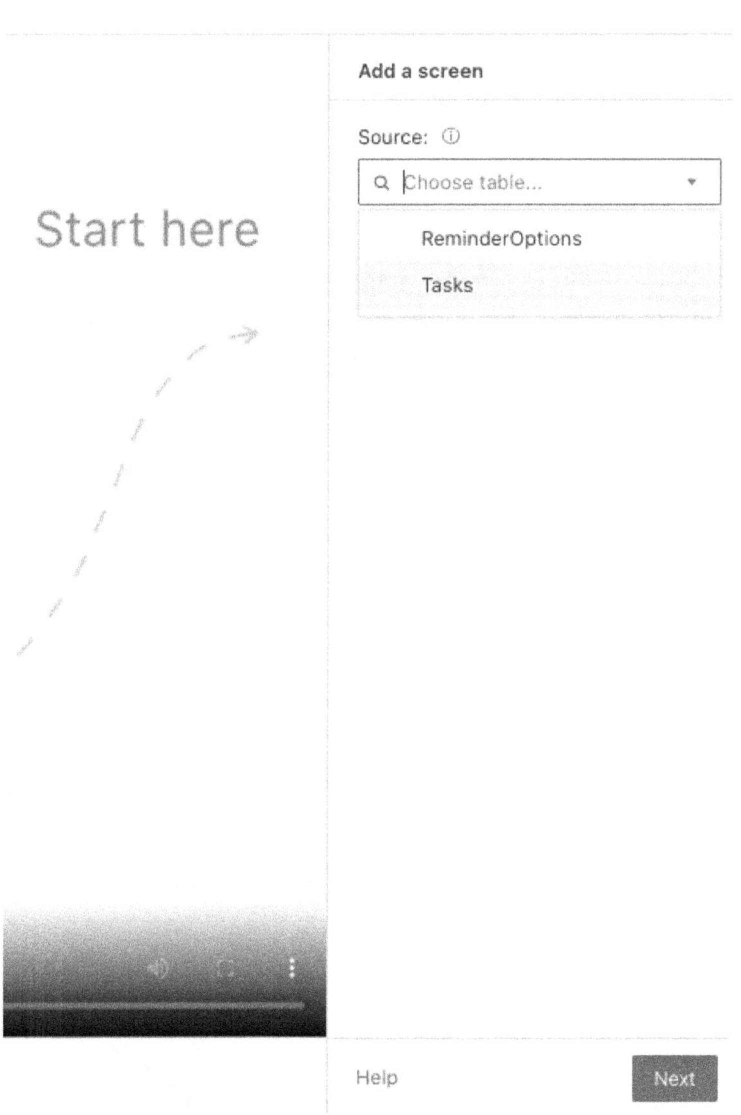

Figure 3.38 – The Select Tasks table as the source on the wizard screen

This creates our first screen, named **Tasks**, as the wizard defaults the name of the screen to be the same as that of the source table. The screen displays the columnar list view of table fields, which is similar to what we created for the **My Tasks** and **Completed Tasks** screens previously.

> **Note**
> App Wizard only supports creating a screen with the entire table as a data source.

3. Now, let's configure the screen that is presented by the wizard (see *Figure 3.39*):

    I. Rename the screen to `My Tasks` by updating it in the **List screen name** field.

    II. Remove the **Reminder preference** field using the delete icon (**X**) next to the column names listed under the **DATA** tab:

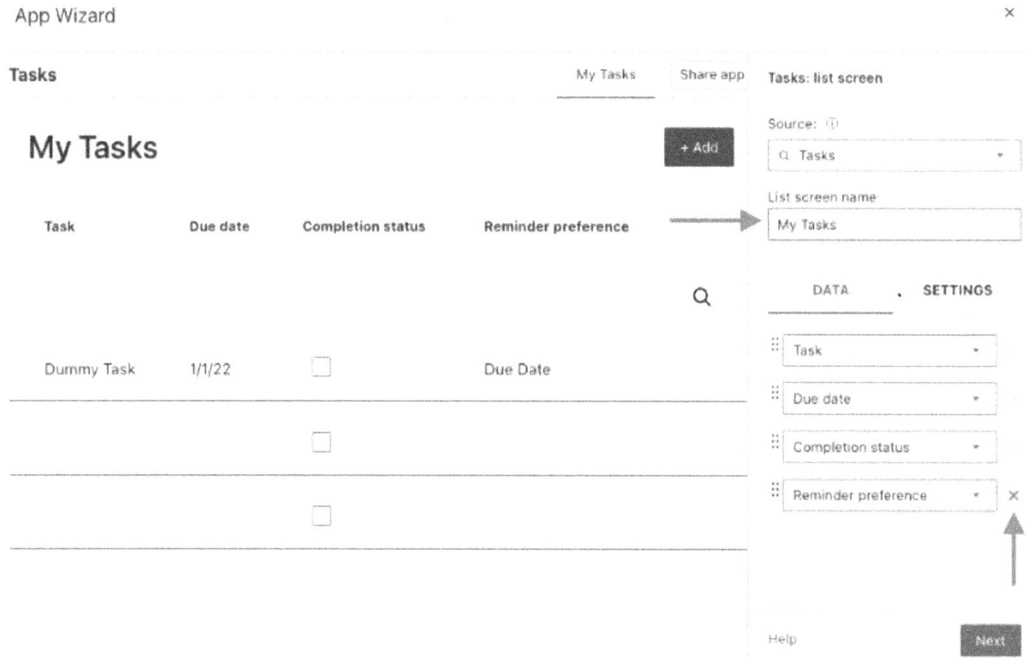

Figure 3.39 – Configuring the wizard-created Tasks screen

III. Under the **SETTINGS** tab, note that the **Add a detail screen** and **Add a form screen** options are selected:

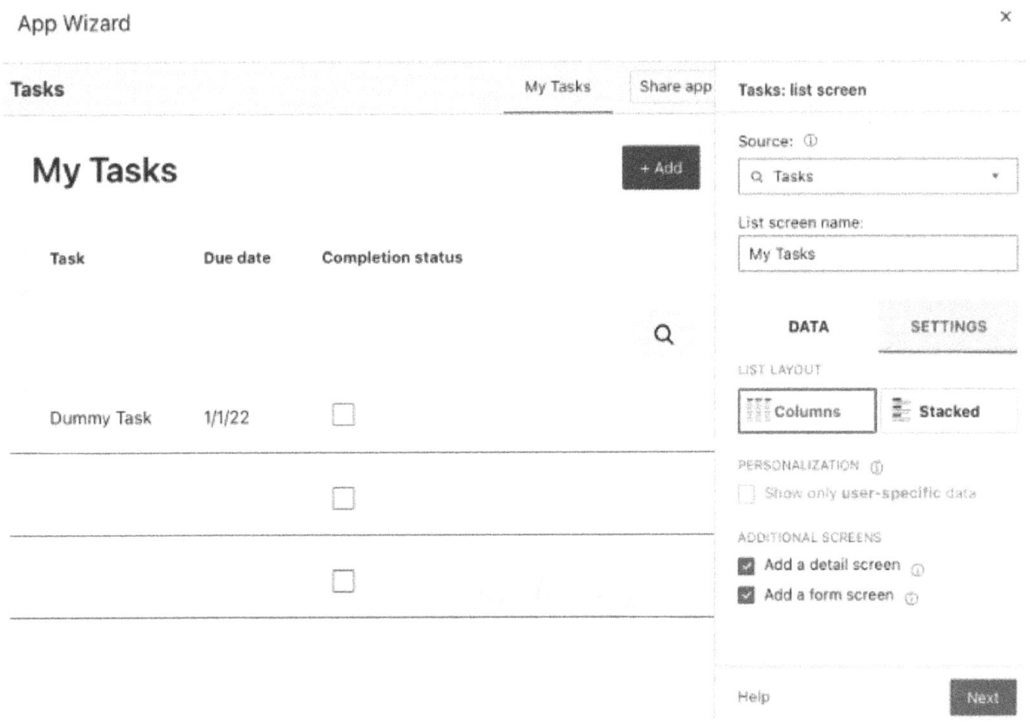

Figure 3.40 – Configuring the wizard-created Tasks screen

IV. Press **Next**, which loads the wizard-created **Detail** screen to configure.

5. Note how the **Detail** screen looks similar to our **Edit** screen, and we'll configure it as following:

    I. Rename the screen to `Edit task`.

    II. Remove the **Completion status** field using the delete icon (**X**) next to the list of fields under the **Detail data** section on the right.

    III. Make **Due date** and **Reminder preference** editable using the pencil icon next to the field.

> **Note**
> The **Delete Row** button, which we did not include in our basic app. We'll revisit the app and add this button to it in *Chapter 5, Powering the Honeycode Apps with Automations,* :

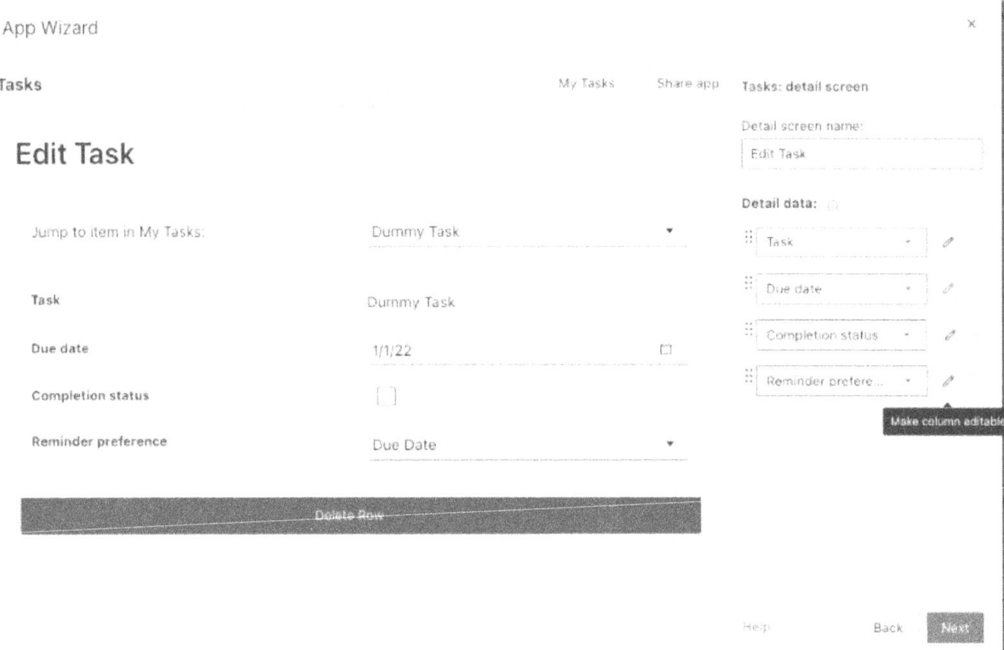

Figure 3.41 – Configuring the wizard-created Tasks screen

IV. Press **Next**, which loads the wizard-created **Form** screen to configure.

5. Recall that we used the **Form** object to create our **Add new task** screen, which is similar to this wizard-created screen. So, let's configure it as following:

    I. Rename the screen `Add new task`.

    II. Remove the **Completion status** field.

    III. Press **Done**:

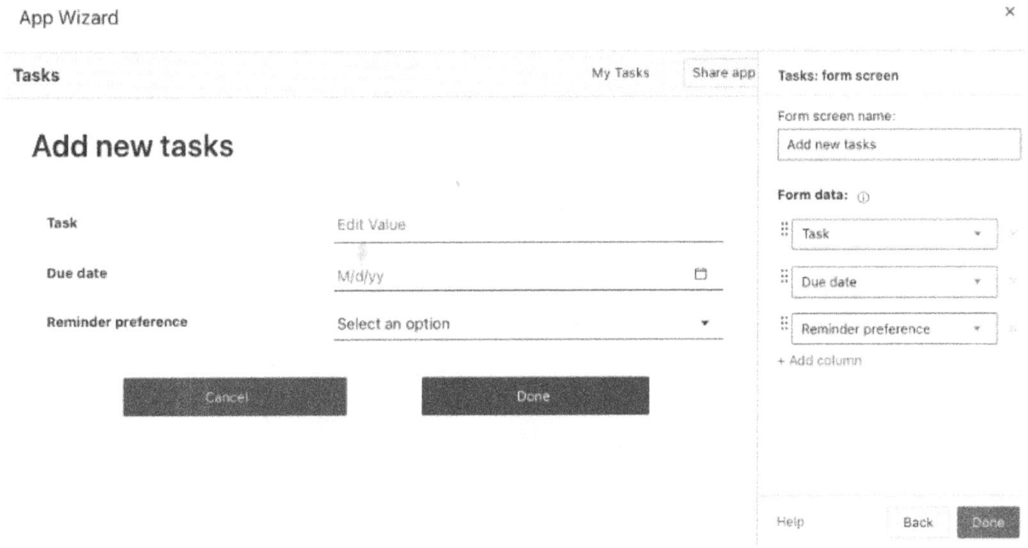

Figure 3.42 – Configuring the wizard-created Tasks screen

4. The next screen that loads allows you to go through another set of steps to create more screens using App Wizard for the other tables in the workbook. It also has a button to view the created web app in a new tab:

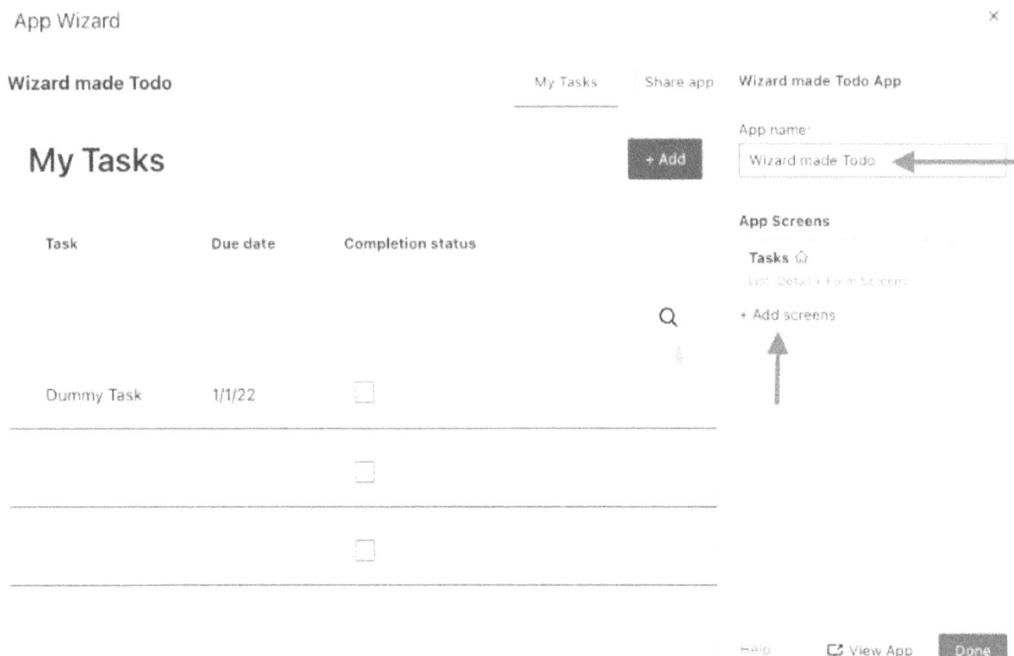

Figure 3.43 – Configuring the wizard-created Tasks screen

> **Note**
>
> App Wizard does not allow us to create another set of screens for the same source table, implying that we cannot build our **Completed tasks** screen using App Wizard.

Since we do not have another table that we want to build screens for, let's just rename our app `Wizard made Todo` and press **Done**. The newly created app will show in the **Apps** list.

5. With the **My Tasks** screen selected, select the list control on the screen and update the **Set source** field with the same `Filter` formula we used for the **My Tasks** screen in the earlier section:

```
Filter(Tasks, "Tasks[Completion Status] = FALSE")
```

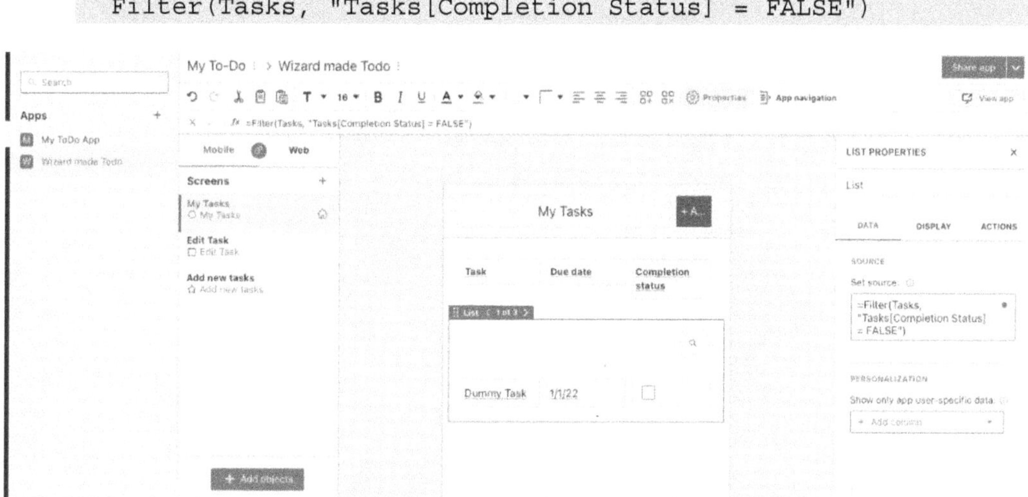

Figure 3.44 – The wizard-created To-Do app

6. Add the `Completed tasks` screen, similar to how we created it before.

That's it! Our basic To-Do app created using App Wizard is ready.

Summary

In this chapter, we built on top of what we learned from *Chapter 2, Introduction to Honeycode,* and put it into action. We used the app builder to create our application interface and tables to create the necessary tables needed for modeling our To-Do data. We also learned how to use App Wizard to create a basic app that can serve as a starting point for building more advanced apps.

Now that we know the basic operations and steps to take to build a basic Honeycode app, we are ready to dive into some advanced builder concepts to further the functionality of our apps, as well as power our app actions and data processing with automation.

In the next chapter, we'll learn how to enhance our To-Do app with advanced builder functionalities, including search, filter, sort, conditional styling, and personalization.

4
Advanced Builder Tools

In the previous chapter, the **To-Do** app we built was functional, but it was very basic and left much to be desired when compared with any app we use in our day-to-day lives. The good news is that Honeycode offers a lot more tools for customization and improving the presentation and functionality of our apps. In this chapter, we'll explore some of the advanced functionality that enables app builders to improve the presentation with conditional styling and controlling the visibility of different components, increase functionality by adding controls for filtering and sorting views, and provide customized views for each app user using personalization.

In this chapter, we're going to cover the following main topics:

- Applying styles to app components
- Controlling the component visibility with conditions
- Filtering and sorting data views on the fly
- Restricting data access per user using personalized views

Technical requirements

To follow this chapter, you'll need to have access to Amazon Honeycode, which requires a laptop with a web browser, preferably Google Chrome, and optionally a mobile device running either a Honeycode-supported version of Android (it currently requires Android 8.0 and up) or iOS (it currently requires iOS 11 or later).

Furthermore, we will use Honeycode terminology and refer to the components that we covered in *Chapter 2, Introduction to Honeycode*, and continue to build upon the To-Do app we created in *Chapter, Building your first Honeycode Application* and therefore, we recommend you complete those first.

Defining the app requirements

Similar to how we identified our app's requirements in *Chapter 3, Building Your First Honeycode Application*, let's list down the requirements or the use cases that we'd like to build to enhance our To-Do app. Recall that we left a couple of requirements in the additional list in *Chapter 3, Building Your First Honeycode Application*, to be covered in the following chapters. We are picking up two of those requirements here and adding a couple more as we learn the advanced builder functionality that can be enabled in our Honeycode apps.

So, here are the use cases that I'd like to see added to my To-Do list app that we will cover in this chapter:

- I'd like my team to use my app, but every user must be able to see only their incomplete tasks while completed tasks are visible to all.
- I'd like to see my overdue tasks highlighted in red.
- I'd like to be able to search for tasks.
- I'd like to be able to filter and sort tasks.
- I'd like to have an additional field to write optional notes for each task that I can show or hide using a control.

These changes are not just on the app interface but will also require minor updates to our data model. Note that we are adding requirements to identify the owner/assignee of the task in the first use case and require an additional field for capturing notes for each task in the fifth use case. Both requirements are essentially adding further information to the task, implying the need for additional table columns in the **Tasks** table.

Alright, so now that we have defined what we want to build and established some idea of the required data model change, let's get started and update our app.

Applying styles to app components

Let's take another look at the app we created in the previous chapter (as shown in *Figure 4.1*). What we built was a very simple but functional To-Do app, devoid of colors, layouts such as presenting each task as a card, or lists with alternating background colors for better readability:

Figure 4.1 – The My To-Do app created in Chapter 3, Building Your First Honeycode Application

However, what we currently have is not due to the lack of options in Honeycode. In this section, we'll learn what Honeycode offers us to stylize our apps and then apply that to our To-Do app.

Style controls in Honeycode

In Honeycode, controls for adding custom styles to apps are provided in the **toolbar** (see *Figure 4.2*). These controls can be categorized under the following four categories:

A. **Text configuration**: This group includes the **Font** and **Font size** controls for selecting the display font and the text size.

B. **Text formatting**: This group contains the standard text formatting controls of **Bold**, **Italic**, and **Underline**, along with the control to set the **Font** color.

C. **Object styles**: This group includes a control to choose the fill color of the objects along with the **Border** style and **Corner** style controls to stylize the objects by selecting which borders to show, choosing the border color, and deciding whether the borders will have angular or rounded corners.

D. **Text alignment**: This group contains the controls to customize whether the text within the container object should be aligned **Left**, **Center**, or **Right**:

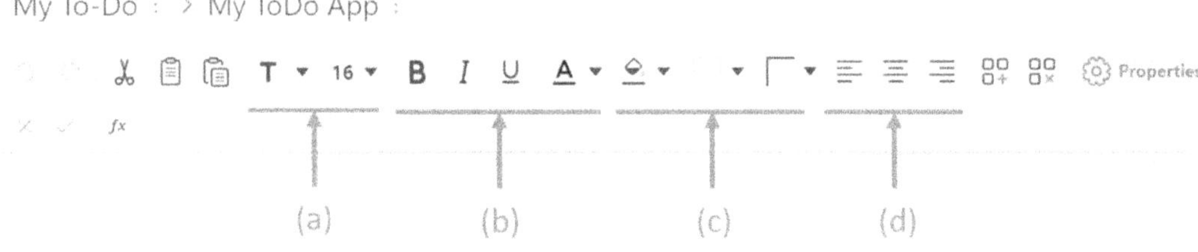

Figure 4.2 – Style controls in Honeycode

Those who have experience building websites or apps or other interfaces for user interaction might find this control set basic, and that won't be an inaccurate statement. But where Honeycode loses on the number of controls available, it makes up by making it very simple for app builders to set conditional logic and customize the app presentation.

Conditional-styling app components

The **conditional styling** feature of Honeycode allows you to stylize objects on the screen based on a set of conditions that you preconfigure as part of building the app. This feature is available for most of the objects that we can add to the app screens and is accessible under the **DISPLAY** tab in the properties panel (see *Figure 4.3*):

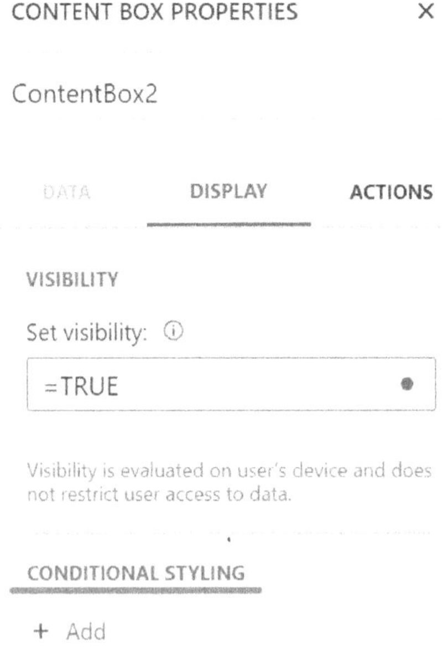

Figure 4.3 – A conditional styling property for a content box

As noted earlier, setting these conditions on the controls is fairly simple and requires three simple steps:

1. Click on the + **Add** control under the **CONDITIONAL STYLING** feature option, as displayed in *Figure 4.3*.
2. Under the **WHEN** block (as shown in *Figure 4.4*), set the condition by providing a formula that evaluates to return either `true` or `false` as output.
3. Lastly, under the **THEN STYLE AS** block (as shown in *Figure 4.4*), click to open the drop-down list, and either choose from one of the defaults provided or select **Custom** to manually set the styles using the *Text formatting* and *Object styles* controls:

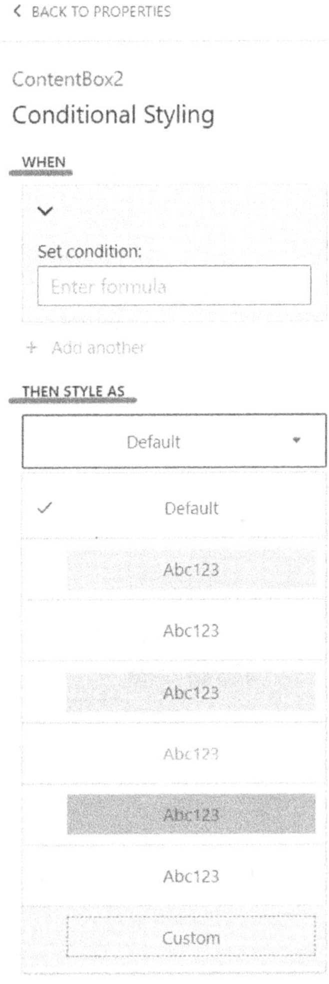

Figure 4.4 – Applying conditional styling on a content box

Now that we know the different styling controls available in Honeycode for styling the apps, and the feature to apply a style on the fly using pre-set conditions, let's apply them to our To-Do app.

Styling the To-Do App

While defining the app requirements in this chapter, we made note of a requirement of a visual indicator for overdue tasks by highlighting them with a red color. So, let's build that, and let's also improve the default presentation by removing the single border at the bottom of each task and setting the default display of each task as green-colored boxes instead.

Setting a default style for each task

We can set the default display of each task as green-colored boxes by following these steps:

1. Open the builder for the **My To-Do** app that we created in *Chapter 3, Building your first Honeycode Application,* and select the **My Tasks** screen.
2. Select the segment inside the list and set the fill color to be a light shade of green:

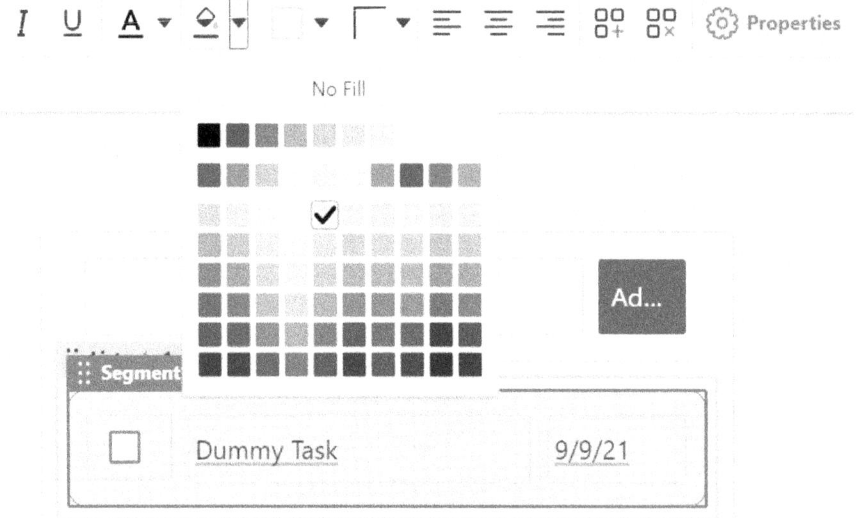

Figure 4.5 – Set a light green shade as the fill color for the segment

3. Next, using **Border** styles, we'll set the border to be a shade of dark green to make it stand out with contrast:

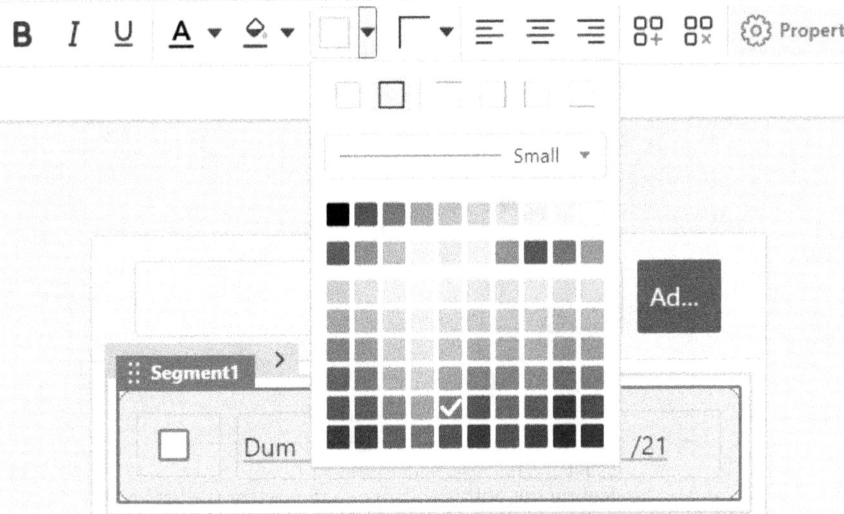

Figure 4.6 – Setting up a dark green border using border styles

4. We will make the corners a little rounded using the **Corner** styles control, as displayed in *Figure 4.7*:

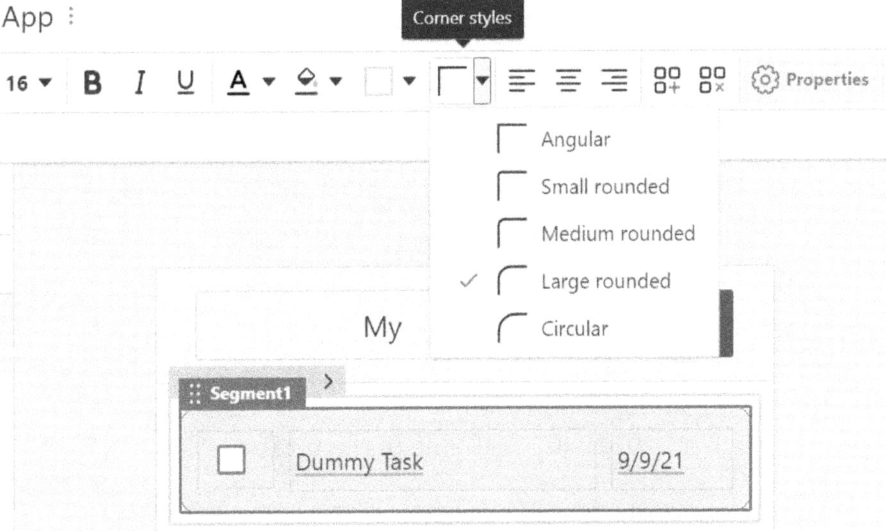

Figure 4.7 – Setting the large rounded corner style

5. Finally, select the list component and remove the border below every task by deselecting the lower border option in the **Border** styles control:

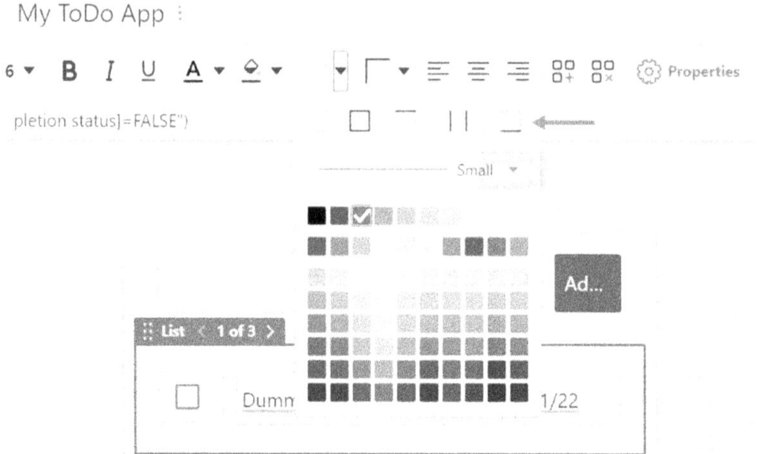

Figure 4.8 – Deselecting the bottom border configured at the list level

And with that, we now have changed the default view of our app's **My Tasks** screen to look like *Figure 4.9*:

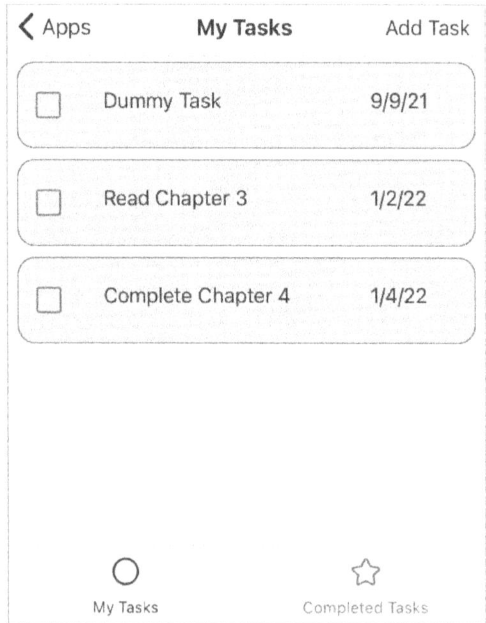

Figure 4.9 – The My Tasks screen styled to display tasks as individual green boxes

Next, let's set up the conditions to highlight the overdue tasks in red.

Setting the conditional style for overdue tasks

As a user of this To-Do app, we want to have a visual indication that a certain task is past its due date. Red is a very commonly used color to highlight things that need immediate attention, and we can set our tasks' background to red instead of green by using conditional styling feature and following these steps:

1. Select the segment inside the list and open the **Conditional Styling** property for it.
2. An overdue task has its due date in the past. We'll therefore use the comparison on the due date with today's date to set the **WHEN** condition, using the following formula: `=Tasks[Due date] < TODAY()`.
3. In the **THEN STYLE AS** block, we'll use the **Custom** style and set the fill color to a lighter shade of red, with dark red borders and large rounded corners to match the default style:

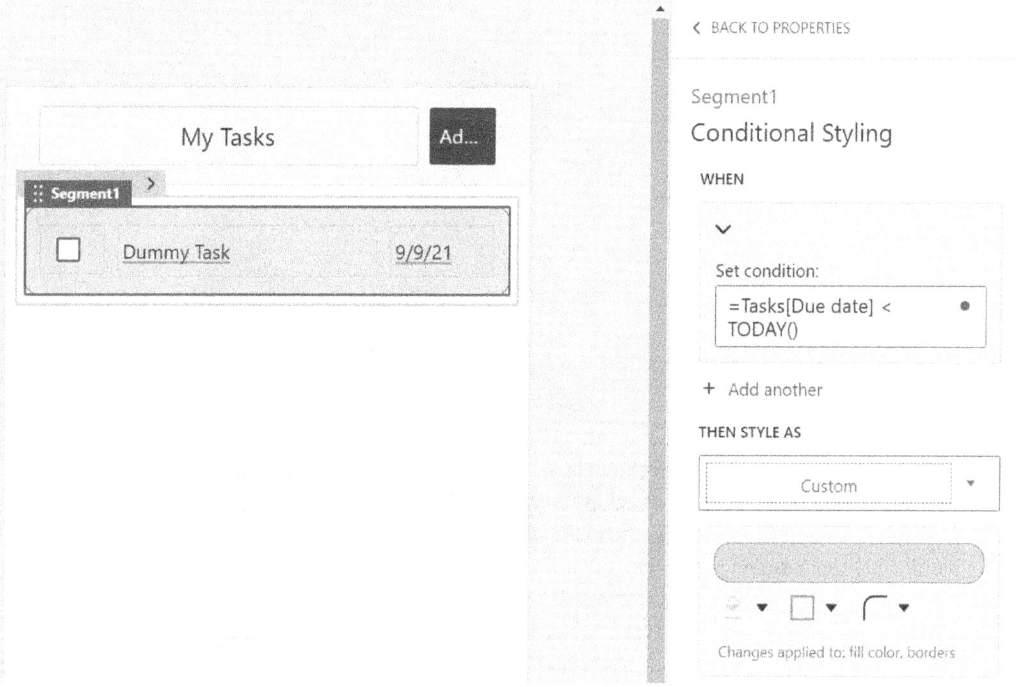

Figure 4.10 – Setting conditional styling to highlight the overdue tasks

4. That's it! Our app is now configured to show the past due items in red, as shown in *Figure 4.11*:

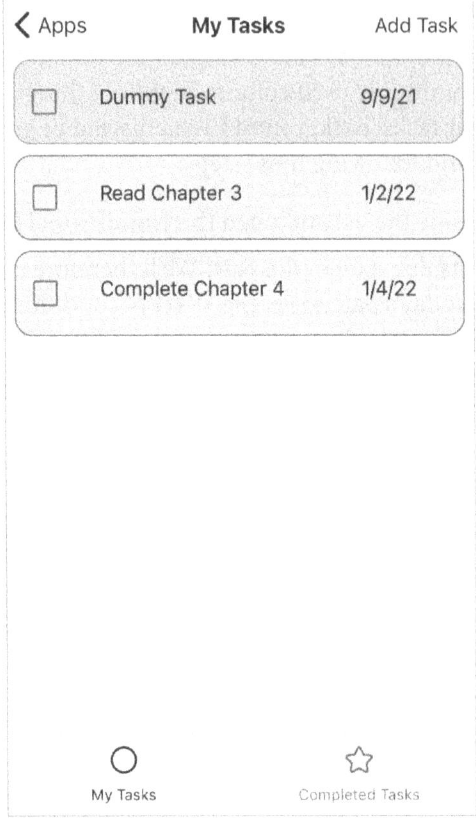

Figure 4.11 – The app with conditional styling applied to show past due items in red

> **Note**
>
> This styling is conditional to the date and time, and therefore it is important to note that *Figure 4.11* was created in October 2021, making **9/9/21** a past due date and shown in red, while the tasks with due dates in 2022 are not yet due and are therefore in green.

> **Exercise 1**
>
> Use conditional styling to change the set color to yellow when the task is coming close to its due date – let's say, 1 day before the due date.

Controlling the component visibility with conditions

In the previous section, we learned how to use conditions to stylize our app. We can similarly affect the layout of our app using conditions to show or hide a component. In the requirements section, we listed a requirement for an optional notes field to capture additional details of a task. However, we do not want that field to be always visible, as notes can be detailed and will affect the utility of the app if they are always visible in the view. So, let's see how we can use conditions to control the onscreen component and implement these requirements:

1. We learned in *Chapter 3, Building your first Honeycode Application*, that for us to store data entered in the app, we need to have a corresponding field in our table. So first, we add a new column to the `Tasks` table and rename it `Notes`. Add some dummy data for the first column.
2. In the **MyTasks** screen, add a new block and move it above the existing lists.
3. Add a **Switch** control to the new block and rename it `ShowNotes`. Under the **Display** tab for this control, set the value of the **Display Data from** field to `Show Notes`.
4. Ensure that under the **DATA** tab, the **Source** type is set as **Variable**:

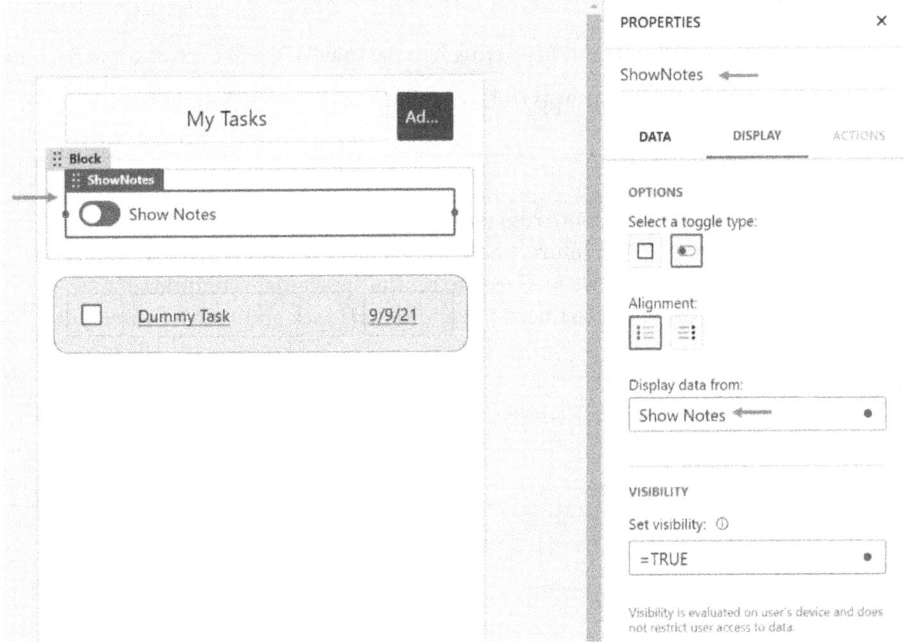

Figure 4.12 – Adding a switch control for toggling the visibility of notes

5. With **List** selected, add a new segment object.
6. Add a data cell to the new segment and bind it to the **Notes** column of the **Tasks** table to display the notes corresponding to the task displayed in the preceding segment.
7. Now, to control the visibility of notes by the **Show Notes** switch control that we added in *step 3*, select the content box containing the data cell, and under the **Display** tab, update its **Set visibility** condition to `=$ [ShowNotes]`:

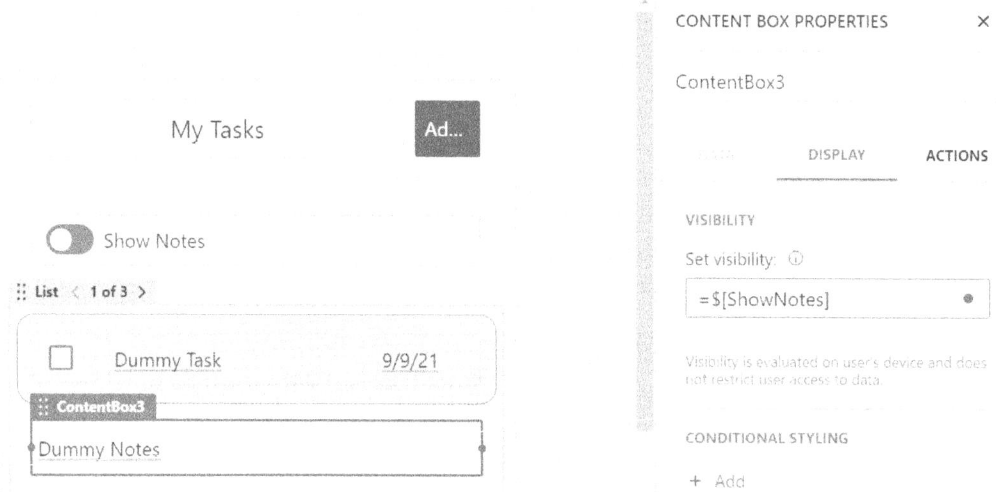

Figure 4.13 – Setting the visibility condition by tying it to the value of the Show Notes switch control

And with that, we have updated our app to have a field with conditional visibility.

> **Note**
>
> While this field is manually controlled using an onscreen toggle switch, if we come across a use case in the future where the visibility needs to be controlled based on the data displayed, we just need to set the appropriate condition for the **visibility** property and won't need the additional block and switch control that we added here:

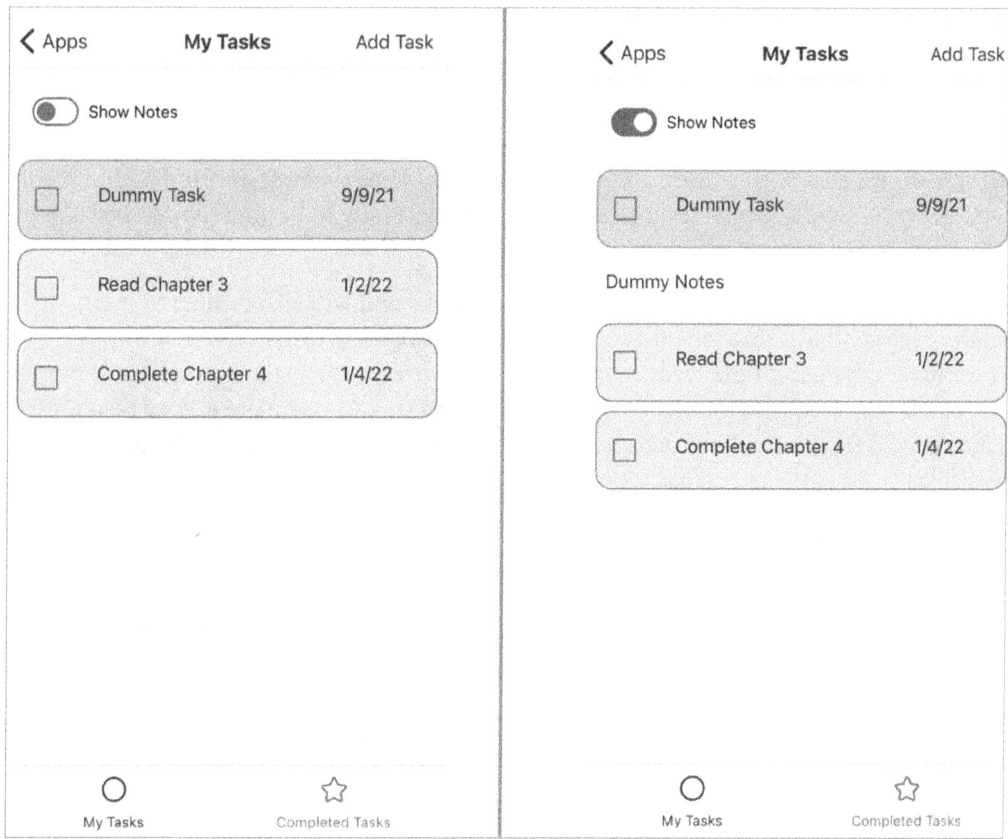

Figure 4.14 – The My Tasks screen with notes hidden on the left and visible on the right, aligned with the state of the Show Notes toggle

> **Exercise 2**
>
> The **Notes** field on the **My Tasks** screen is view-only. For us to make use of the newly added **Notes** field, we should be able to add or update data in it. So, the task is to add the **Notes** field to the **Edit Task** screen. It is important to note that, ideally, we also need to add this field to the **Add Task** screen; however, to be able to do that, we need to know about building automations, which we will cover in the next chapter.

Restricting data access per user using personalized views

Until now, we have been using this app for personal use only. However, with the updates made in the previous two sections improving the visual appeal of our app and adding the functionality for additional notes for each task, I feel it's ready to be shared with friends and family.

But once we share, every user will be able to create tasks and will also be able to mark them done. Potentially, this can add some chaos and confusion, especially in a work environment, where multiple team members can be working on a single project and might find it confusing as to who a task is meant for and who should complete it. As a result, we will also need a functionality to determine the tasks meant for each user of the app and only display those tasks to them.

Let's see how to make this happen:

1. To capture the assignee information for each task, we need a field for storing this information. So, we start by adding another table column to the **Tasks** table, rename it `Assignee`, and set the format to **Contact**:

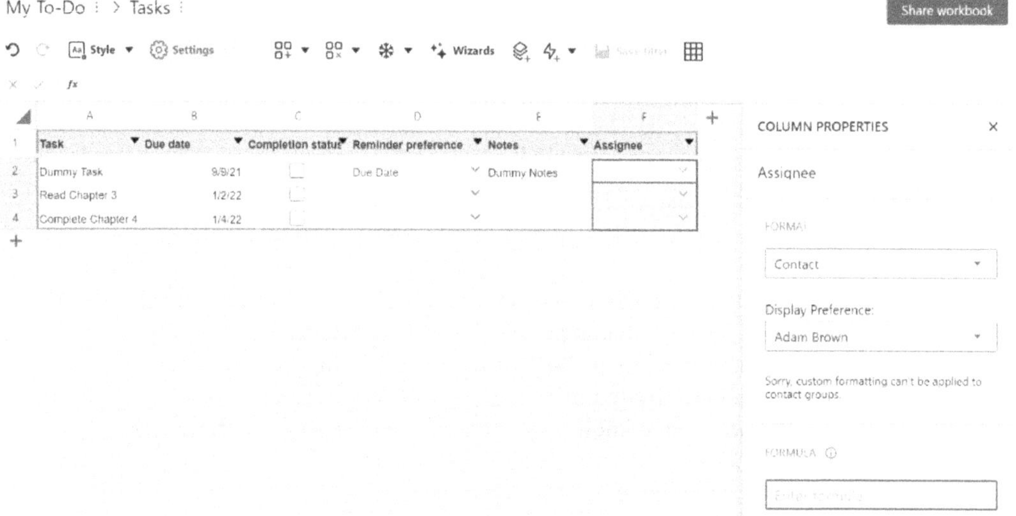

Figure 4.15 – Adding a table column for Assignee and setting the Contact format

> **Note**
>
> **Contact** is a format unique to Honeycode that is used to present your teammate's information in a table cell, app, or automation. A teammate's information is available for use the moment the app is shared with them if they have an existing Honeycode account; otherwise, they'll first have to create an account and then their information will be available.

2. Next, in the **My Tasks** screen, select the task list. In the **LIST PROPERTIES** panel, locate the **PERSONALIZATION** property under the **DATA** tab and set the value of **Show only app user-specific data** to the **Assignee** column:

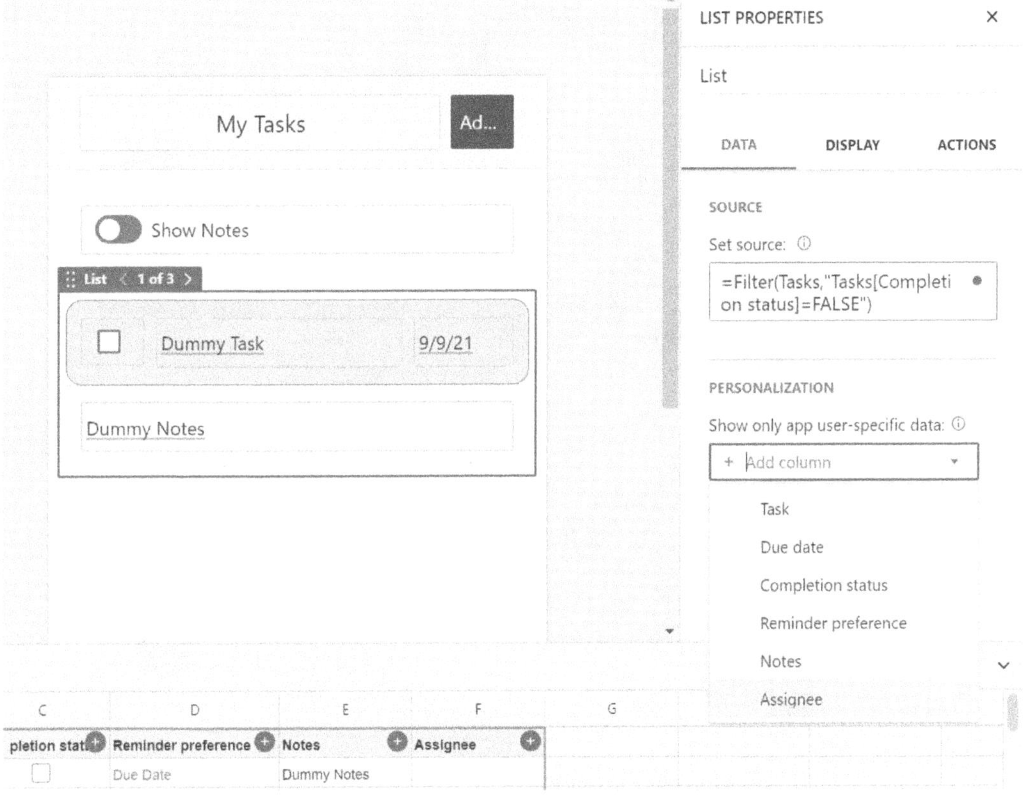

Figure 4.16 – Personalizing the task list to show only the tasks assigned to the app user

And it's done! This is another powerful feature of Honeycode, allowing screen customization in literally one step without having to write a single line of code. *Figure 4.17* shows the updated app as seen from two different user accounts:

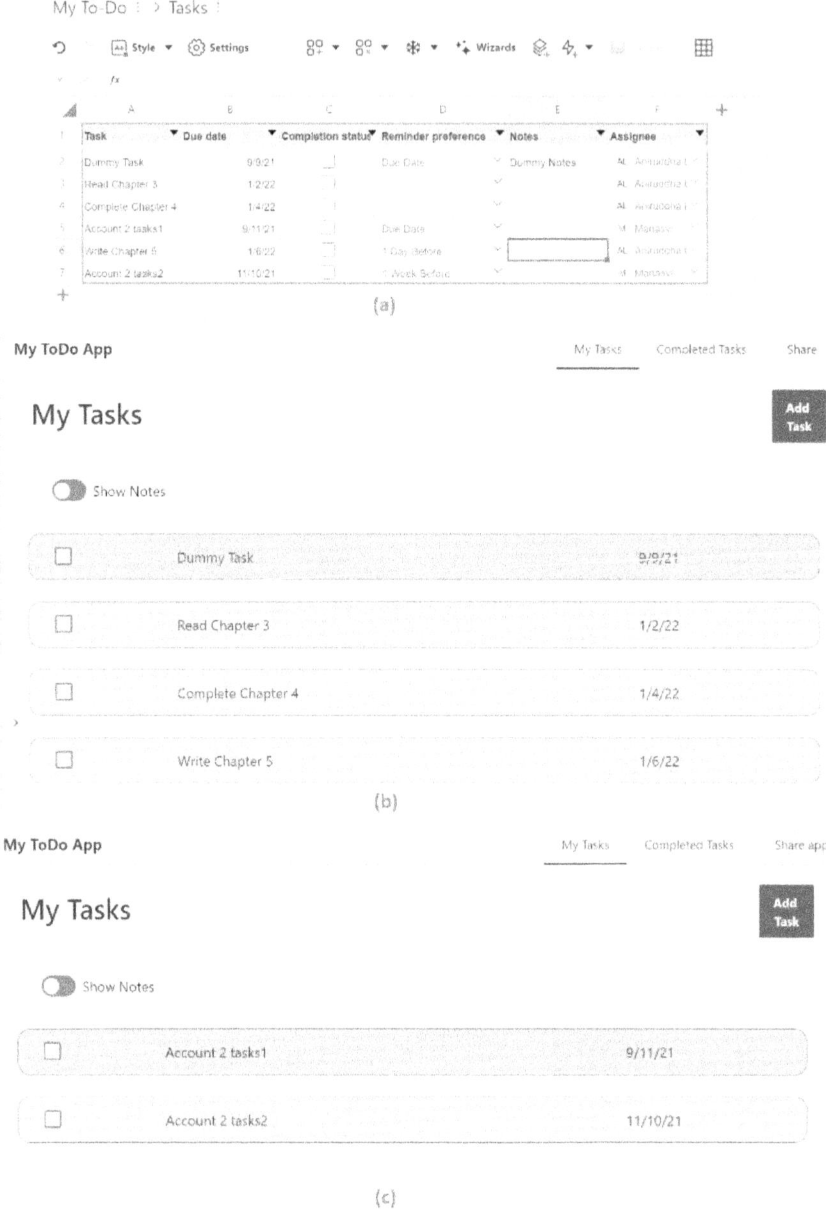

Figure 4.17 – (a) Table data showing assignments, (b) a personalized app view from the primary account, and (c) a personalized app view from Account 2

Exercise 3

We added the **Assignee** table column to the Tasks table, but there is no screen that sets or displays the value of this field. So, the task is to do the following:

1. Add the **Assignee** field to the **Edit Task** screen.
2. Add the **Assignee** field in the **Completed Tasks** screen for displaying the person who completed the listed task. Note that like the **Notes** field, the **Assignee** field also needs to be added to the **Add New Task** screen and will require an understanding of building automations.

Searching, filtering, and sorting data views on the fly

In a list of items, the requirement to be able to search, filter, and sort the data becomes a necessity as the list begins to grow. Like the creation of personalized views, Honeycode makes the inclusion of these three features a breeze. Let's check it out:

1. Go to the **My Tasks** screen. Select **List** and open the properties panel if not already open.
2. Under the **DISPLAY** tab, locate the **APP USER CONTROLS** property and the three checkboxes for **Search**, **Filter**, and **Sort**. Select these checkboxes, and voila – the feature to search, filter, and sort in the tasks list is now available in the **My Tasks** screen of our app:

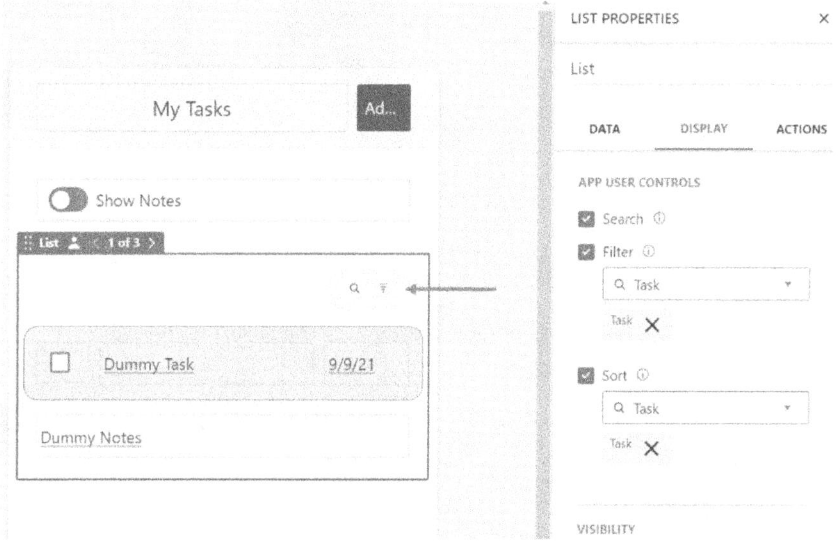

Figure 4.18 – Fields to enable searching, filtering, and sorting in the list

> **Note**
>
> In *Figure 4.18*, the new icons appear over the segment containing task details as you select the fields in the panel. Also, note that, by default, the first column of the table is selected for **Filter** and **Sort** fields. However, you can use the dropdown to select additional fields where filtering and sorting should be enabled.

For example, let's see *Figure 4.19*:

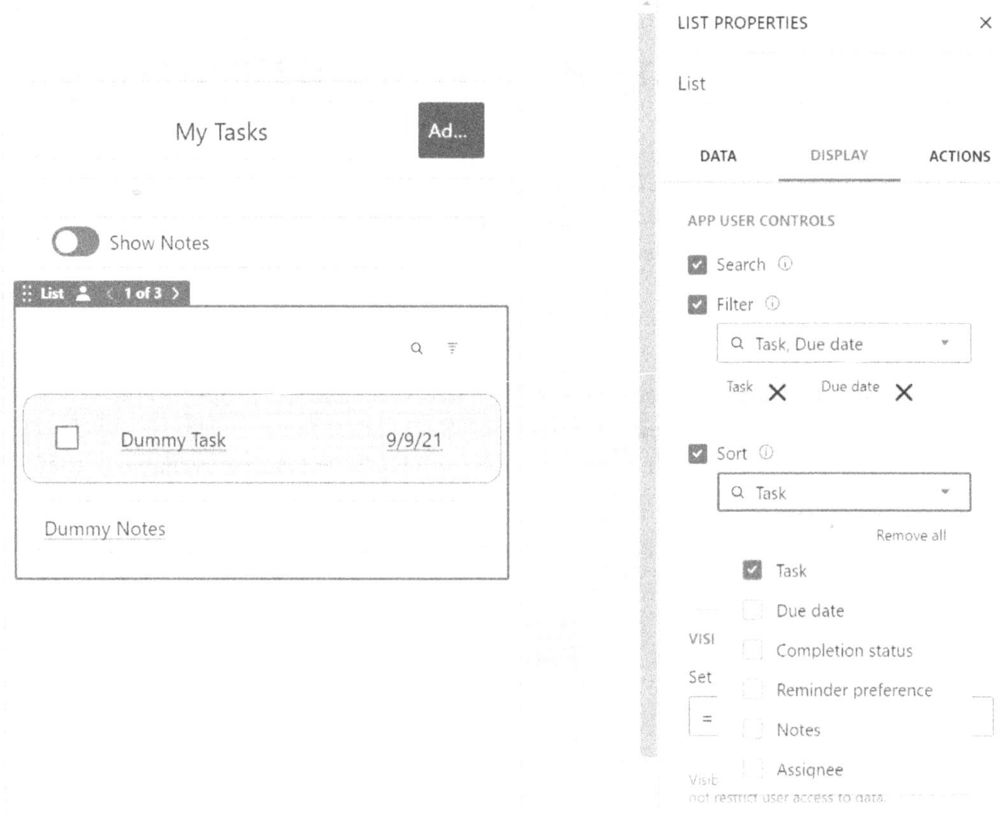

Figure 4.19 – Enabling additional fields to filter and sort in the list

The outcome on the web app looks like *Figure 4.20* when you click on the filter icon:

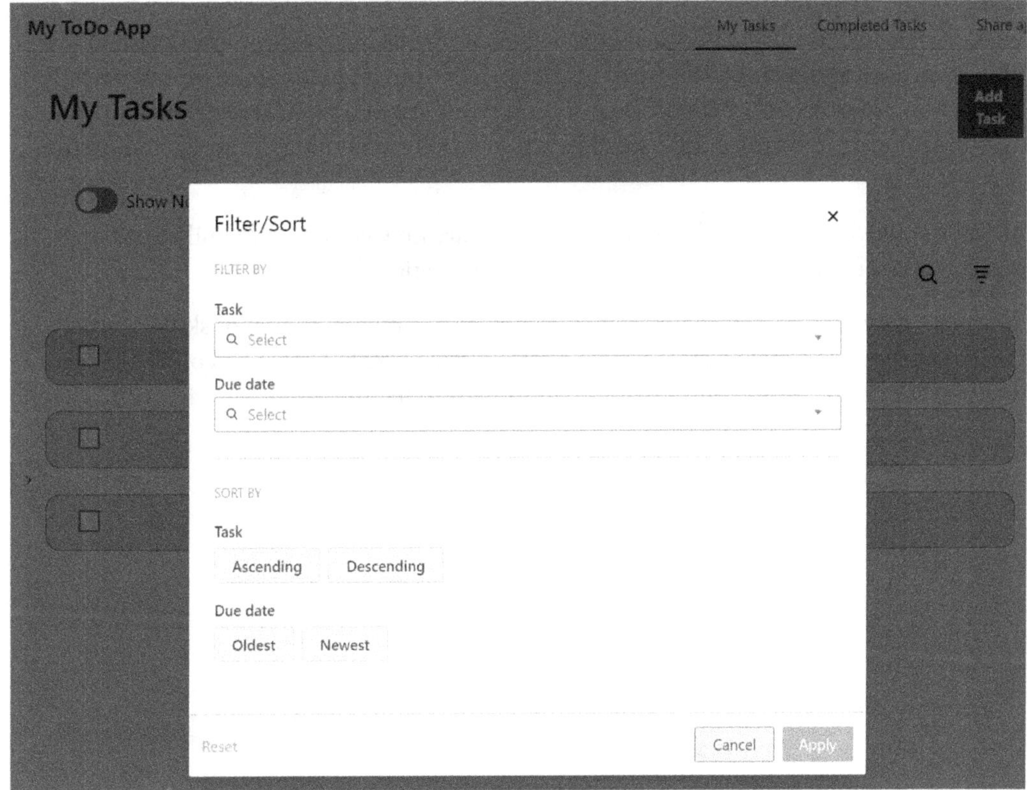

Figure 4.20 – The web app displaying the filter/sort model

We now know how to search filter and sort on the fly.

> **Exercise 4**
>
> Add the capability to search, filter, and sort on the **Completed Tasks** screen.

Summary

In this chapter, we build on top of our learning from *Chapter 3, Building your first Honeycode Application,* and upgraded our To-Do app while learning the advanced builder features. We used styles and visibility to improve the layout and added visual indicators for overdue tasks with the use of conditional styling. We learned how to create personalized views for each user and added additional functionality for searching, filtering, and sorting in our tasks list.

With the knowledge of these advanced features, we are now equipped to build apps with more flair, more enhanced interactions, and personalization for each user.

In the next chapter, we'll learn about **automations** and complete the set of skills needed to build or understand any given Honeycode application. We'll do that by continuing our journey with our To-Do app and further upgrading it to its final version.

5
Powering Apps with Automations

In the previous chapters, we learned about building the data model and the presentation layer of our applications along with different customization controls available in Honeycode. This helped us to improve the presentability of our applications. Although the **To-Do** app that we created while learning these concepts may seem powerful and complete, it is not. It may surprise you to learn that it is still incomplete. Have you been receiving the reminders based on the selected preference?

Moreover, recall that in the exercises in *Chapter 4, Advanced Builder Tools in Honeycode*, we added the new `Notes` and `Assignee` fields only to the **Edit Task** screen, whereas they should also be present on the **New Task** screen. This was because, while we could have added those fields to the screen with what we have learned so far, saving the information added to those fields would require a change to automation to process it. Now, you may be thinking that we never wrote any automation in the first place, so what change is being referred to here? And guess what, you are not wrong! It was Honeycode's out-of-the-box objects that did that part for us and until now, we used it as a black box.

In Honeycode, automations can be configured on App objects that get triggered through user actions, and on Table objects that are triggered through pre-defined conditions or indirectly because of a user's action. In this chapter, we'll continue on our learning journey to cover the final pillar of Honeycode, and with it, we will also update our **ToDo** app and complete it. We will also cover another important concept of Honeycode called **variables**, which are essential for understanding and building automation.

In this chapter, we're going to cover the following main topics:

- Understanding variables in Honeycode
- Making apps more powerful with automations
- Processing data with automation based on triggers
- Debugging automations

Technical requirements

To follow this chapter, you'll need to have access to Amazon Honeycode, and that requires a laptop with a web browser, preferably Google Chrome, and optionally a mobile device running either a Honeycode-supported version of Android (currently requires *Android 8.0* or later) or iOS (currently requires *iOS 11* or later).

Furthermore, we'll use the Honeycode terminology and refer to the components that we covered in *Chapter 2, Introduction to Honeycode*, and continue to build upon the **ToDo** app we have from *Chapter 4, Advanced Builder Tools in Honeycode*; therefore, I recommend that you complete those first.

Defining the app requirements

Continuing with our practice of defining our app's requirements upfront, let's list the requirements or the use cases that we would like to build to complete our **ToDo** app. So, here are the use cases that I would like to see added in my **ToDo** app that we will cover in this chapter:

1. I'd like to be able to add notes when creating a new task.
2. I'd like to be able to select an assignee when creating a new task.
3. I'd like to be able to delete tasks.
4. I'd like the app users to receive a *Task-Due* reminder based on the preference set for each task.

5. I'd like the app users to receive a notification whenever a task is assigned to them by someone else.
6. I'd like the creator of the task to be notified of its completion.

While the focus of this chapter is on **automations**, by now you will have realized that these app requirements will warrant changes to the app interface, as well as minor updates to our data model. So, let's build it up!

However, before we get to building our app, we need to know another key concept of Honeycode called **variables**. In the next section, we will learn about variables in Honeycode and how they are used. This will enable us to use them to build our automation in the remainder of the chapter.

Understanding variables in Honeycode

Variables, in general, are placeholders that can assume or be assigned different values based on the context. In Honeycode too, they serve a similar feature and while we did come across and set variables in the previous chapters, we did not spend time detailing their importance, which we will cover in this section.

In Honeycode, there are two types of variables, distinguished by how they are made available: **user-defined variables** and **system variables**. They will both be discussed in the following subsections. In terms of representation, both types of variables are represented the same way: `$[Variable_Name]`.

User-defined variables

In Honeycode, user-defined variables are created and configured only through a **data cell**. As noted in *Chapter 2, Introduction to Honeycode*, a data cell allows us to display data from a table by creating direct mapping to it. At the same time, it can also be used to pass data from one app screen to another, as well as a temporary container to intake user data and update the table through automation. To distinguish between these two modes of data cell, they have been categorized under two types: **shared** and **variable**.

Shared data cell

As the name suggests, this variable is shared across the app users, which means all users will see the same value. The only way that this can be made possible across different user sessions is by the value being derived from a shared source, which, in the case of Honeycode applications, is the underlying tables. Thereby, the shared data cell type has the **set shared source** field, which gets its value assigned through a formula creating a binding to the underlying tables. When made editable, any change made to this value will reflect for all active and future app sessions.

Variable data cell

This variable data cell type is local to the session, which means each user can update its value independently from what is being done in other sessions of the same or a different user. The value of these variables is referred to by using the name of the data cell. Additionally, Honeycode provides an optional property for these variable types that can be used to set an initial value of this variable; a feature typically useful in scenarios where there will be a specific value assigned to the variable in most cases, and we want to provide that as a default option to users.

System variables

As the name suggests, these are system- (Honeycode-) provided variables and are available for use across the product. Honeycode provides three such variables:

- `SYS_USER`: This variable indicates the name of the person using Honeycode and works across the board in **Tables**, **Builder**, and **Automations**. The value returned is of the **Contact** data format and therefore, you can dereference the value similar to a contact cell and request only First name or only Last name or only Email, if desired.

> **Note**
>
> Unless you want the value of a cell to change to display the name of the last user saving data, you should avoid adding this formula to a table cell and only use it in App and Automations.

- `SYS_USER_GROUPS`: This variable resolves to the list of groups that the current user is a member of. This variable is only useful in the Plus and Pro tiers as they only support **single sign-on (SSO)** integration, and consequently allow the leverage of existing groups at the identity provider.

- `PREVIOUS`: This variable is only useful in Automations, and specifically to retrieve the previous value for a column or a cell that is configured as a trigger for the automation. Using this in other places will have undefined behavior and is, therefore, not recommended.

If this section feels a little sparse, simply make a note of the two variable types and their subtypes for now. We will see the use of both types of variables and their subtypes as we continue along with this chapter and update our app with automation, which should provide clarity to the concept of variables.

Making apps more powerful with automations

In *Chapter 2, Introduction to Honeycode*, we learned that Honeycode categorizes automation into two types – **Workbook automations** and **App automations**, also referred to as **Actions**. In this section, we will learn to add some of these actions to our **ToDo** app and provide the foundation for using actions.

Using actions for data input through forms

Data input into the system is one of the key use cases for most applications, except a few that are meant to be used as read-only. In *Chapter 3, Building Your First Honeycode Application*, we used the **Form** control provided in Honeycode to create the **Add New Task** screen for adding new tasks to our **ToDo** list. By now, you will have used that screen many times to create new tasks and found that they were being added to your list seamlessly.

However, when in *Chapter 4, Advanced Builder Tools in Honeycode,* we added two new fields to our list, namely `Notes` and `Assignee`, we added them to the **Edit Task** screen only. This essentially meant that we had not provided us or other App users with any way to add the notes and assign the task at the time of creating it. This was a deliberate choice made in the chapter because, while adding the controls to display those new fields to the **Add New Task** screen would have been like what we did for the **Edit Task** screen, that, by itself, would not have resulted in the value of those fields being saved to our list. Why?

Because of the simplification that was offered to us by Honeycode. When we added the **Form** control to the screen, it also created the required actions under the hood to save the data provided in the form. At that moment, we did not dive into how that was happening, but now we are ready to deconstruct this set of actions. In the next two subsections, we will first add the new fields to the screen and then extend the existing action definition to apply to our new fields and persist with them.

Adding the new fields to the Add New Task screen

Let's see how we can add new fields to the **Add New Task** screen:

1. Go to the **Add New Task** screen. Select **Form** and, using the **Add Objects** control, add two new segments to the screen, one for the `Assignee` field and the other for the `Notes` field. Drag these newly added segments and put them above the **Cancel** and **Done** buttons, as shown in *Figure 5.1*:

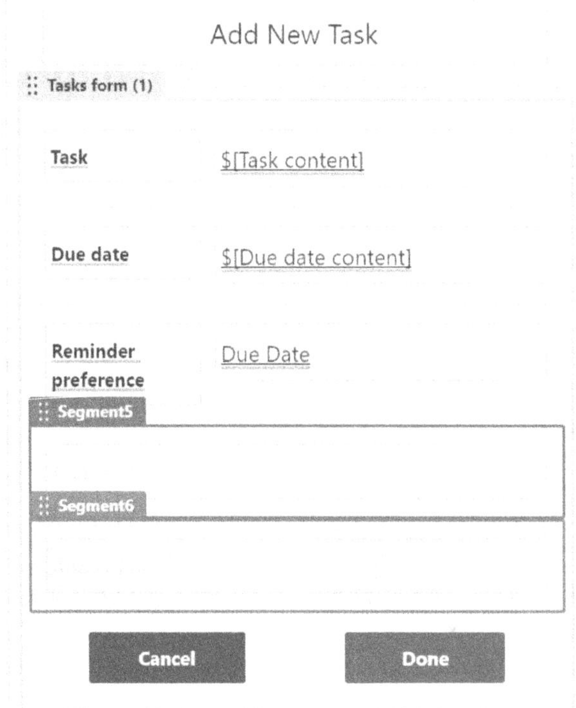

Figure 5.1 – Add New Task screen with the two new segments

2. Within these new segments, resize the existing content box to the left of the screen, such that the content box aligns under the existing ones with the field names. Now, add a data cell inside these content boxes.

   Set the shared source field of the data cell as follows:

   `=Tasks[[#Headers], [Assignee]]` and
   `=Tasks[[#Headers],[Notes]]`

3. These formulas will set the data cells to display the values from the **header row** of the **table column**. Also, with the content boxes selected, format the text as **bold** to keep it consistent with the previous fields, as shown in *Figure 5.2*:

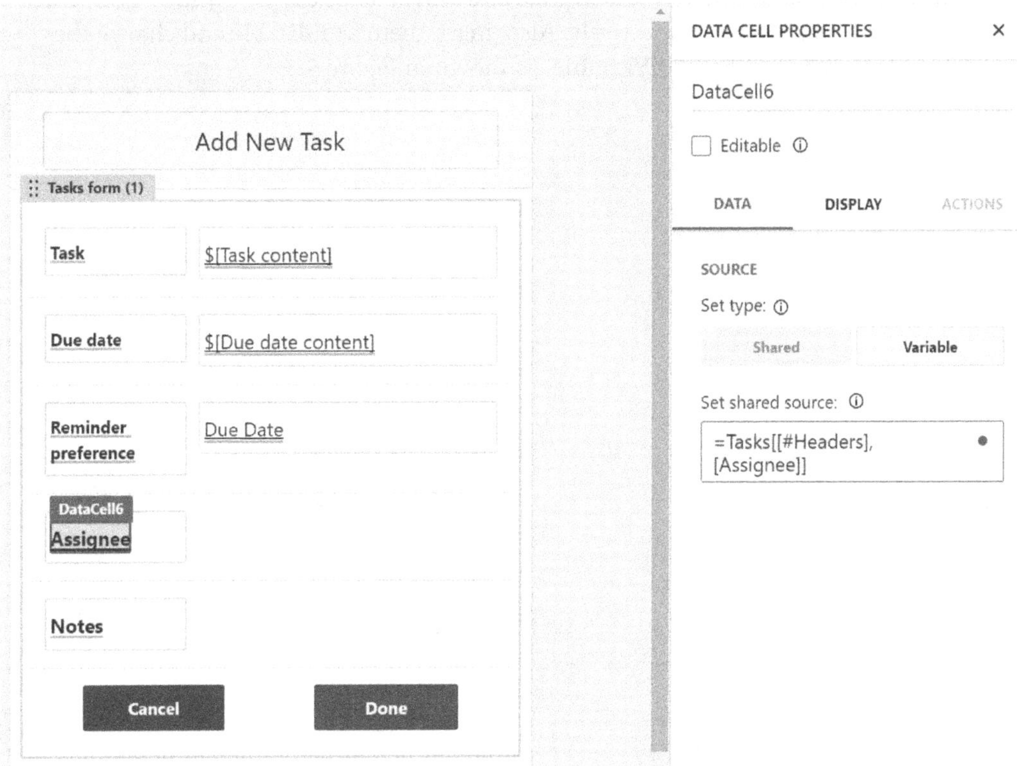

Figure 5.2 – Creating label data cells using the formula to display the table column header

Why Did We Use the Formula for Labels Compared to Writing the Label Values Approach We Had in the Edit Screen?

This is because the formula is a reference to the value of the header row of the column itself, which allows us to maintain a direct reference to the binding column. If, in the future, we change the column headers and want to reflect the same on the app, we would not have to make an edit to the app. It is important to note that each approach has its merit. While the use of formulas allows us to keep the screen labels consistent with those of the underlying column labels, on occasions, we do not want to expose these values outside to our users to keep our internal representation decoupled from the external, and for those scenarios, the approach used in the Edit screen is the way to go.

4. Now, right next to the content boxes in the two segments, add **Data cells** such that they align under the input fields.
5. Following the convention from these existing data cells, rename the newly added data cells corresponding to the **Assignee** and **Notes** fields as `Assignee content` and `Notes content`, respectively. Also, mark them as **Editable** and change the value of the **Set type** field to **Variable**, as shown in *Figure 5.3*:

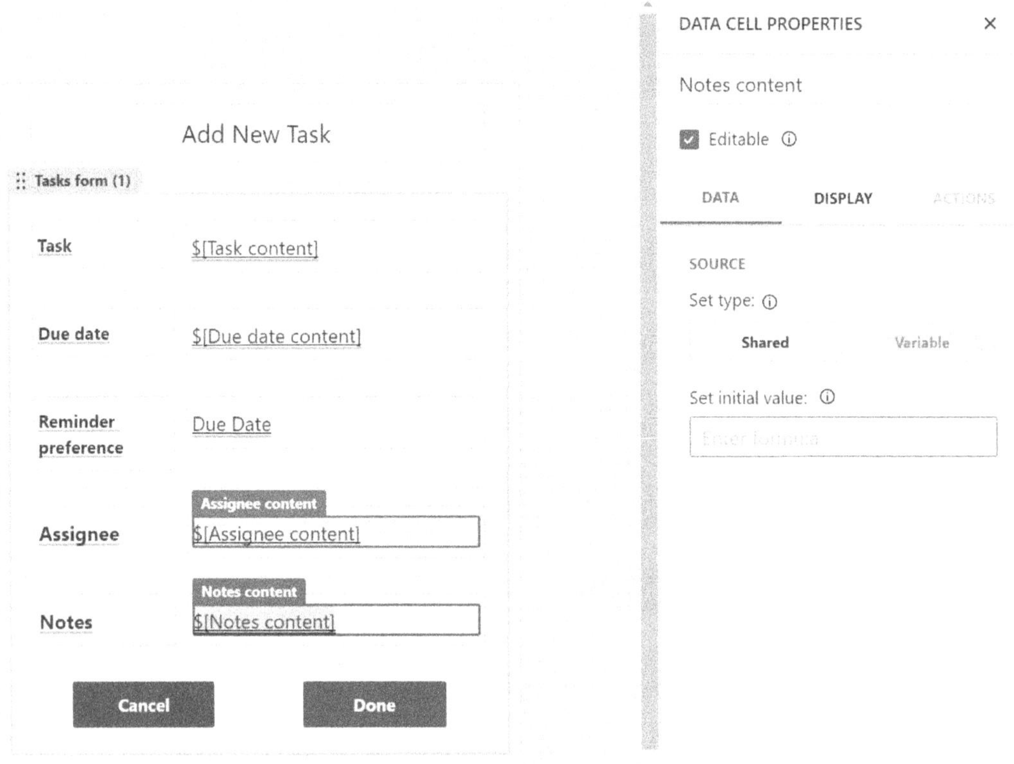

Figure 5.3 – Adding the data cells corresponding to the Assignee and Notes fields for user input content

6. While we are building functionality to be able to create and assign a task to anyone in the team, typically everyone most likely will create their own tasks. So, we simplify the task creation for them by selecting the default assignee to be the creator by setting the initial value of the **Assignee content** data cell as `=$[SYS_USER]`, as shown in *Figure 5.4*:

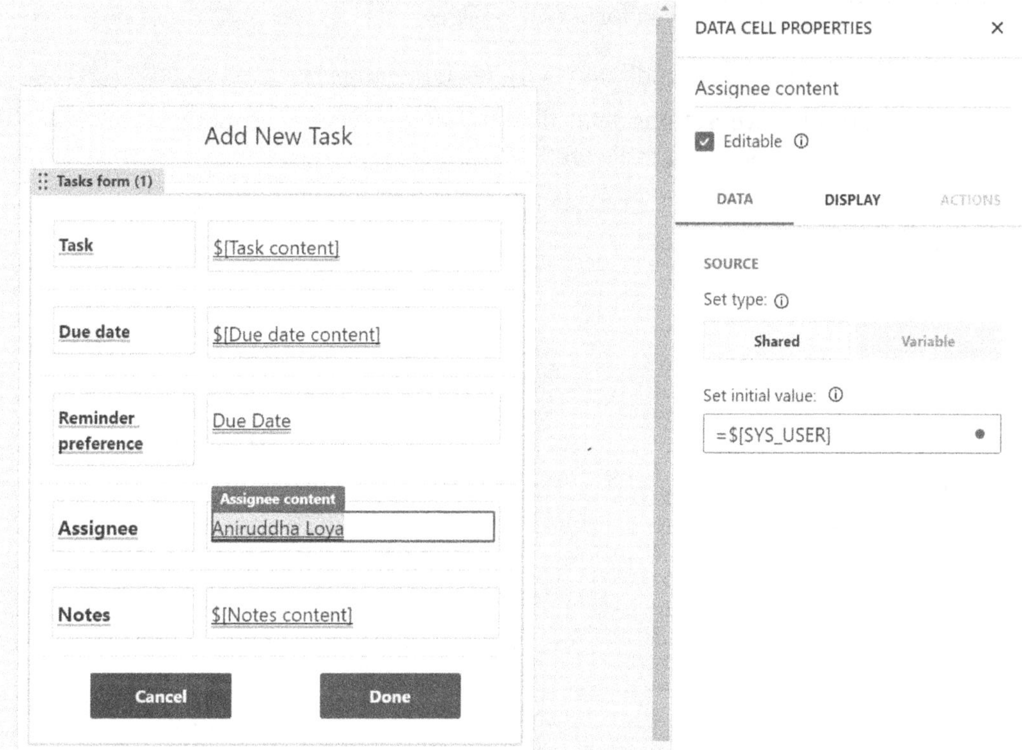

Figure 5.4 – Setting the default Assignee value to be the user

> **Note**
>
> Notice the difference between *Figure 5.3* and *Figure 5.4*. After I added the formula to the **Set initial value** property on the **Assignee content** data cell, the value inside the cell changed to my name, `Aniruddha Loya`, as I am the current user of the system.

Extending the existing Actions to persist the values for new fields

In the previous subsection, we added two new fields on the **Add New Task** screen. However, as we had noted earlier, if you try to add a new task with these fields filled, the values will not be retained, and the fields would show up empty in the **My Tasks** list view. So, now let's update the automation defined on the **Done** button and learn how we can persist the information from a given form to our tables:

1. Click on the **Done** button and select **Actions** in the property panel.
2. Since we already have **Actions** defined on this button, we see an option that reads **Edit automation**, as shown in *Figure 5.5*:

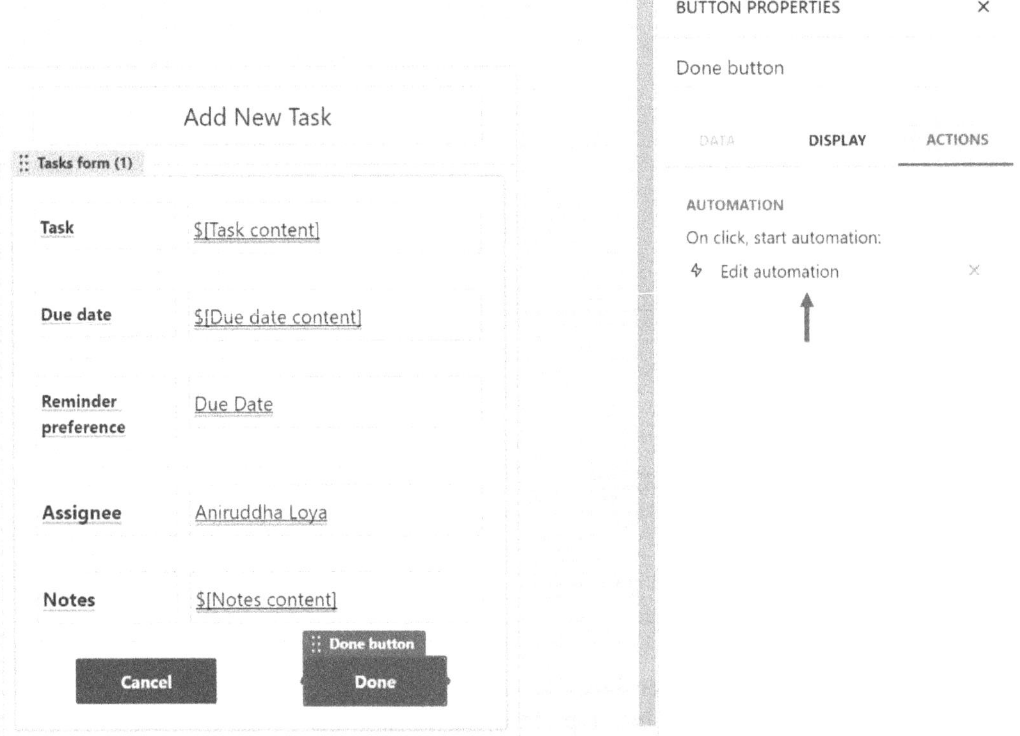

Figure 5.5 – Editing the existing automation defined on the Done button

> **Note**
>
> Had there not been an automation defined here, we would have seen an option for **Create an automation**, along with a couple of other common actions, under the **QUICK ACTIONS** property that Honeycode has made it even more straightforward to configure. Refer to *Figure 5.6* for my example:

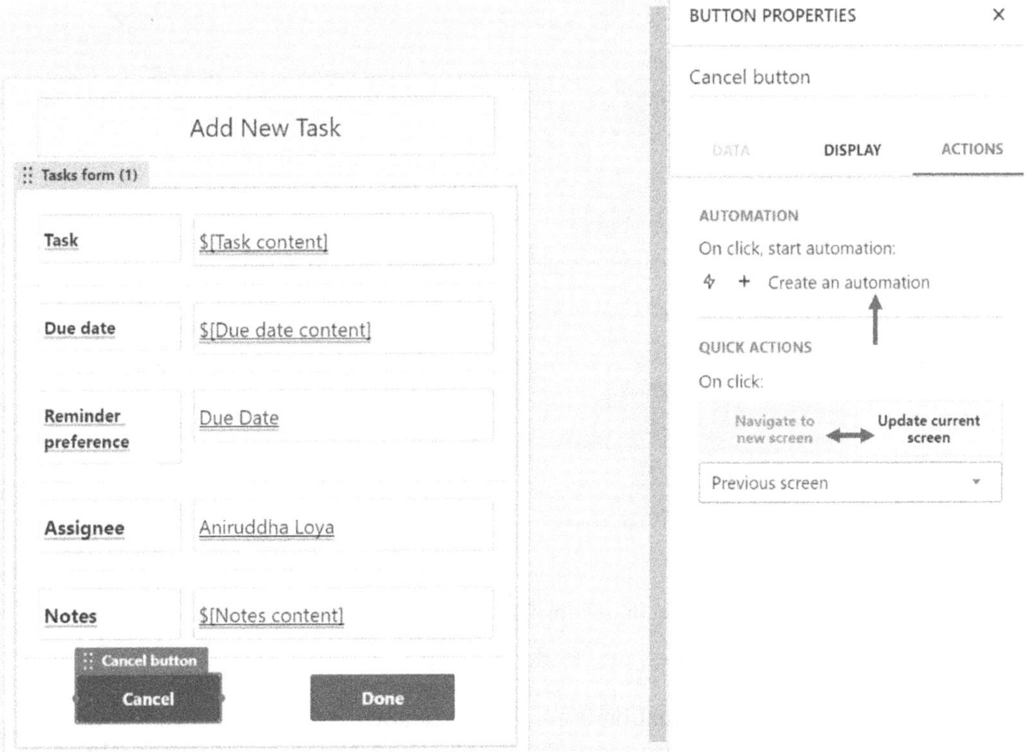

Figure 5.6 – Adding the AUTOMATION option along with QUICK ACTIONS

3. On clicking **Edit automation**, the right panel updates to display the configured automation in the edit mode with the option to add more action blocks or update the existing ones, as shown in *Figure 5.7*. There are two actions defined for this automation – the first block is configured to add the new task to the table and the second block defines the navigation upon completion of the first action.

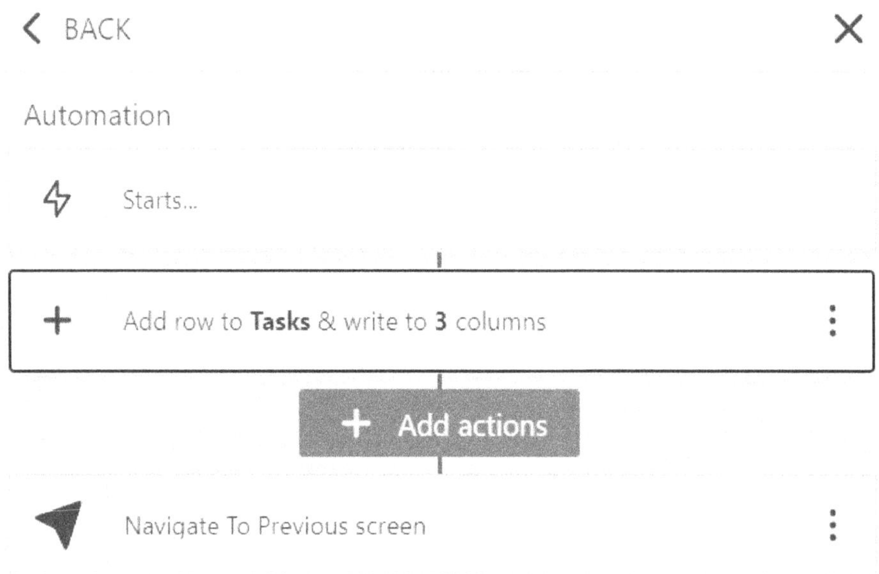

Figure 5.7 – Editing the automation defined on the Done button

4. Let's take a moment to review and understand these blocks a little more:

   I. The first block of automation is configured to add a new row to the tasks table using the **Add row to:** property. Below that, three sections are configured to take the data from the variables that were defined for taking the inputs from the existing fields, and write it to the cell of the newly added row identified by the chosen table column. And finally, there is the **Add another** link to add more sections, which is what we will use to define the mapping for our new data cells and the corresponding columns in the table, as shown in *Figure 5.8*:

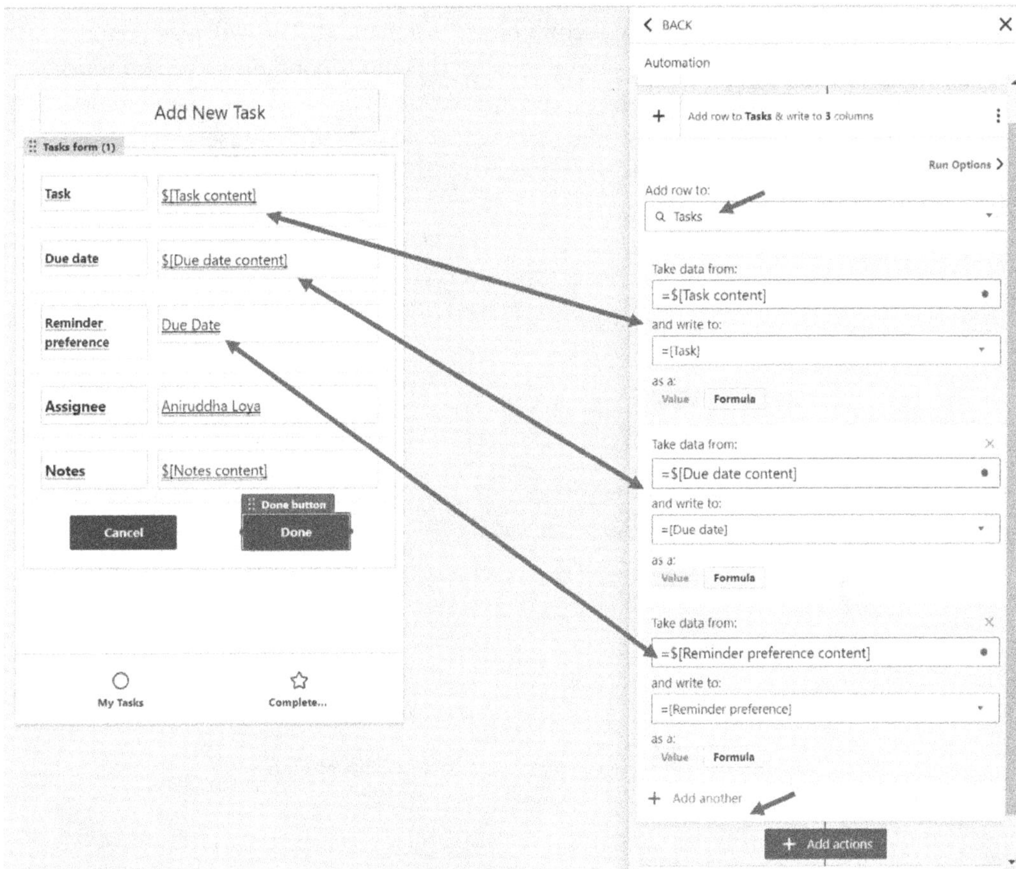

Figure 5.8 – Editing more automation defined on the Done button

> **Note**
>
> You can see how the variables defined on the data cells in the screen are referenced in the **Take data from:** field using the variable representation syntax.

II. Now, click on the row of the second block to expand it. This will also result in collapsing the first block automatically. This block configures the action to navigate back to the previous screen, the action to be performed on completion of the previous block, and adding a new task, which, in this case, is to navigate back to the previous screen. This is shown in *Figure 5.9*:

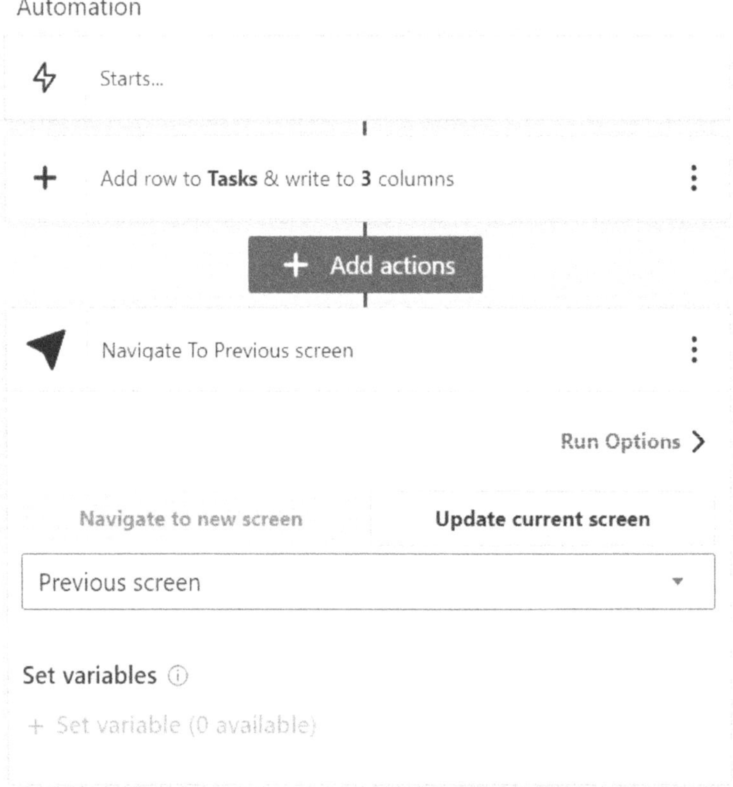

Figure 5.9 – Editing the automation defined on the Done button

> **Note**
>
> This is an App automation and can only take place on a click event on the screen object, a button. In this case, the start blocks of automation do not allow any configuration conditions and we can only configure the actions to be performed.

5. Now, let's expand the first block again and click on the **Add another** link. This will add another section similar to the previous three. In this section, fill **Take data from:** with `=$[Assignee content]` and for the **and write to:** property, select the `=[Assignee]` column from the dropdown, as shown in *Figure 5.10*:

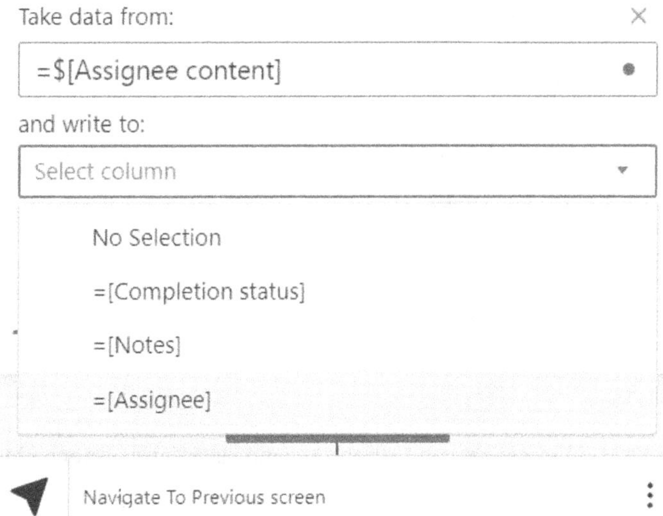

Figure 5.10 – Mapping the Assignee content variable value to the Assignee table column

6. Repeat the action to create the mapping for notes by setting `=$ [Notes content]` and selecting the `= [Notes]` column for mapping. The automation block one will now have five sections with the last two newly added ones looking as shown in *Figure 5.11*:

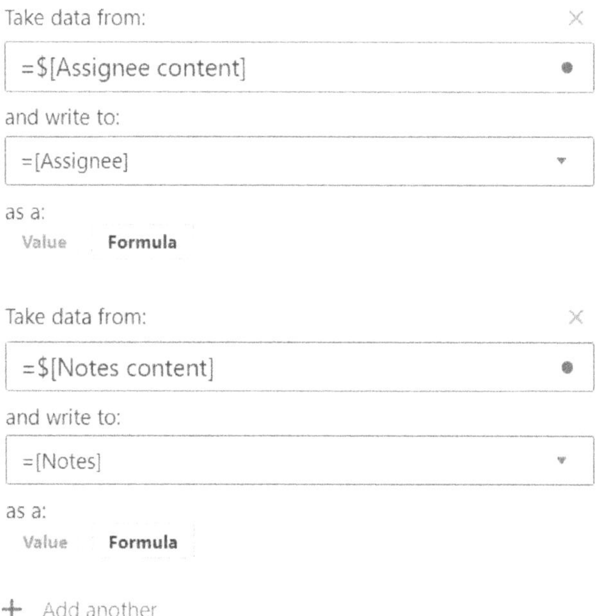

Figure 5.11 – Automation with both the Assignee and Notes sections added with mapping to a table column

7. Finally, click on the **BACK** link at the top of the panel, as shown in *Figures 5.7* and *5.8*. With that, the automation, and thereby the app, is updated to save the value of the assignee and notes if any were added at the time of task creation.

Adding functionality to delete a task

Like adding a new task row, we can also delete a row using actions defined on a control. For building this functionality, we will define the action on the **Edit Task** screen with a new **Delete** button using the following steps:

1. Click on the **Edit Task** screen and select the segment containing the **Done** button. Using the **Add Objects** pane, select and add a button in the same segment. Rearrange to align the two buttons side by side.

2. With the newly added button selected, double-click on the button to change its text. Update the value to **Delete**.
3. In the next step, set the text format to **Bold** and using the fill color tool, set the background color of the button to **Red**, as shown in *Figure 5.12*:

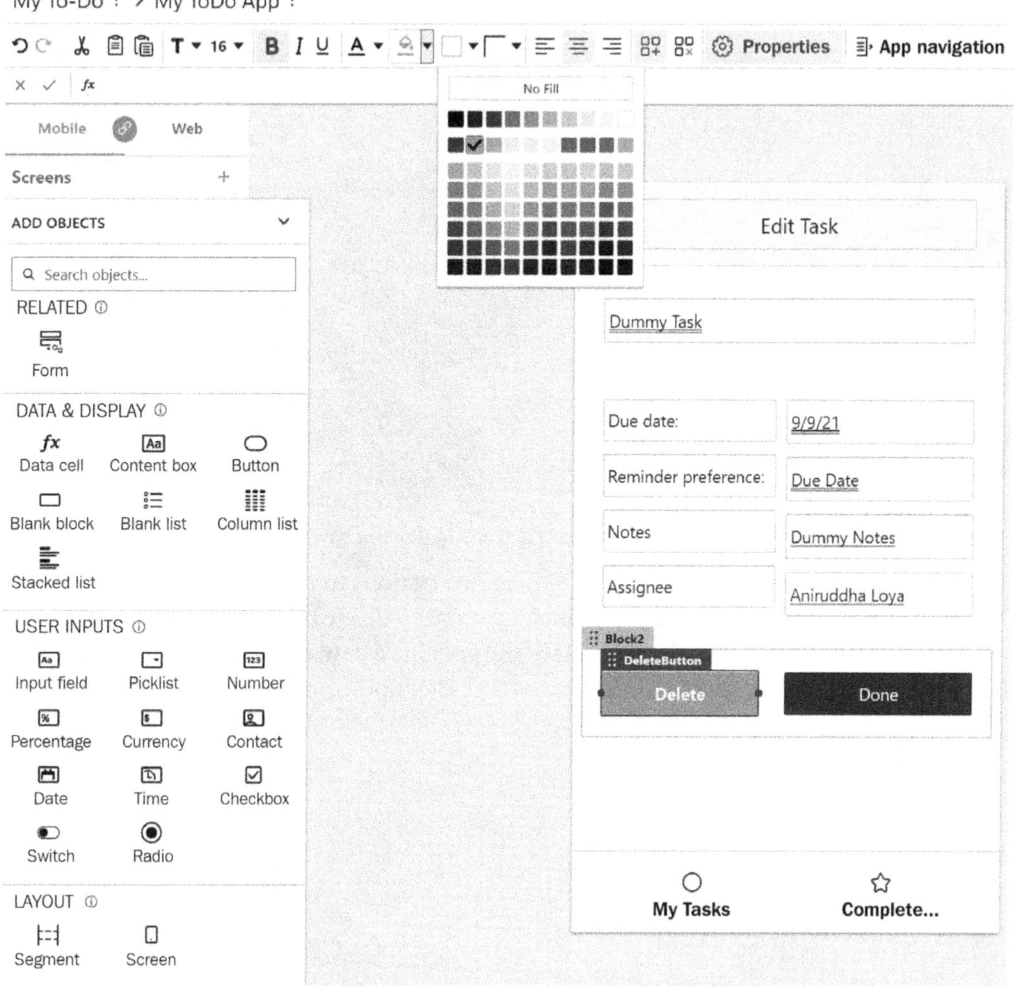

Figure 5.12 – Creating and stylizing the Delete button

4. Now, click on **Create an automation** under the **Actions** tab for this **Delete** button and bring up the **App automation** editor.

5. Click **Add actions** and select **Delete a row** from the list of possible actions, as shown in *Figure 5.13*:

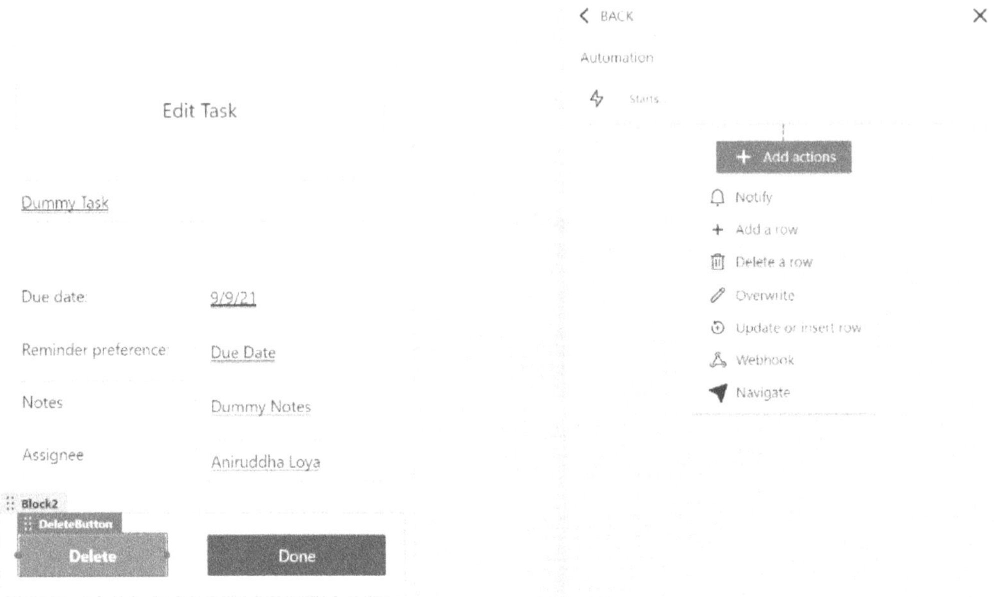

Figure 5.13 – Adding actions to delete a row

6. In the action block that gets added, we see two options to choose from for determining which row to delete. Since we want to delete the task we were editing, we leave the default selection to **Delete the specified row** and set the formula in the next field as `=$[InputRow]`. Unsure where the inputrow value is coming from? Refer to *Chapter 3*, *Building Your First Honeycode Application*, subsection *Creating the Edit Task screen.*

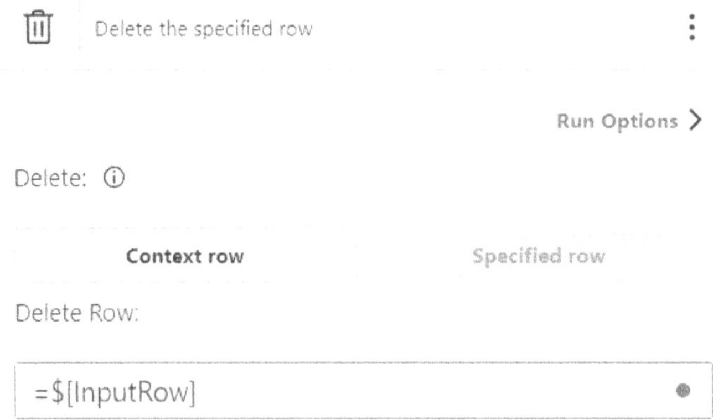

Figure 5.14 – Delete Row block

7. Lastly, we add another action to navigate and choose the **Previous screen** option as the destination where we would like to return once the task is deleted.

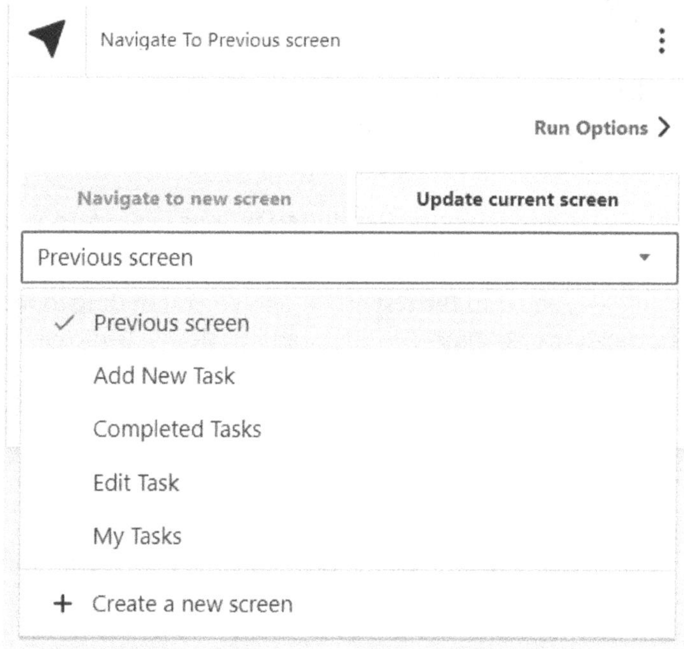

Figure 5.15 – Configuring the app to navigate to the previous screen after deleting the task

And, with that, we have added the functionality for deleting a task. Now, wasn't that easy!

Processing data with automation based on triggers

In the previous section, we learned how different automations are configured on the objects on the app screen. But not all automation needs can be fulfilled by them. Sometimes, we need to start processing an automation based on triggers that may not be a direct consequence of user action. To address such cases, Honeycode provides users with the option to create automation at the workbook level and configure them to be triggered on events, such as reaching a specific date and time and changing the value of a cell in a specific column. In the following subsections, we will learn how to configure such automations as we work toward building the use cases we defined at the beginning for updating our **ToDo** app.

Sending reminders based on set preference

So far, we have enabled our **ToDo** app user to set the preference but haven't made use of that information for anything. Let's fix that by creating automation using the **Date & Time** reached trigger.

Currently, we have taken the user preference for a reminder in text format. However, for a trigger to work, we need to translate that to a specific date and time. So, let's do the translation using the following steps:

1. Go to the **Tables** view and select the **ReminderOptions** table. Add a column to the table and set its name as **Number of days**.
2. Fill the values of `0`, `1`, and `7` in the respective cells corresponding to the **Remind On** values of `Due Date`, `1 Day Before`, and `1 Week Before`, as shown in *Figure 5.16*:

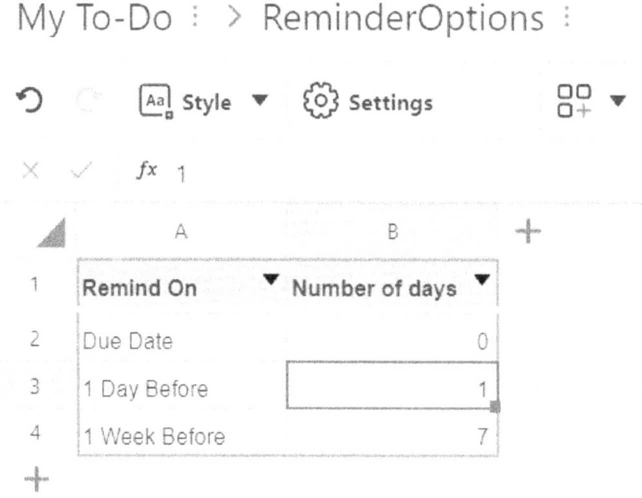

Figure 5.16 – Adding a Number of days column to the ReminderOptions table and filling its values

3. Next, select the **Tasks** table and add a column with the name set as **Notify on**.
4. In the table column properties, set `formula = [Due date] - [Reminder preference] [Number of days]`, as shown in *Figure 5.17*:

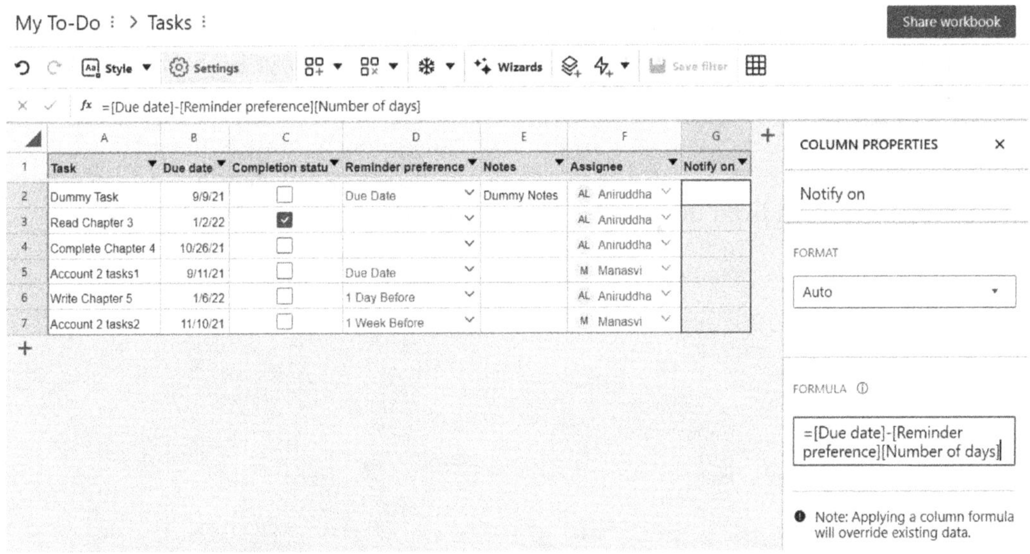

Figure 5.17 – Setting the table column formula to compute the date to send a reminder on

Once you hit *Enter* or click outside of the formula field, the formula will be copied to every cell in that column and compute the date (see *Figure 5.18*) to send a notification that we can now use for configuring our automation trigger.

My To-Do > Tasks

Style | Settings | Wizards | Save filter

fx 1 Week Before

	Task	Due date	Completion status	Reminder preference	Notes	Assignee	Notify on
2	Dummy Task	9/9/21	☐	Due Date	Dummy Notes	AL Aniruddha I	9/9/21
3	Read Chapter 3	1/2/22	☑	1 Day Before		AL Aniruddha I	1/1/22
4	Complete Chapter 4	10/26/21	☐	1 Week Before		AL Aniruddha I	10/19/21
5	Account 2 tasks1	9/11/21	☐	Due Date		M Manasvi	9/11/21
6	Write Chapter 5	1/6/22	☐	1 Day Before		AL Aniruddha I	1/5/22
7	Account 2 tasks2	11/10/21	☐	1 Week Before		M Manasvi	11/3/21

Figure 5.18 – Table column applied to all cells resulting in computed dates for sending reminder notifications

5. Navigate to **Automations** using the left nav bar. Create a new automation and rename it `Reminder Automation`.

6. Select **Date & Time Reached** in the **Automation trigger** options. Select the **Tasks** table using the dropdown for **Automation will start once for each row in this table**.
7. Next, set the number of days to `0` in the first field under the **Date & Time (UTC)** label, and instead of **this date**, use the dropdown to select **row date**. Then, select the **Notify on** column from the **Select column** dropdown, as shown in *Figure 5.19*:

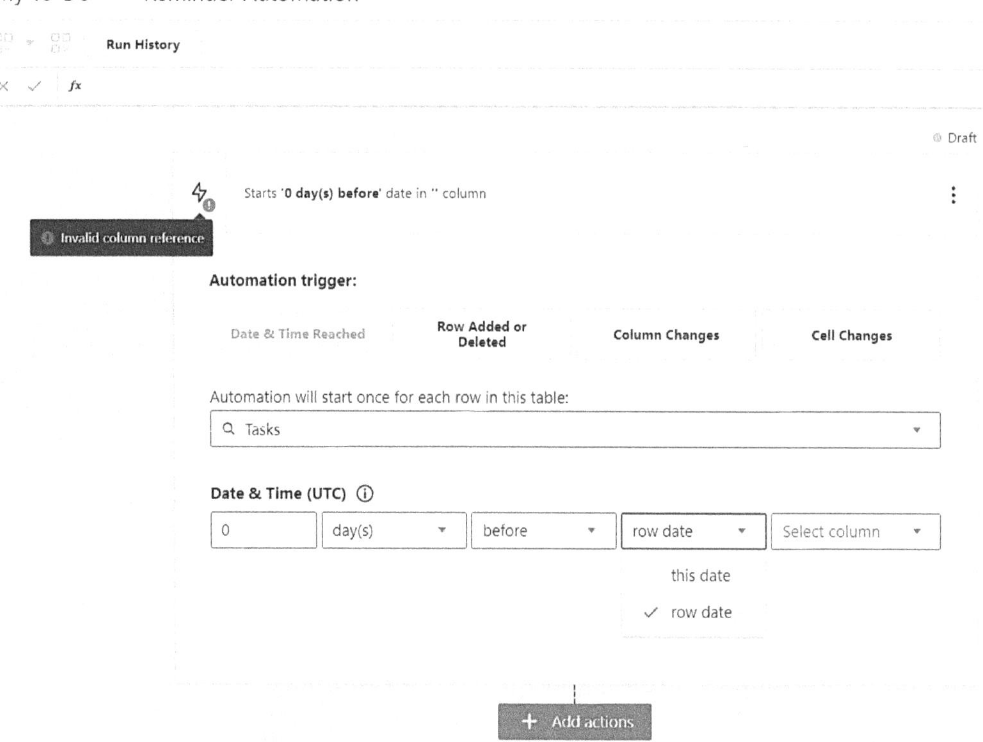

Figure 5.19 – Configuring the start condition block

8. Now, using **Add actions**, add a **Notify** block.
9. Fill this action block as follows:

   I. Set the To field to =[Assignee].

   II. Set **Subject** to `Reminder: =[Task] due on =[Due date].`

   III. Set **Message** to `This is a reminder to complete =[Task].`

   IV. There is no point in sending a reminder for completed tasks, so let's exclude the completed tasks. Click on the **Run** options and set `=[Completion status] = FALSE` as the condition to run this step.

10. Refer to *Figure 5.20* to see the result.

> **Note**
> The formulas will be resolved by Honeycode and will replace those with values from the first row qualifying the criteria.

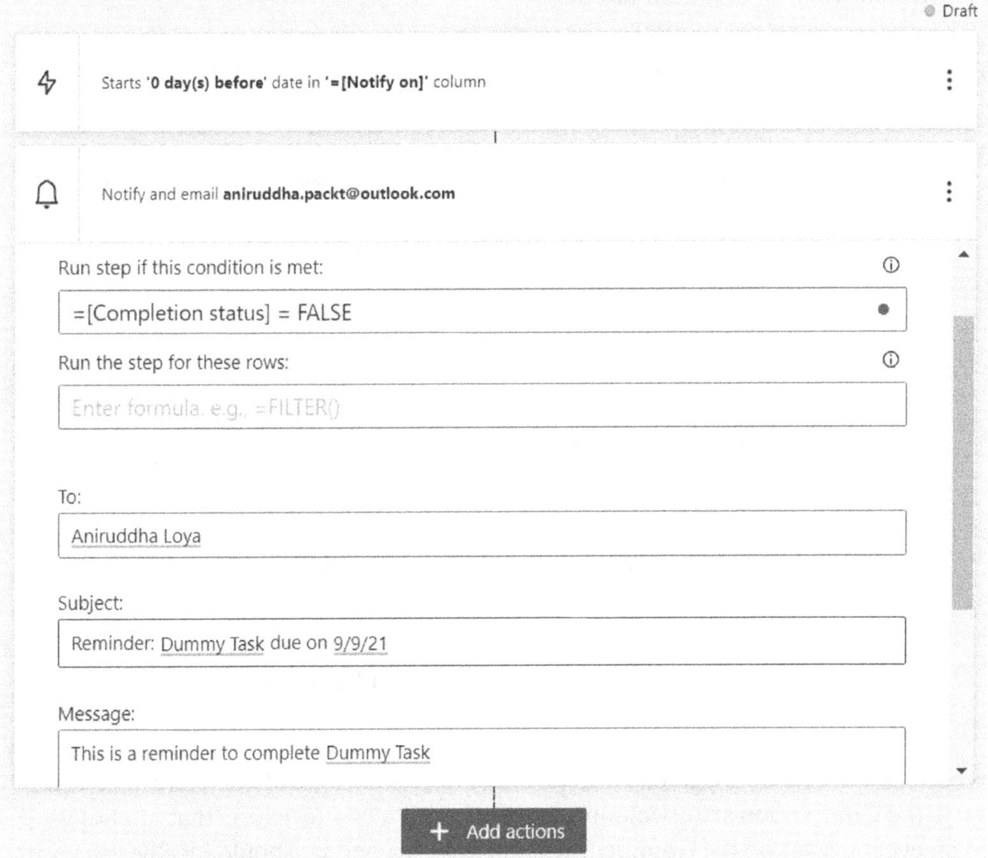

Figure 5.20 – Configured notification block

11. You may choose to add the link to the app in the notification message using the **Attach link to app** dropdown.
12. Finally, hit the **Publish** button on the top right, and with that our automation is set to remind our app users when a task is due.

> **Note**
>
> The toast at the bottom of the screen confirms the publishing of the automation and also that the orange dot symbolizing a draft automation goes away.

Honeycode sends notifications both as an email and as a push notification on the Honeycode app on the mobile, as shown in *Figure 5.21*. The notification also shows up under the notifications on the left nav bar.

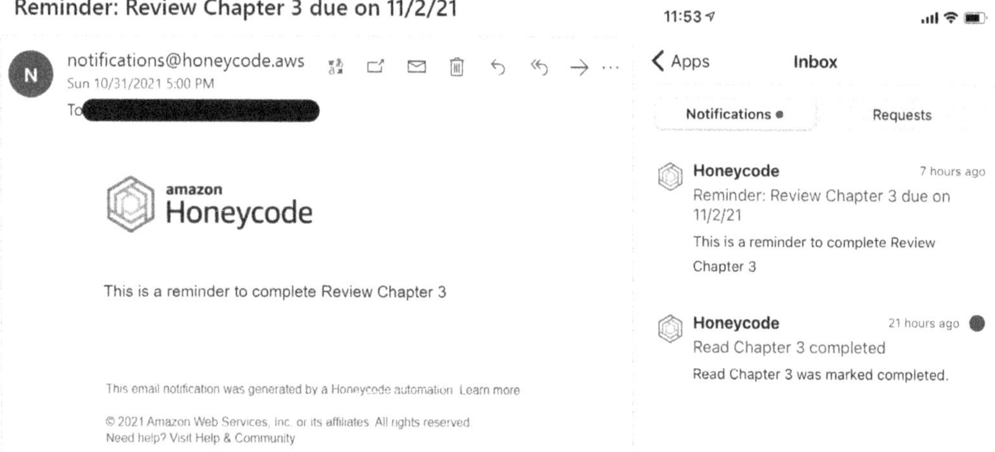

Figure 5.21 – Sample notifications, with email on the left and the Honeycode mobile iOS app on the right

Sending a task completion notification to the task creator

This automation is to be triggered when any task's status changes to complete, that is, its value in the **Completion** status column changes from `False` to `True`. That implies we need an event trigger on the **Completion** status column, and we should ideally also ensure that the previous value was not the same as the current one.

> **Exercise 1**
>
> To notify the creator of the task, we need to know who created the task in the first place. So, the exercise is to update the app to save the information of the user who created the task. Hint – you will need to use the `SYS_USER` system variable to capture the user who created the task.

After completing the preceding exercise, follow the next steps to build this automation:

1. Create a new automation and rename it to **Completion** notification.
2. In the **Start** block, select the **Column Changes** option under **Automation trigger**. Choose the **Tasks** table and the **Completion status** column, and set the condition to run automation as `=AND([Completion status] = TRUE, $[PREVIOUS] = FALSE)`, as shown in *Figure 5.22*:

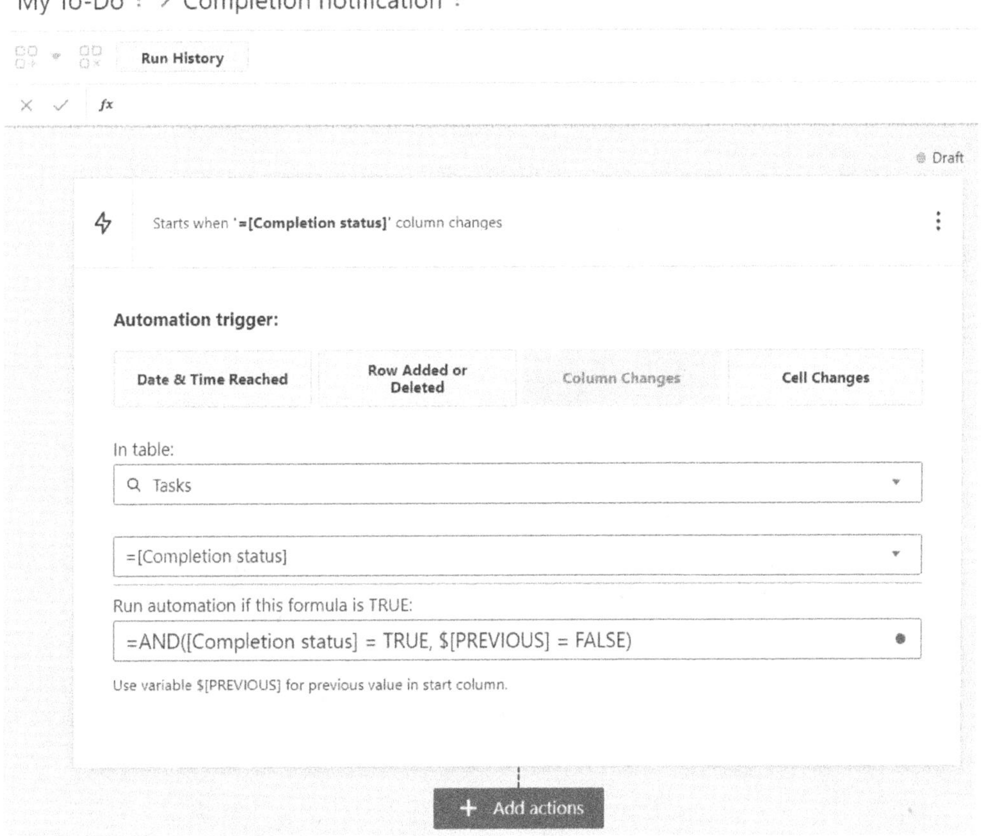

Figure 5.22 – A configured Start block

3. Now, using **Add action**, add a **Notify** block.
4. Fill this automation block with the following points. Refer to *Figure 5.23* for the completed block:

   I. Set the **To** field to `= [Created by]`.

   II. Set **Subject** to `= [Task] completed`.

III. Set **Message** to = `[Task] was marked completed.`

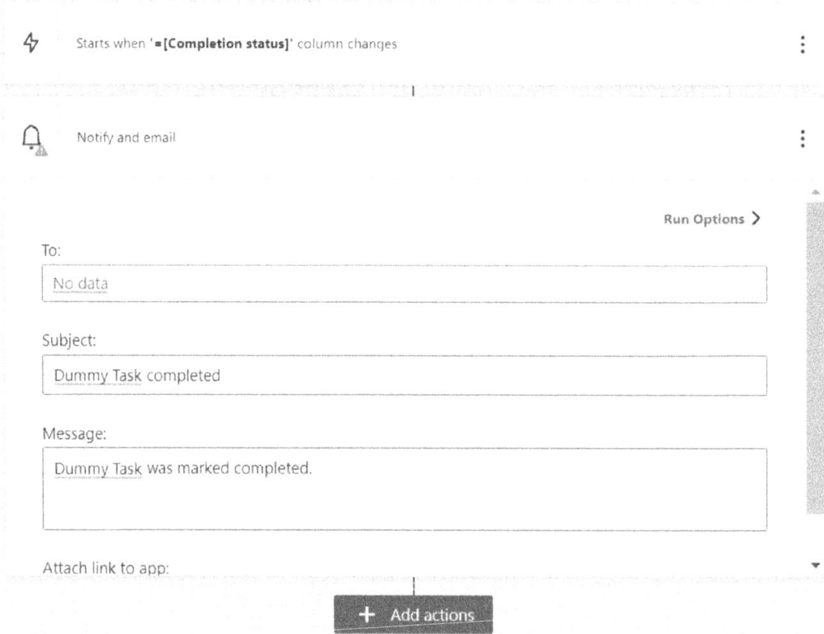

Figure 5.23 – A completed notification block

5. Publish the automation and we are done!

   Figure 5.24 shows a sample notification that was triggered when I marked one of the tasks I had created as completed:

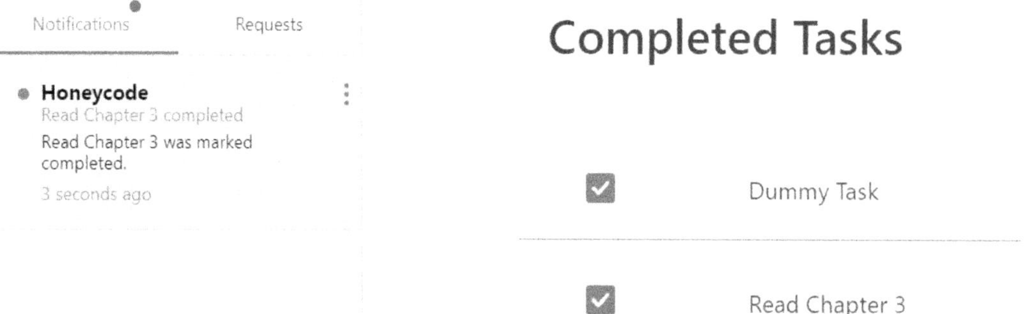

Figure 5.24 – Example notification sent to the creator on task completion

> **Exercise 2**
>
> Send a notification to the new and old assignees of the task.

Debugging automations

Debugging is an exercise to identify what went wrong in the execution of a scenario that resulted in unexpected behavior. This is a constant friend of any application built by anyone, and Honeycode is no exception to it. Debugging can be simple when you can see and test what was wrong, but is difficult when a scenario failed due to some external triggers beyond your control. For such cases, you need some information (logs) about the execution. For Honeycode automation, this information is available through the **Run History** button on the toolbar. An example of run history is shown in *Figure 5.25*:

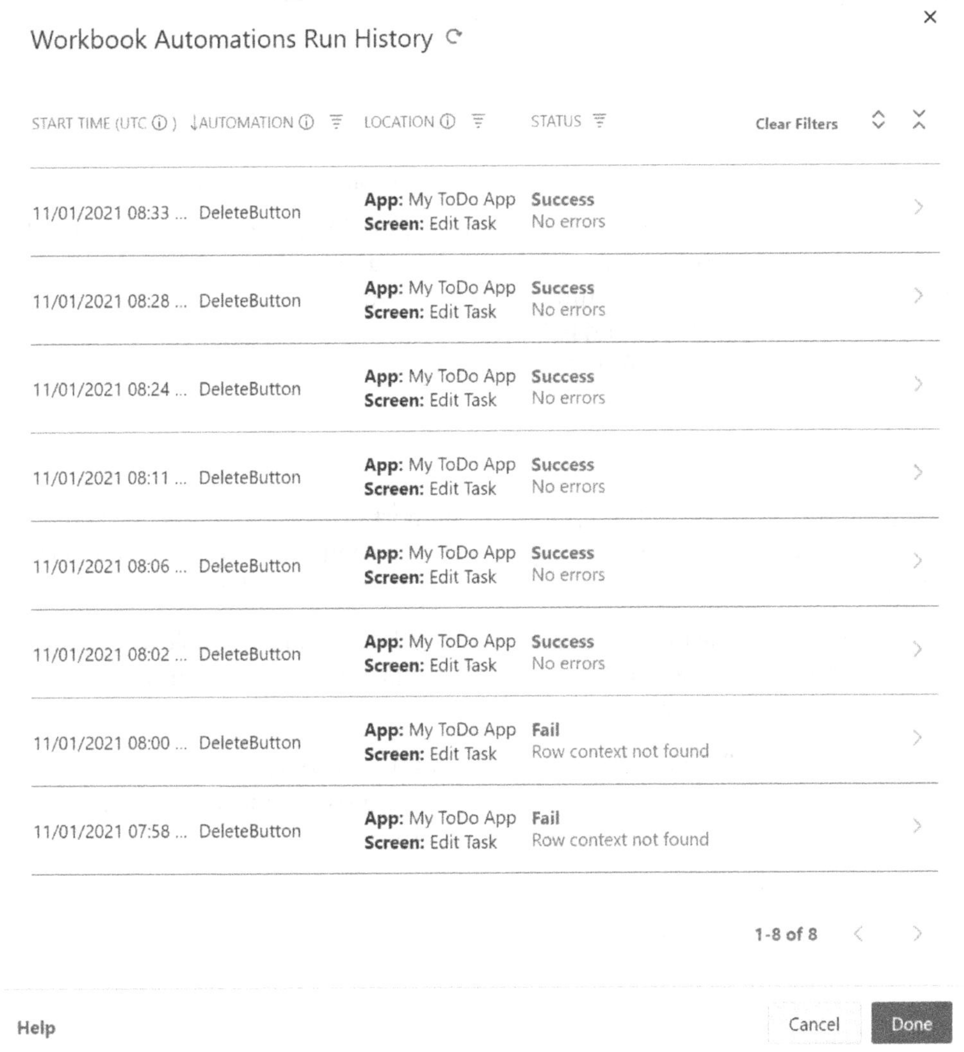

Workbook Automations Run History

START TIME (UTC)	AUTOMATION	LOCATION	STATUS
11/01/2021 08:33 ...	DeleteButton	**App:** My ToDo App **Screen:** Edit Task	**Success** No errors
11/01/2021 08:28 ...	DeleteButton	**App:** My ToDo App **Screen:** Edit Task	**Success** No errors
11/01/2021 08:24 ...	DeleteButton	**App:** My ToDo App **Screen:** Edit Task	**Success** No errors
11/01/2021 08:11 ...	DeleteButton	**App:** My ToDo App **Screen:** Edit Task	**Success** No errors
11/01/2021 08:06 ...	DeleteButton	**App:** My ToDo App **Screen:** Edit Task	**Success** No errors
11/01/2021 08:02 ...	DeleteButton	**App:** My ToDo App **Screen:** Edit Task	**Success** No errors
11/01/2021 08:00 ...	DeleteButton	**App:** My ToDo App **Screen:** Edit Task	**Fail** Row context not found
11/01/2021 07:58 ...	DeleteButton	**App:** My ToDo App **Screen:** Edit Task	**Fail** Row context not found

Clear Filters

1-8 of 8

Help

Cancel Done

Figure 5.25 – Run history of an automation

Another useful trick for debugging or temporarily disabling some of the automation blocks is to set the `run` condition to `False()`, in which case the entire block is skipped during execution, as shown in *Figure 5.26*:

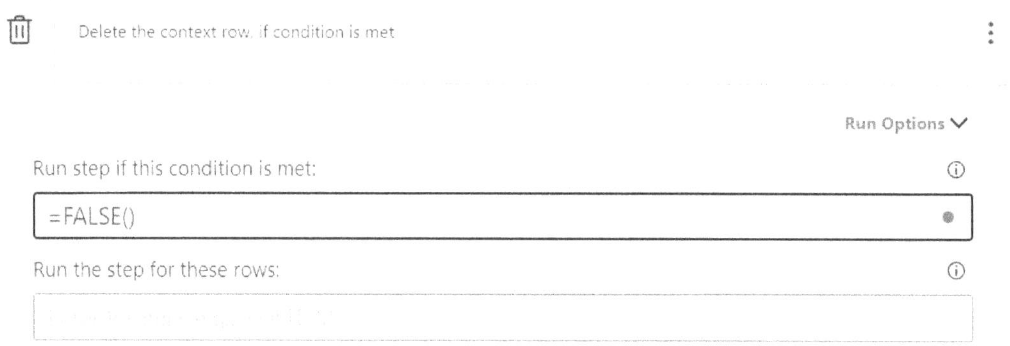

Figure 5.26 – Debugging automation by temporarily disabling the execution of a block

Now, while the debugging options for automations are minimal, the two ways listed previously are good to get us through a wide variety of automations, and there can be other tricks that we can develop over time. But, editing and testing an automation for a workbook or an app that is shared and being used is risky and can create side effects, and, therefore, requires better support from Honeycode.

Summary

In this chapter, we learned about the automation support systems in Honeycode and how to use them. We did so while updating our **ToDo** app to completion by adding the missing functionalities.

We have now covered all the major pillars of Honeycode and should be able to understand any application built using this platform, as well as build applications of varying levels of complexities. This concludes the first part of the book.

In the next chapter, we will learn about the **templates** that are offered by Honeycode, and in the following chapters, we will deconstruct some of those to supplement our learning by looking at different use cases and building patterns.

Part 2: Deep-Dive into Honeycode Templates

In this part, readers will get to know what Honeycode templates are, and what use cases are available out of the box with these templates. We then dive into four of those templates with increasing levels of complexity to further understand how they are made and how they work, enabling readers to modify and adapt them for their own usage. Furthermore, the template deep-dives are structured to not just provide a framework for analyzing other templates but also to enable the reader to structure their ideas for building their own apps from the ground up.

This section comprises the following chapters:

- *Chapter 6, Introduction to Honeycode Templates*
- *Chapter 7, Simple Survey Template*
- *Chapter 8, Instant Polls Template*
- *Chapter 9, Event Management Template*
- *Chapter 10, Inventory Management Template*

6
Introduction to Honeycode Templates

A **template** can be defined as something that can serve as a model, used by others to replicate various outcomes, which, in return, will provide consistency, save time, lower effort, and much more.

This set of Honeycode templates aims at helping users to jump-start their journey with the product as well as provide a showcase of the features and capabilities of Honeycode. This chapter is a bird's-eye view of all the templates that Honeycode currently provides and the use cases that are currently enabled out of the box through them.

After reading the first five chapters, you will be confident with your skills and will contemplate the use case of templates, since you can now build everything yourself. While you may be able to develop everything from the ground up, as you go through the templates in the current and subsequent chapters, you will realize that there is much to be gained with this additional knowledge.

You are not just enabling yourself to get a head start in building your next great application; you'll also be learning various patterns and best practices in Honeycode by reviewing these template workbooks and apps.

In this chapter, we're going to cover the following main topics:

- Getting to know Honeycode templates
- Learning how the templates work

Technical requirements

In order to follow this chapter, you'll need to have access to Amazon Honeycode, which requires a laptop with a web browser, preferably Google Chrome, and optionally a mobile device running either a Honeycode-supported version of Android (which currently requires Android 8.0 and up) or iOS (which currently requires iOS 11 or later).

Getting to know Honeycode templates

Honeycode is Amazon's offering in the space of low-code/no-code platforms. But, as we browsed through the last three chapters trying to build our **To-Do** app, you must have realized that it is not a no-code platform in the literal sense and does require some understanding of data modeling (databases), as well as the occasional need for Boolean logic and formulas. And therefore, to simplify the platform for citizen developers, Honeycode provides a decent set of templates, covering a variety of everyday use cases, allowing for easier adoption. These templates also serve as samples for learning various constructs and features of the product.

> **Note**
>
> As of *November 2021*, Honeycode is available with 19 templates, as shown in *Figure 6.1*:

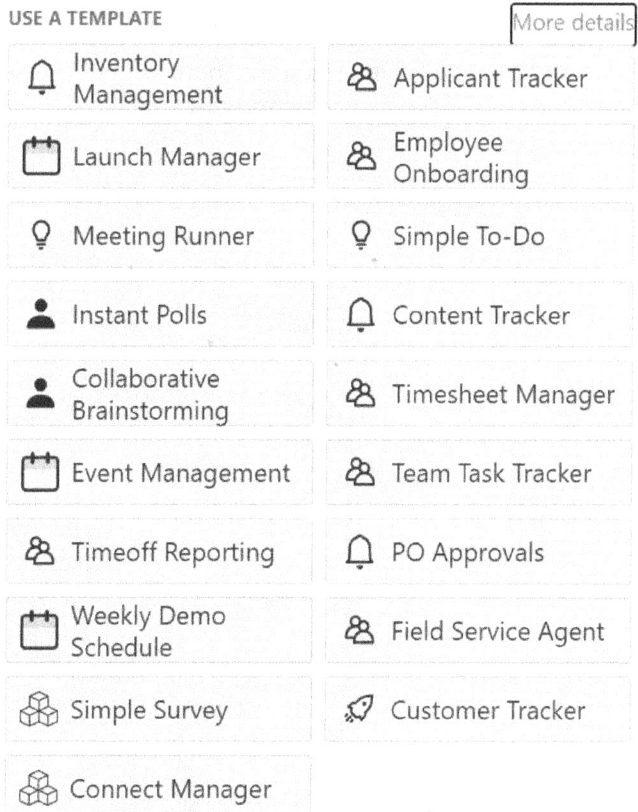

Figure 6.1 – The templates available in Honeycode

In the following subsections, we will take a brief look at each of these, and in the next set of chapters, we will also do a deep dive into some of them. Details for these can also be found on the official website at `https://www.honeycode.aws/templates`.

Applicant Tracker

Have you found yourself or your team using emails or some form of spreadsheets to manage an internal hiring process? If so, you may find this template is ideal for you. This template comes with two apps:

- One for job postings with applicant tracking
- One for managing an interview process

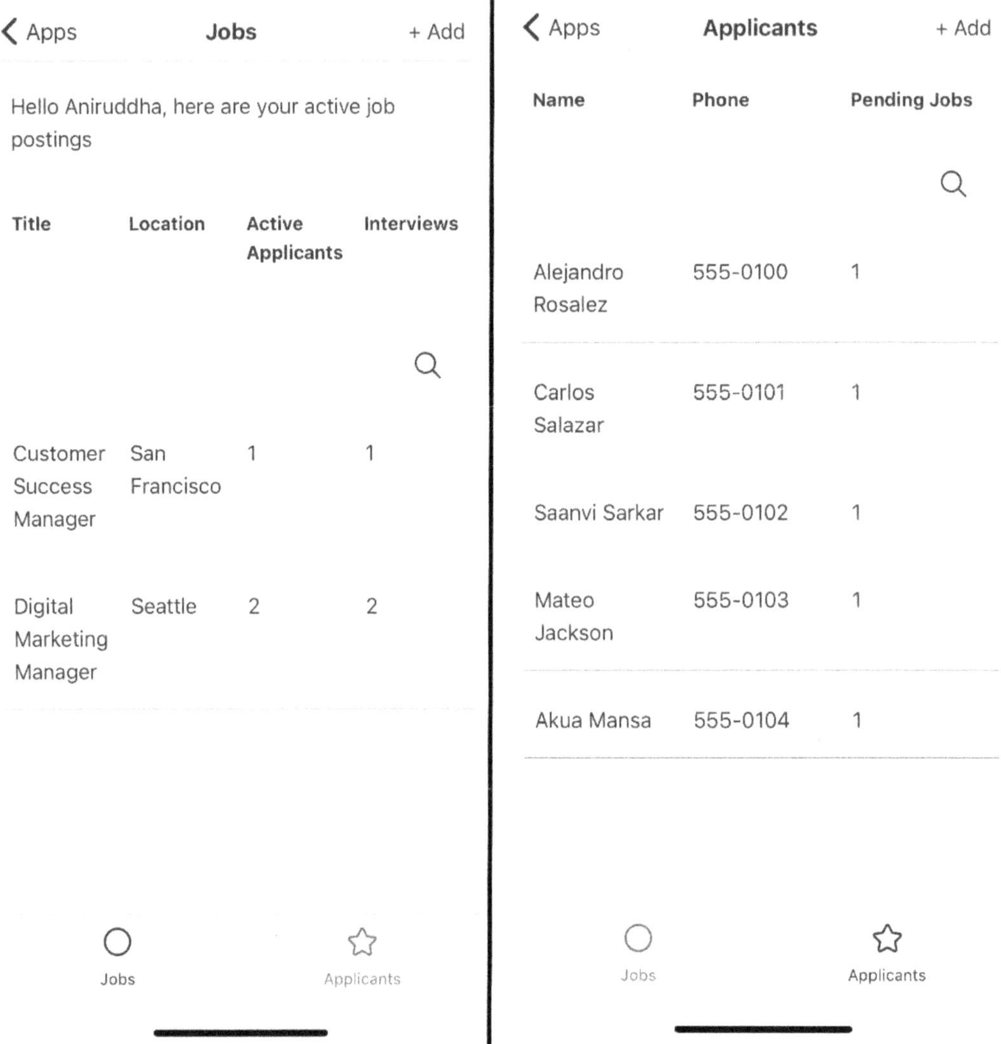

Figure 6.2 – Screenshots of the Jobs and Applicants screens from the Jobs app, captured on an iOS device

Figure 6.2 shows the two primary screens from the **Jobs** app:

- One that lists the active job postings along with their details and is personalized to only show listings for which you are the hiring manager
- The other listing all the applicants in the system.

The hiring manager or the recruiter can add an applicant to a specific job and then use the app to schedule the interview, which will then show up in the other app. It also allows the hiring manager to provide feedback and the final hire/no-hire vote for the candidate.

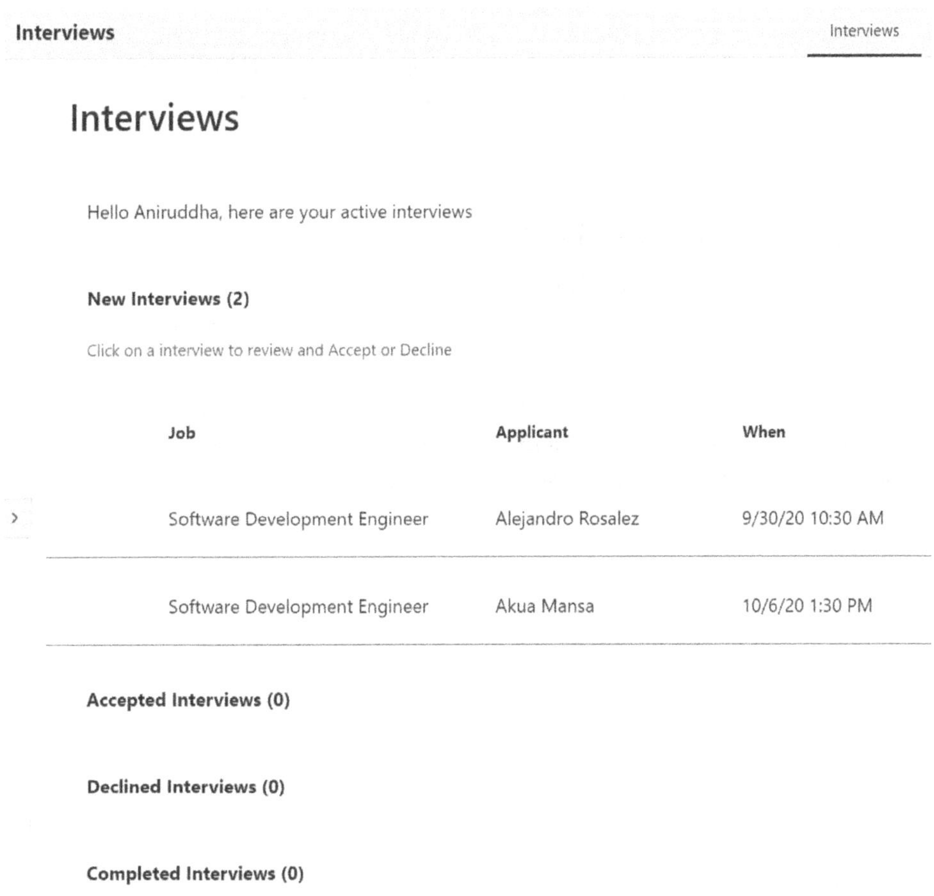

Figure 6.3 – A screenshot of the Interviews web app

Figure 6.3 shows the primary screen of the **Interviews** app. This allows for coordination of the interview where the interviewer can accept or decline an assigned interview. For the accepted interview, the interviewer then can provide their feedback on the interview along with their hire/no-hire vote.

> **Note**
>
> In the description of the template, I mentioned, *manage the internal hiring process*. This is because, if you recall, for anyone to be able to use your application, they must first be a part of the team, which is typically not something you'd do for external candidates, is it?
>
> And this may be one big downside for this template's real-life usability if your organization is already using some form of external-facing recruitment system, as most of those will typically come with an interface to handle your internal process too. Also, generally having a process split across two systems is more pain than gain.
>
> Can you connect the two systems? Possibly. Would you want to do it? Probably not, unless the recruitment system used in the organization is really bad and you have some technical chops to dive into setting up integration between the two systems.

Collaborative brainstorming

This template is listed on Honeycode's website under the title **Ideation**. It is designed to enable the asynchronous collection of ideas from a group and even vote and rank them if the intended outcome is to be chosen by a popular vote.

The template comes with a single app, with the home screen set to display all the questions currently open and requesting input (*Figure 6.4*). Two other screens let users review the resolved questions and track the questions they have asked.

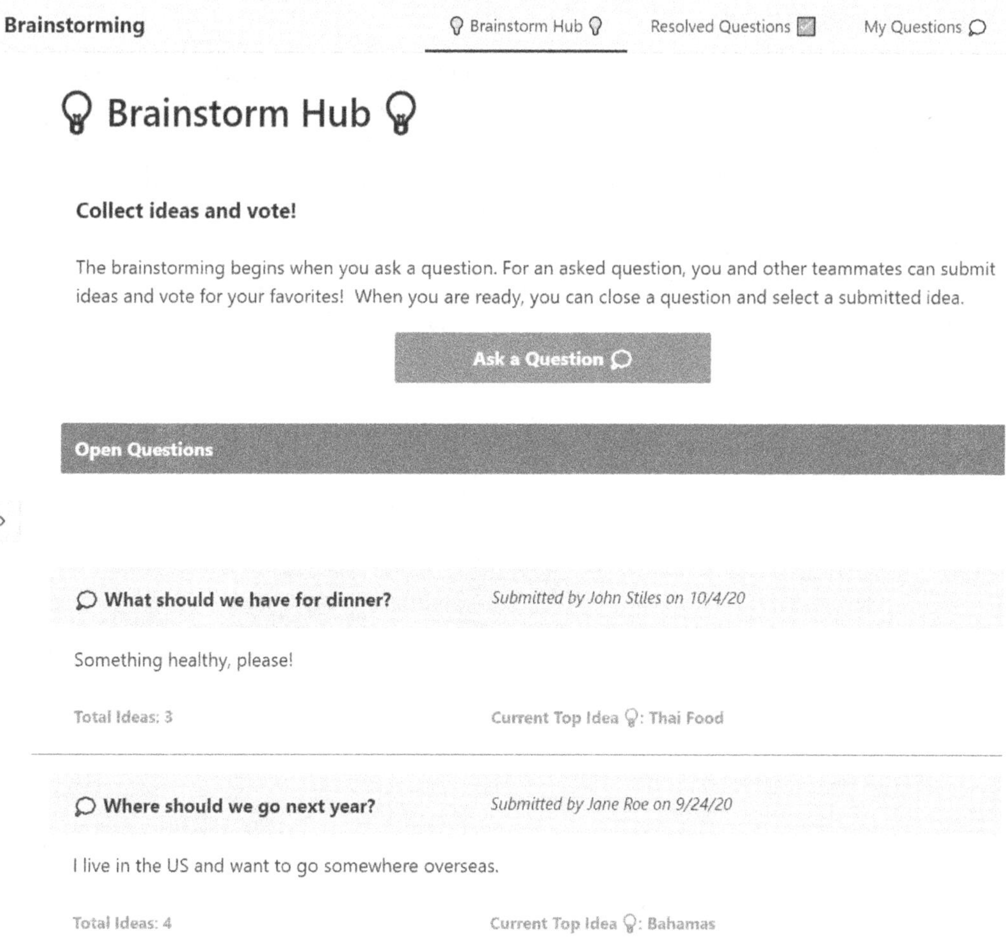

Figure 6.4 – The home screen of the Brainstorming app

Need a quick solution for deciding the next office party or a game or movie night with friends? This app will get you set up in a matter of minutes.

Connect manager

This template is a showcase of how you can integrate Honeycode and other Amazon services. The template uses **Amazon Connect** as an example to demonstrate this. However, even if you are not keen on using Honeycode with other services, this template provides a very useful and commonly used pattern of privileges applied with the hierarchy. The template comes with three separate apps to be made available as per the role of the users, and each app comes with an increasing set of privileges available to the user. With this template, you can learn how to set up an audit trail and build a simple approval flow.

Shift Supervisor
Messages
He
Messages
As a Shift Supervisor you can choose which of the available messages is live. If you need to create, edit or approve messages use the Call Center Manager or Customer Support Director apps. To make this application work with Amazon Connect you may need help from your IT or engineering team - see Help screen for more information.
Choose a message type:
EmergencyMessage
Choose a version to view or make live:
Storm message
Live
Earthquake message
Available
Storm message - copy
Pending
Earthquake message - copy
Pending

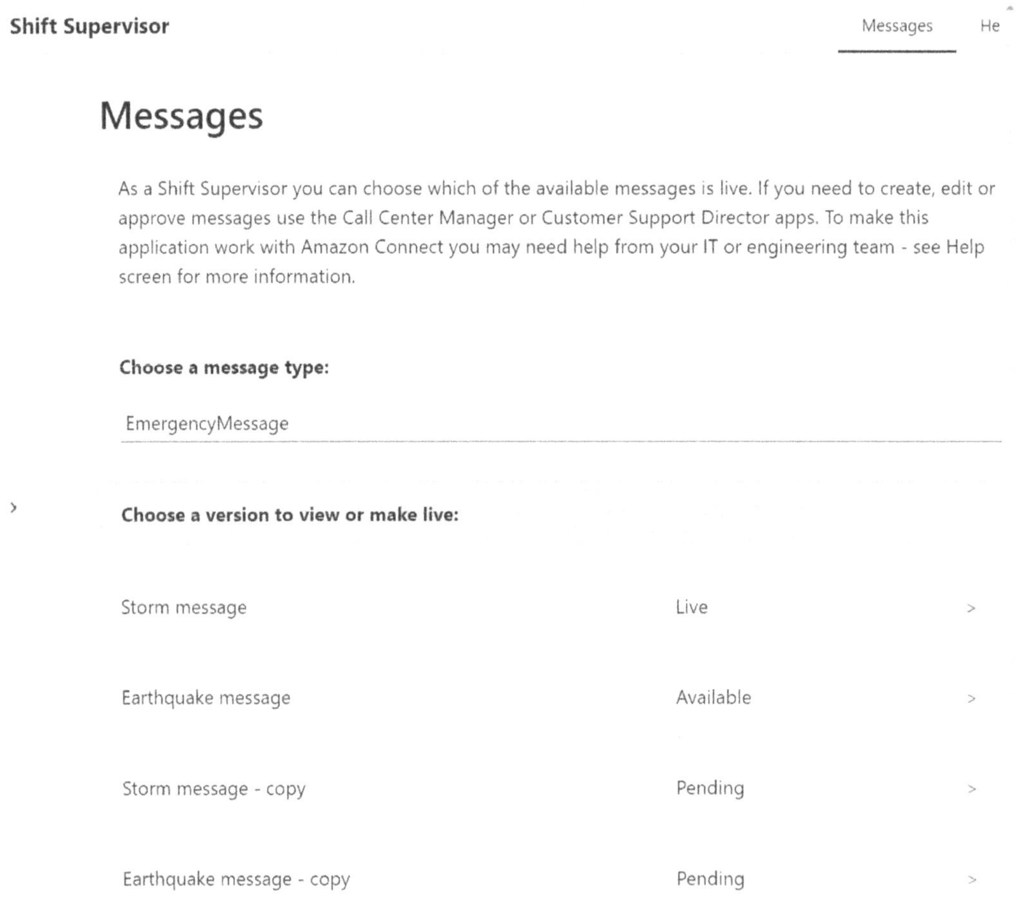

Figure 6.5 – A screenshot of the Shift Supervisor app with the least set of privileges to only set the status of the existing approved message

As per *Figure 6.5*, the app is for **Shift Supervisor**, and users of this app will only be able to set the status of any existing approved message to **Live**.

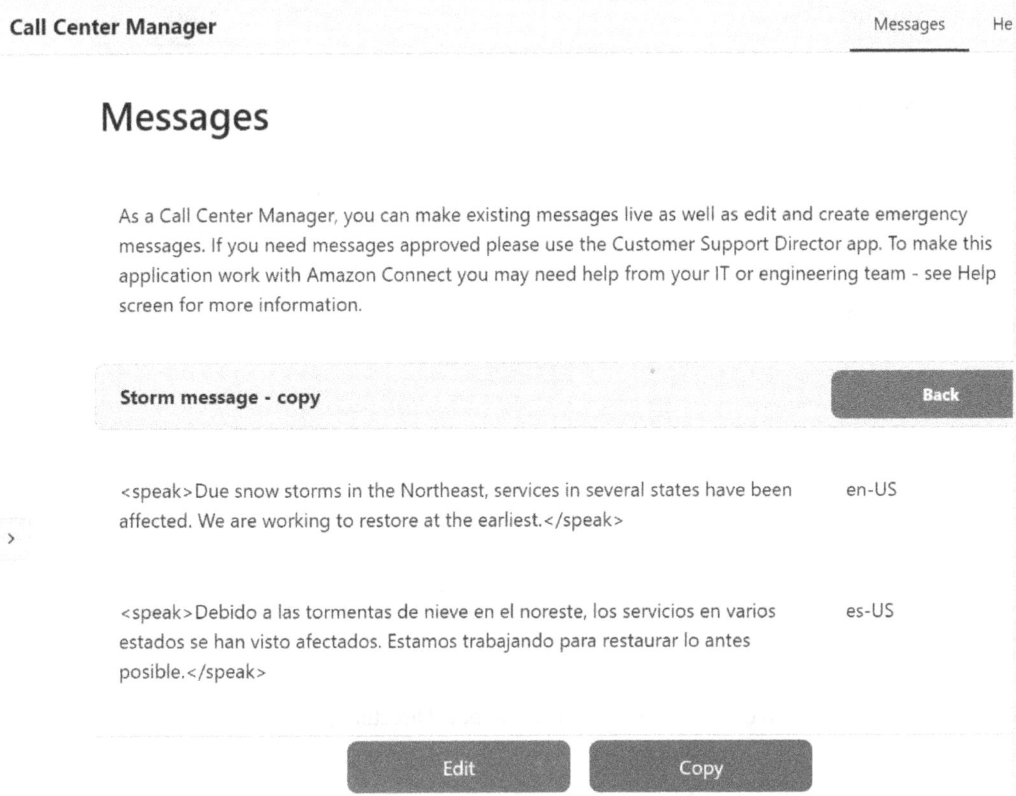

Figure 6.6 – A screenshot of the Call Center Manager app, displaying a message that can be edited or copied over to create a new message and seek approval on it

As shown in *Figure 6.6*, the second app is for **Call Center Manager**. Along with the ability to change message status, this app allows you to edit and create new messages.

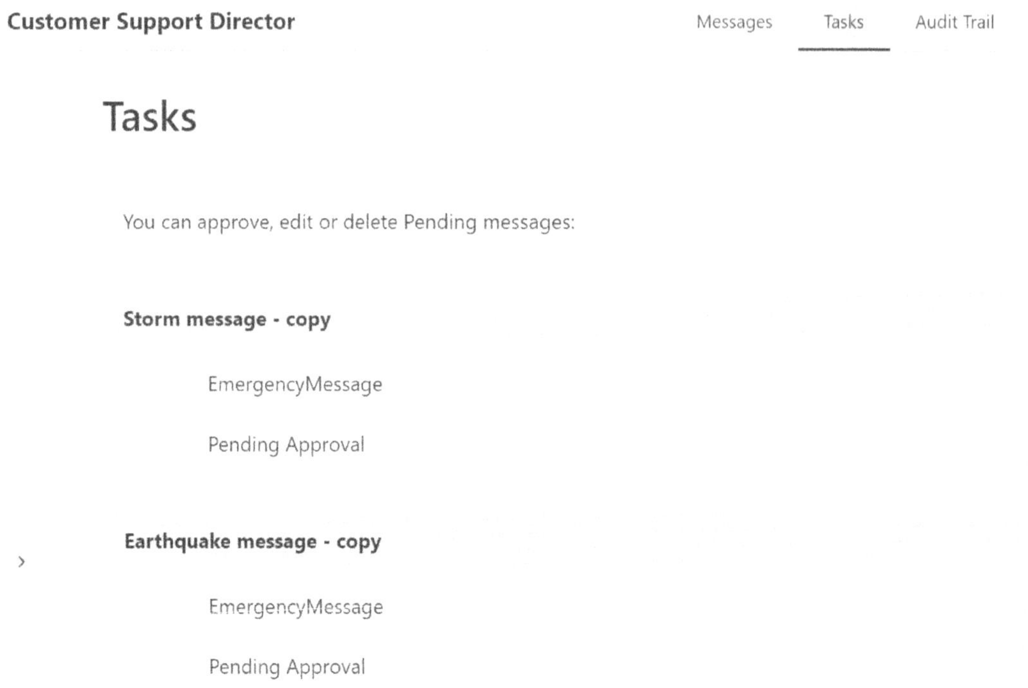

Customer Support Director   Messages   Tasks   Audit Trail

Tasks

You can approve, edit or delete Pending messages:

Storm message - copy

EmergencyMessage

Pending Approval

Earthquake message - copy

EmergencyMessage

Pending Approval

Figure 6.7 – A screenshot of the Customer Support Director app in the Tasks screen for the approval or rejection of messages

The third app is for **Customer Support Director**. This app adds to the **Call Center Manager** app the ability to approve or deny any newly created messages (see *Figure 6.7*) and also has a screen to display the audit trail of any messages (see *Figure 6.8*):

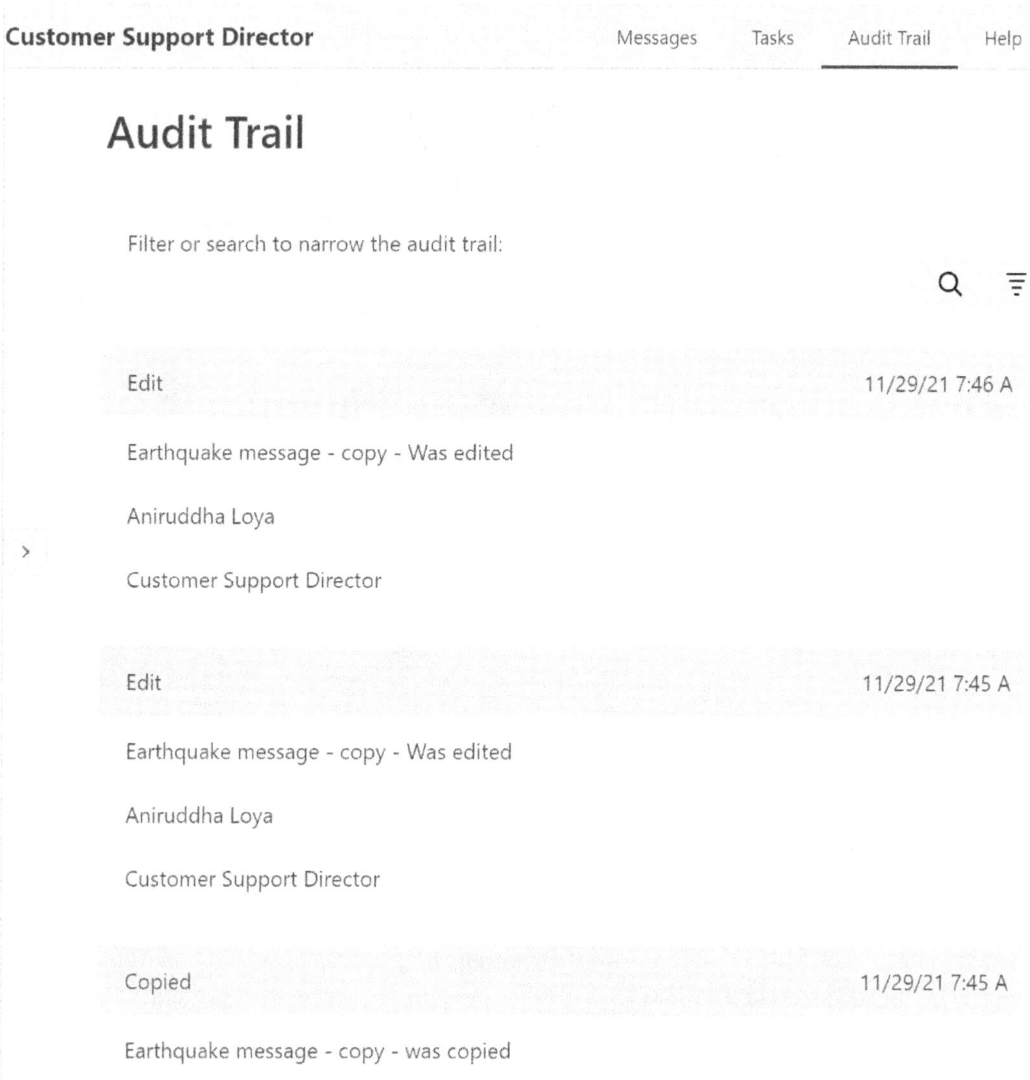

Customer Support Director Messages Tasks Audit Trail Help

Audit Trail

Filter or search to narrow the audit trail:

Edit 11/29/21 7:46 A

Earthquake message - copy - Was edited

Aniruddha Loya

Customer Support Director

Edit 11/29/21 7:45 A

Earthquake message - copy - Was edited

Aniruddha Loya

Customer Support Director

Copied 11/29/21 7:45 A

Earthquake message - copy - was copied

Figure 6.8 – A screenshot of the Customer Support Director app, displaying the message audit trail

Further to the app, there is a **GitHub repository** (`https://github.com/aws-samples/amazon-honeycode-connect-integration-sample`) that provides the code to set up and integrate with an instance of Amazon Connect.

Content tracker

The Content tracker template uses the example of a team of content creators managing different projects, with details on their owners, cost, status, and so on. However, this template is essentially a blueprint for building any kind of project tracker. It can be a tracker at the c-suite level, tracking the progress of high-level initiatives, or an organization's tracker for various features of a single product being worked across the different teams.

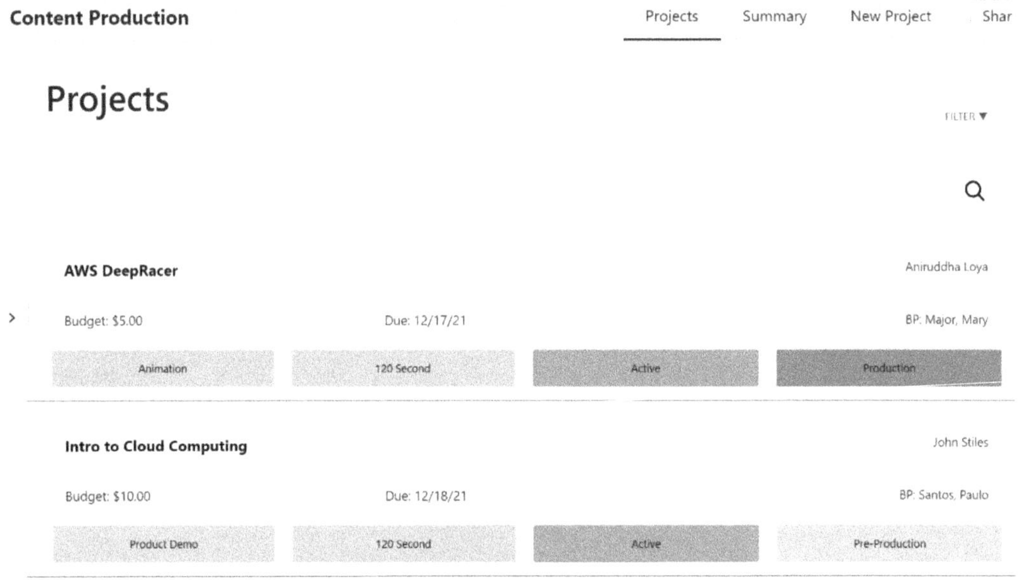

Figure 6.9 – A screenshot of a Content Production app

The template comes with a single **Content Production** app that shows a list of different projects being tracked (see *Figure 6.9*). It also has a summary view with groups by different criteria and a screen to add a new project.

Customer tracker

The template provides a basic CRM to manage customer relationships by maintaining the history of customer engagement by the team. This out-of-the-box template can be useful for smaller organizations (not using a fully fledged CRM solution) and customized as per the need to build additional features until the organization has grown to a substantial size and demands a complete CRM solution.

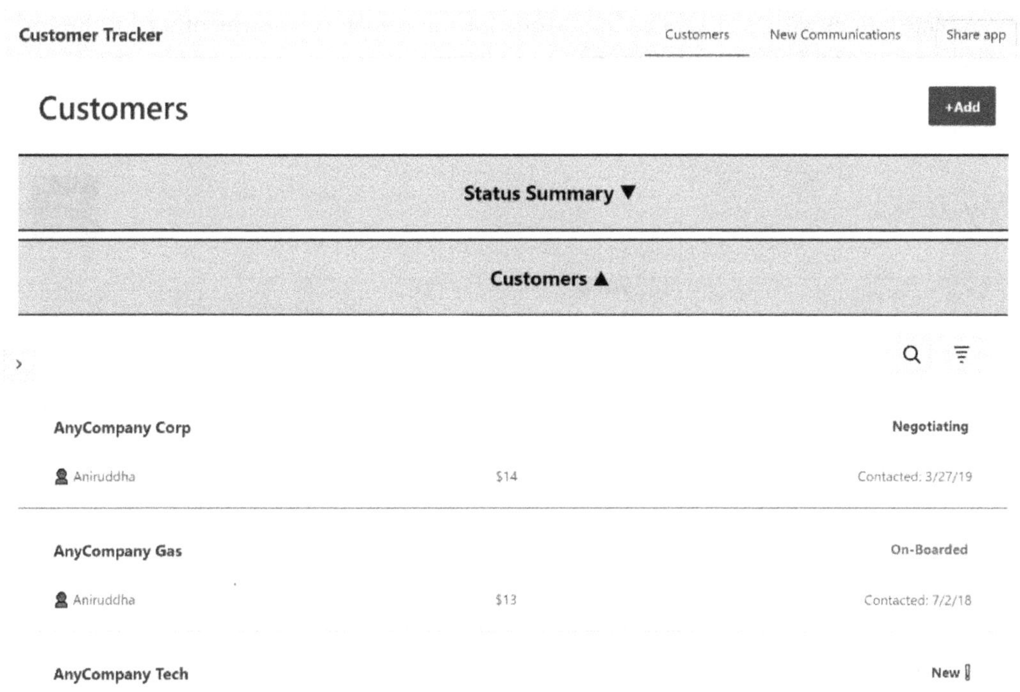

Figure 6.10 – A screenshot of the Customer Tracker app with the collapsed Status Summary block

The template comes with a single **Customer Tracker** app that shows the summary followed by the list of customers, with some key data as part of the list (see *Figure 6.10*). The template is also useful collapsible sections on the app screen. Another pattern to learn here is how to create different navigation based on different areas of clicking on a single list.

Employee onboarding

Onboarding is the first experience your new employee will have after joining, and therefore, in my opinion, it is very crucial to have a well-defined onboarding plan. So, if you do not have a defined template for onboarding or have been using some ad hoc text document or spreadsheet, this template is ideal for you and your organization.

The template comes with two applications – one for the employee and the other for the manager to create and assign the plan to the employee.

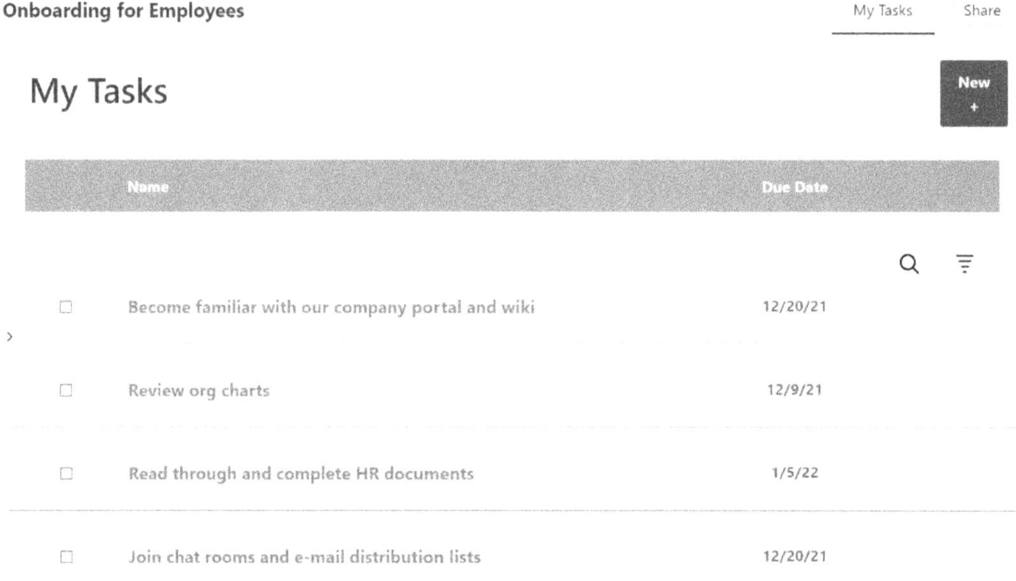

Figure 6.11 – A screenshot of the Onboarding for Employees app

As shown in *Figure 6.11*, the **Onboarding for Employees** app is a simple list of tasks assigned with due dates and controls to add new tasks for the employee. Does this remind you of something we built in earlier chapters? Yes, it's indeed a variant of the **To-Do** list in its most basic form.

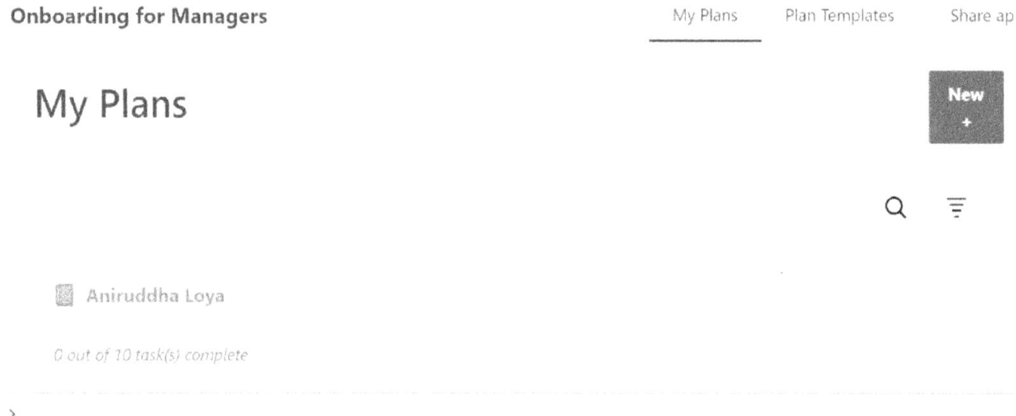

Figure 6.12 – A screenshot of the Onboarding for Managers app

The **Onboarding for Managers** app (see *Figure 6.12*) allows a manager to create and assign onboarding plans to employees. Furthermore, it allows managers to build a template that can be applied over and over without having to add new tasks for each plan manually. So, build once and keep using. Of course, edit and update as time passes by. Finally, there is a detailed screen for each plan, which lists the tasks that the manager needs to perform as part of onboarding and also provides a view of how the employee is progressing along with their plan (see *Figure 6.13*):

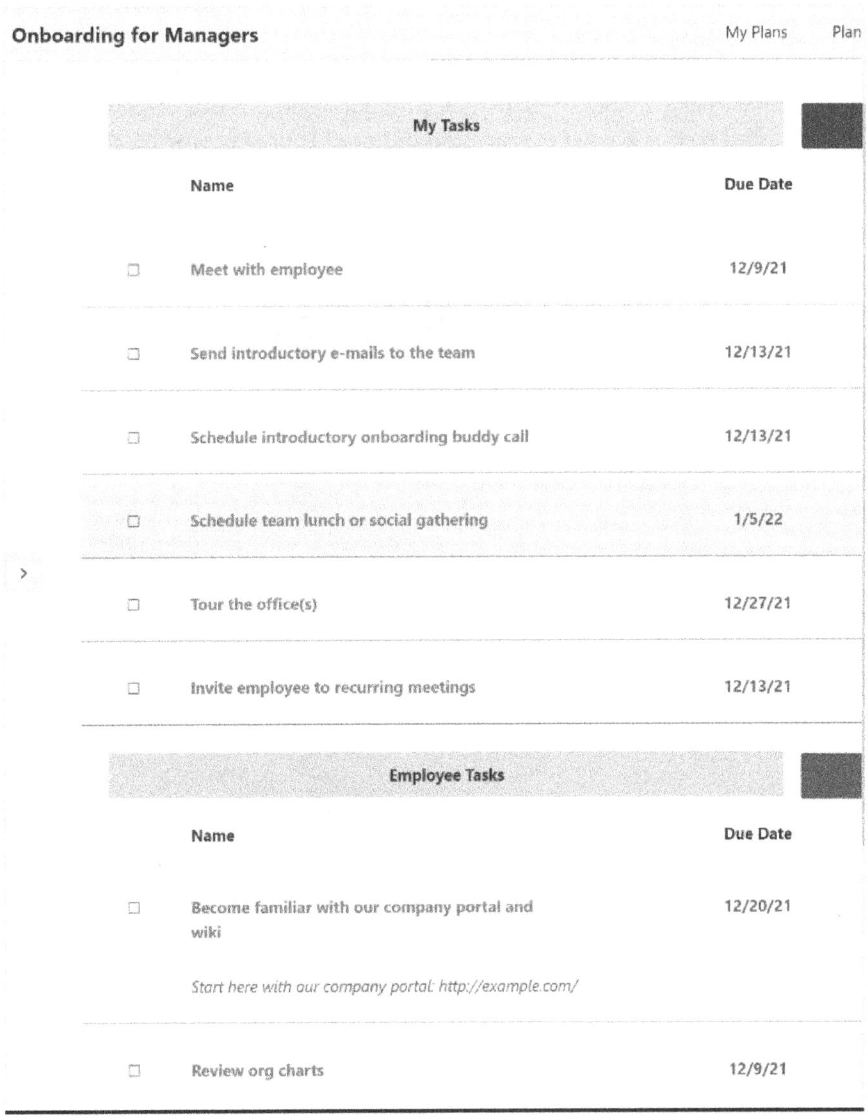

Figure 6.13 – A screenshot of the Onboarding for Managers app, displaying details of the plan

Applying templates to create a new plan is a new pattern that can be learned from this template and has many applications in everyday processes. Furthermore, there are other types of plans, such as personal growth plans and performance management plans, that can be built by customizing this template.

Event management

Have you ever been to a conference or a large event and found yourself constantly opening the website to find details of the next session to attend? Maybe you were on the side and hosting such an event and thinking about whether to spend on building a custom mobile app for the guests to easily access the schedules and details of various ongoing events, register for them, and so on. If you have been in either of those categories, this template is for you.

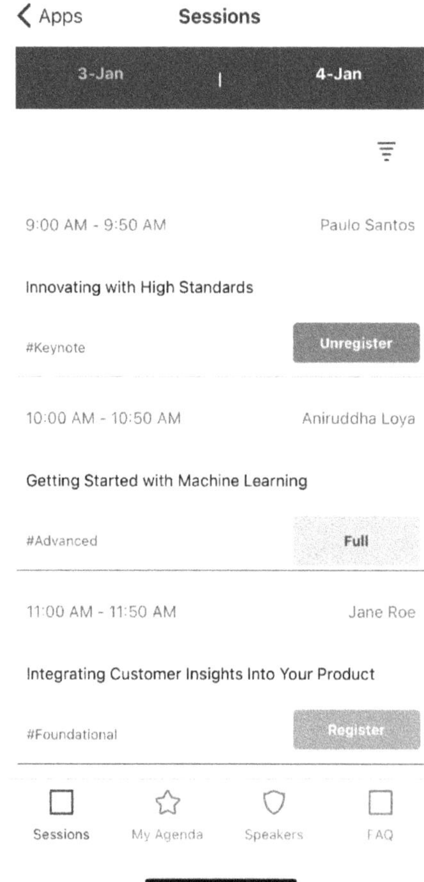

Figure 6.14 – A screenshot of the Product Conf Attendee app, captured on an iOS device

The template comes with a single app titled **Product Conf Attendee** (see *Figure 6.14*). It allows browsing through the various sessions and even supports multi-day events, registering, deregistering, and browsing the speaker bios, and also provides a view for your registered session so that you know exactly what is next, when, and where.

> **Note**
>
> Similar to **Applicant Tracker**, this template also shares the drawback of Honeycode requiring app users to be part of a team. So, it adds a layer of overhead for organizers to add attendees to their team and a layer of friction for attendees to sign up for Honeycode, which will be independent of their signup on the conference website. Also, since, in most cases, the attendees will be external, there is the other challenge of access controls or account separation as you add externals to your team. And finally, the cost of each additional member over the 20 included can very quickly bring Honeycode pricing to be on par with the cost of a more customized and personalized dedicated app for the event.

Field Service Agent

Are you running a business with field service agents on the road for an on-site job? This app is ideal for you to manage your customer information as well as service requests. The template comes with a single app that provides a list of work orders with filtered views for **Today**, **Upcoming**, and **All**. Also, it provides a screen for managing customers' information and the history of their service requests (see *Figure 6.15*):

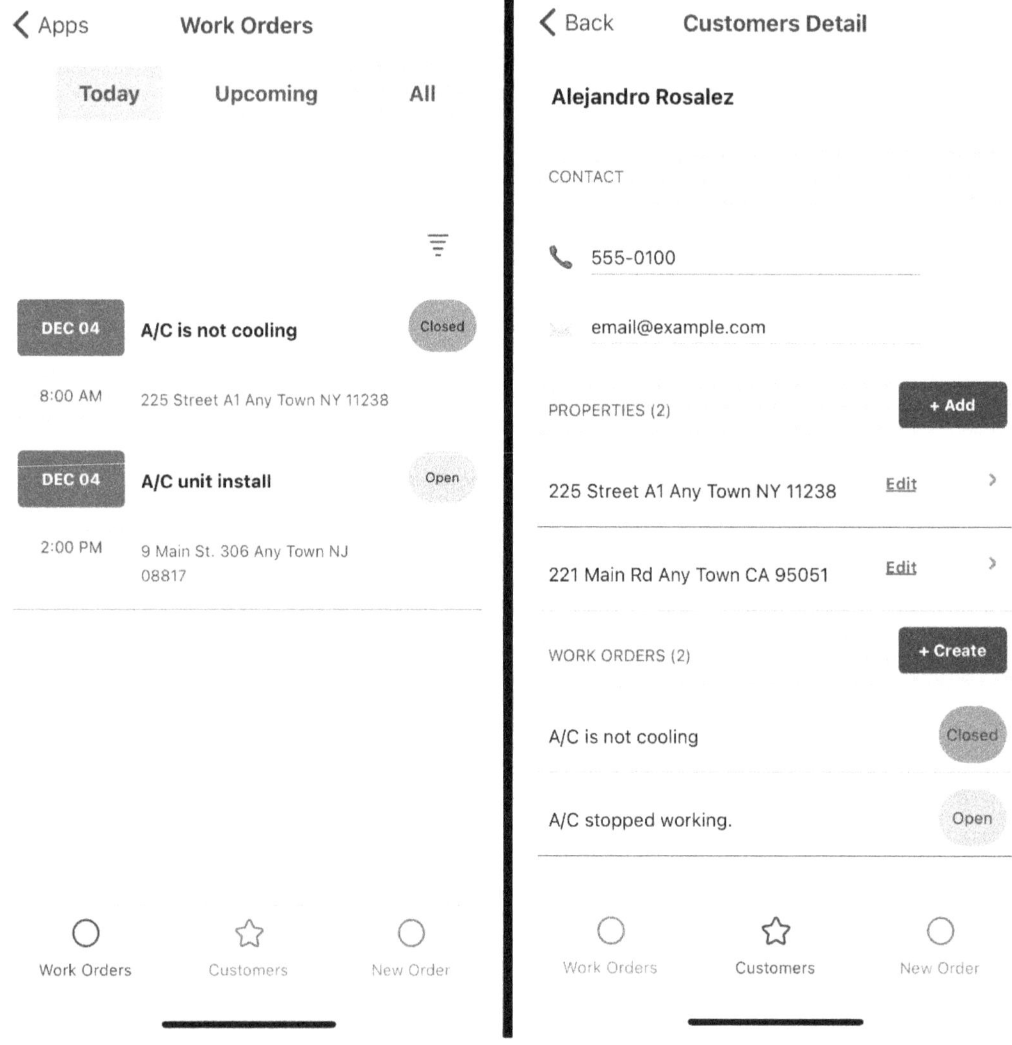

Figure 6.15 – A screenshot of Work Orders for the day on the left and a Customers Detail screen on the right, captured on an iOS device

Field Service Agent is similar to the **Onboarding for Employees** app that we saw earlier, and you may notice that the **Work Orders** screen is also a variant of the **To-Do** list.

Instant Polls

This is another common-use template for a group of people, allowing easy creation and sharing of polls that can be accessed both on mobile and desktop/laptops. The template comes with a single **Instant Polls** app that allows you to create a poll or respond to existing ones:

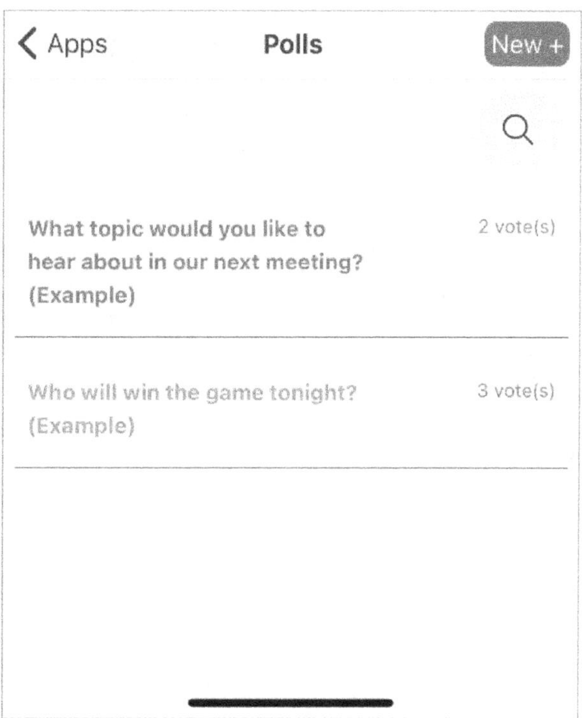

Figure 6.16 – A screenshot of the Instant Polls app, captured on an iOS device

This app can essentially be seen as a restricted functionality variant of the **Brainstorming** app that we reviewed in the earlier subsection, the primary difference being that the answers are pre-entered and need to be selected (voted) instead of being submitted. However, the screen for creating a new poll does provide an example of updating an existing screen on a button press rather than navigating to another screen, which we used while building our **To-Do** application.

Inventory Management

This template makes it a cakewalk to move away from managing your organization's assets (from a spreadsheet or a document) to an app, as well as keeping track of who has been assigned which asset. The template comes with two apps:

- One for reviewing your assigned devices (as well as requesting a new one)
- One for managing your organization's inventory and assignment mapping

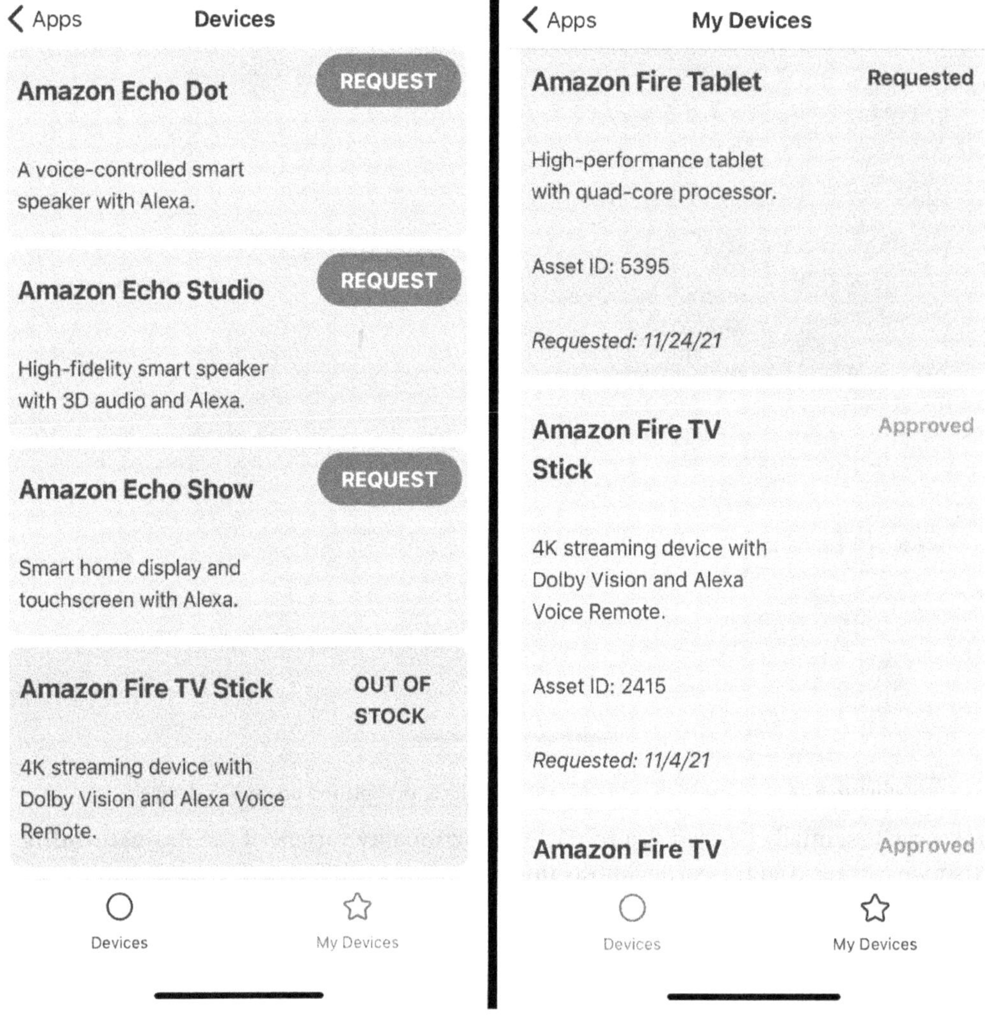

Figure 6.17 – A screenshot of the My Devices app, showing on the left a list of all devices, and on the right, a list of devices requested by or assigned to the app user, captured on an iOS device

The **My Devices** app allows everyone to review and request devices. It also allows users to view the devices that you already have assigned or requested. However, it does lack the feature of being able to return an assigned device and the ability to cancel a request.

My Devices - Manager   All Devices   New Device   Manage Inventory   Share

All Devices

Device	Total	Available	Pending
Echo Dot	3	1	1
Echo Studio	6	4	1
Echo Show	4	3	1
Fire TV Stick	2	0	0
Fire Tablet	3	1	1

Figure 6.18 – A screenshot of the My Devices - Manager app, showing the list of devices with their quantity and availability status

The **My Devices - Manager** app provides the feature to add an entirely new device or a new item of an existing device. It also provides functionality to approve requests and assign devices, as well as keep track of who currently has or has had a given asset but lacks a state to maintain the status of an asset as returned.

Launch Manager

Launching a new product is not easy, especially where you have multiple moving parts requiring coordination. An app that can be at the center of an entire process (a single source of truth) can help with the nerves and also with streamlining all activities. This template enables exactly that. While the template is not listed on the Honeycode website, the app shipped with the template has a **Help** screen that makes up for the missing web page.

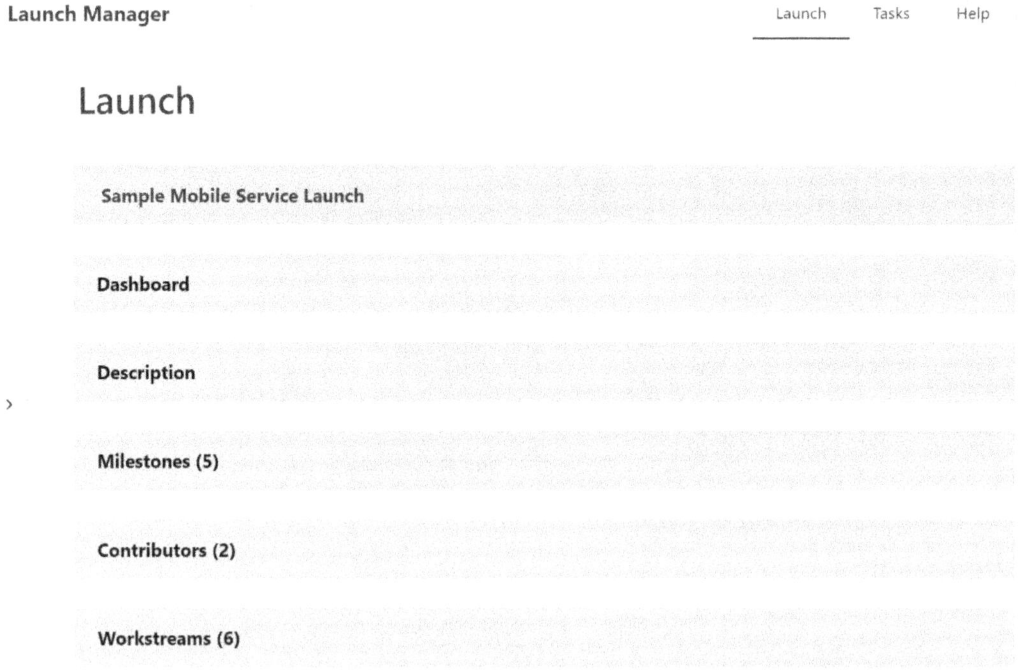

Figure 6.19 – A screenshot displaying the Launch screen of the app

The template comes with a single app, **Launch Manager**. The app is a sample of a service launch setup using this app to help users understand. However, unlike the pattern we have seen so far, where you can add a new item and continue, this app does not provide an option to add a new launch. It instead requires users to edit the sample and customize it to fit the need. Alternatively, you can update the app to allow the creation of new launches and thereby make it possible to have multiple launches tracked in a single app. Either way, this is an advanced app built using Honeycode with a practical use.

Meeting Runner

How is your organization managing meeting notes and derived action items, multiple email threads, shared documents, and no notes? How do you track the **Action Items** (**AIs**) and have a shared common source of truth? This template is here to address these issues

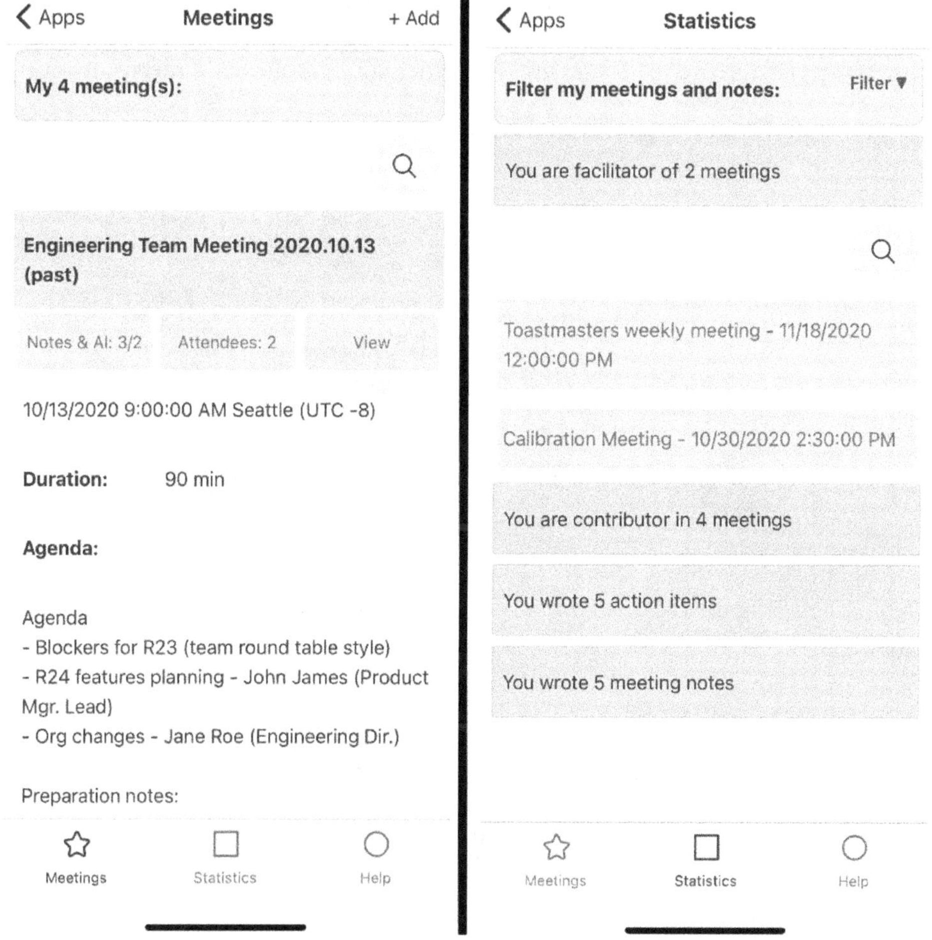

Figure 6.20 – A screenshot of the Meeting Runner app home screen, with Meetings details expanded on the left and a user's Statistics screen on the right, captured on an iOS device

The app allows you to create meeting instances and add the attendees to them. The events in the app are visible only to the attendees. The template also has an interface to add notes and action items, removing the need for a follow-up email thread. Furthermore, it is open for everyone to add notes and action items, thus making the process more collaborative. The app also has a **Statistics** screen to get a quick overview and access to different entities associated with you.

PO Approvals

This template is listed on the Honeycode website under **Budget Approval**. This template implements a very common organizational process of requesting and approving a purchase order. The template comes with two apps:

- One for requesting approval
- One for approving the request along with other configurational screens

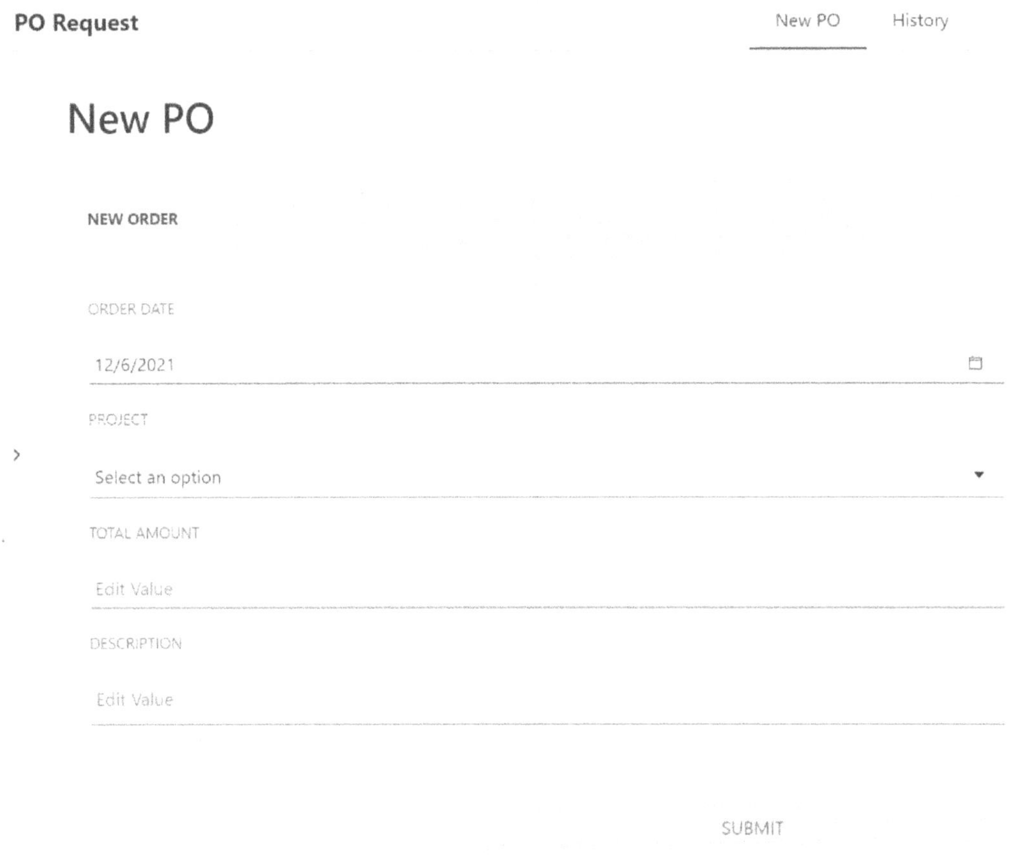

PO Request

New PO History

New PO

NEW ORDER

ORDER DATE

12/6/2021

PROJECT

Select an option

TOTAL AMOUNT

Edit Value

DESCRIPTION

Edit Value

SUBMIT

Figure 6.21 – A screenshot of the PO Request app

As shown in *Figure 6.21*, the **PO Request** app is a simple form to submit a request, which is handy when on the go and used through the mobile version of the app. It also has a history screen to review older requests:

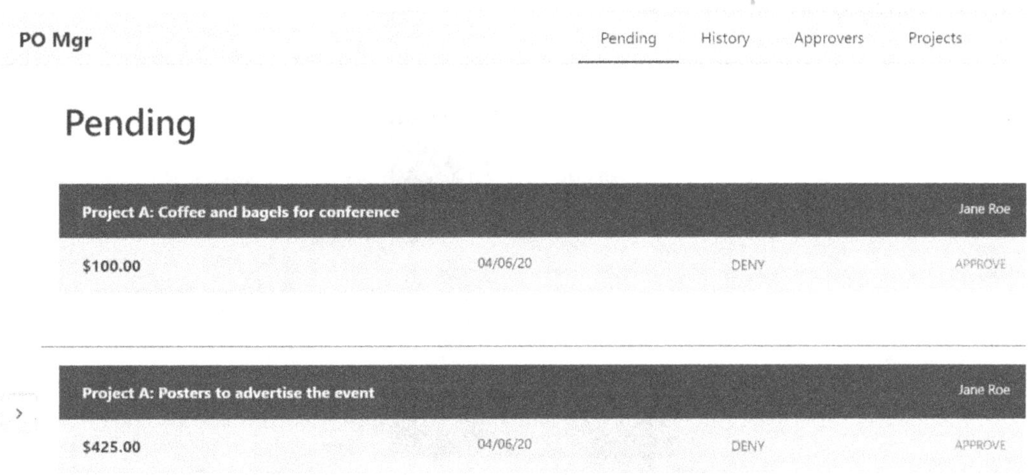

Figure 6.22 – A screenshot of the PO Mgr app

The **PO Mgr** app has a pending approvals list as the home screen. It has another screen for reviewing the request history and decisions. The app also enables the creation of new projects along with the means to set up the approval chain for the project. You can even configure auto-approve limits at each project level, and the automation set up in the template takes care of auto-approving the request and notifying the requester. But the most interesting piece of information to learn about this template is the implementation for multiple levels of approval. The configuration screen for setting the approval flow is also something unique to this template.

Simple Survey

Similar to **Instant Polls**, this is another template that you will find useful even without customization. Just get your questions framed, set up your measuring scale(s), quickly create from the app, and share it out.

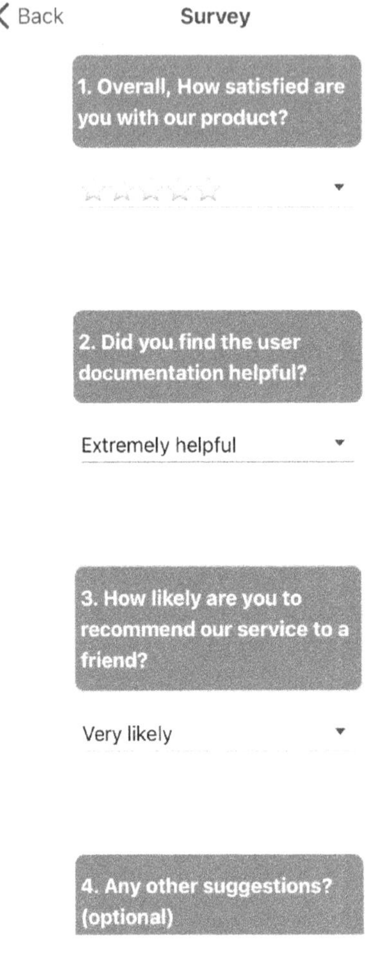

Figure 6.23 – A screenshot of the Simple Survey app, captured on an iOS device

The template has a simple app titled **Survey** that consists of only three screens if you review the builder interface, and for the end user, it's a one-directional flow without any other navigable screen. It's short, sweet, and really simple.

Simple To-Do

We started this book by using this template in *Chapter 1, Amazon Honeycode – Day One*, and then, throughout three chapters, we built a To-Do app ourselves. So, there is not much to add to this template.

Team Task Tracker

Almost all teams, irrespective of their size and distribution, maintain a list of tasks that they are working on and a list of what to work on next. Typically, they will also use a tool such as Jira, Asana, or similar apps to track these tasks and lists. However, if, for some reason, your team is not using any of these tools and practices, this template will get you started.

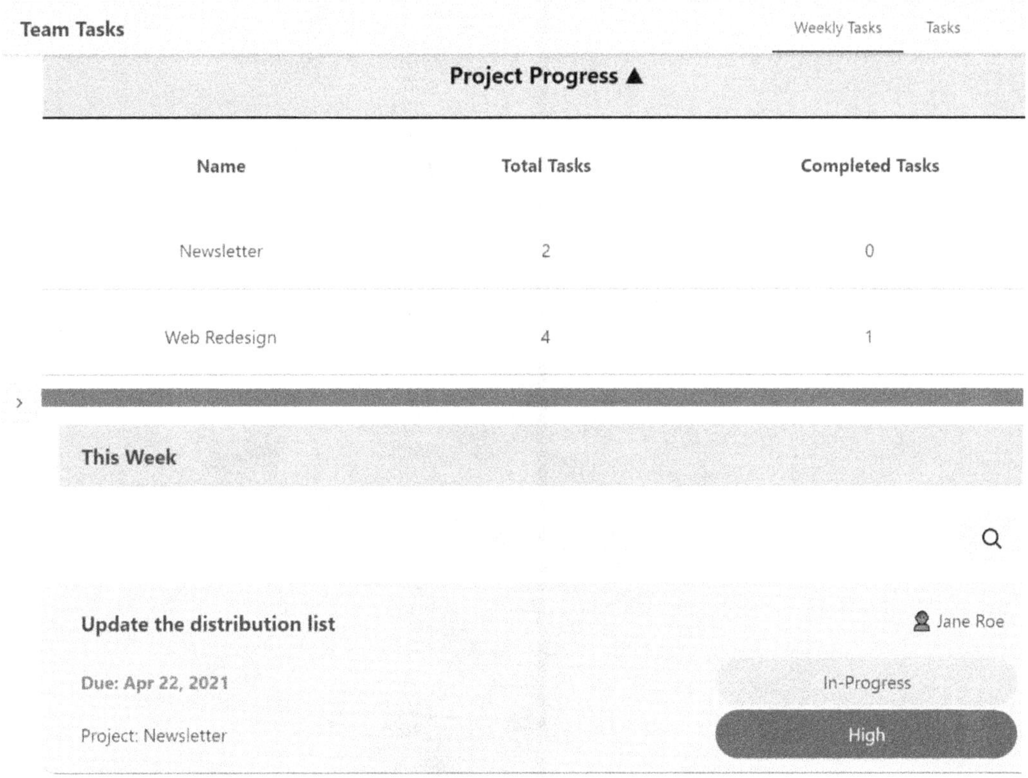

Figure 6.24 – A screenshot of the dashboard of the Team Tasks app

The template provides a single **Team Tasks** app. It has two views – one is a dashboard with the overall status of tasks of the different projects being worked on, along with the tasks of the week, and the second shows a complete list of tasks in the backlog.

Timeoff Reporting

This is another template designed for a common organizational use case of requesting and approving leave. The template comes with two apps – one for requesting and another for approving leave. Both the apps also provide a **History** screen to review previous requests and their statuses

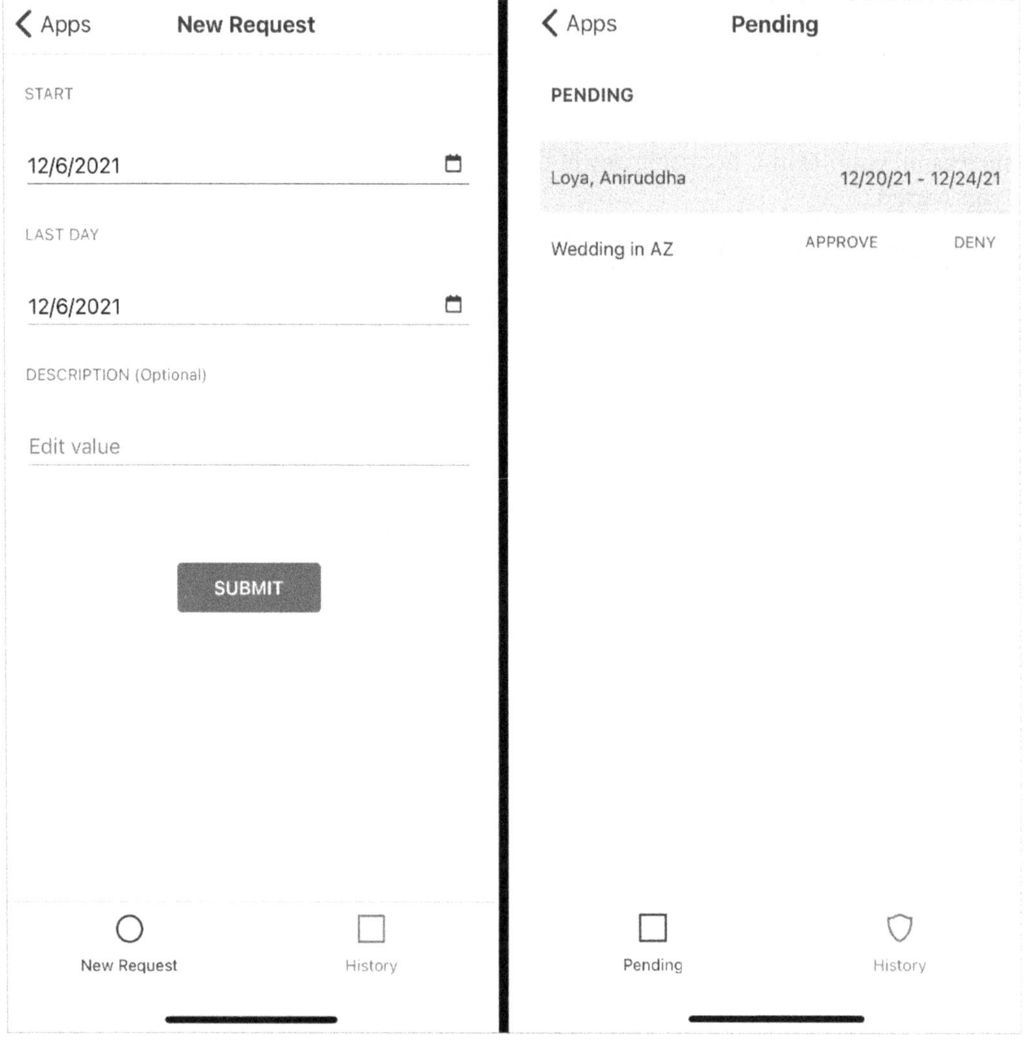

Figure 6.25 – Screenshots of the Time Off app on the left and Time Off – Manager on the right

We saw this request-approve pattern in the **PO Approvals** template, and this template provides the most basic implementation of that pattern. Unlike **PO Approvals**, which had a few more features in terms of being able to set up new projects and define the approval chain, this template offers no such thing. It actually requires a manual entry to the **People** table to add a new employee-manager relationship, as there is no provision for adding new employees or managers, or their mapping, in either of the apps.

Timesheet Manager

This template is functionally a variant of **Time Off Reporting**. While the former applies typically to full-time employees needing only to report their time off, this template is geared toward managing the same flow of hourly employees and, potentially, even contract workers. In line with the request-approve pattern, the template comes with two apps – one for requesting a timesheet and another for approving it.

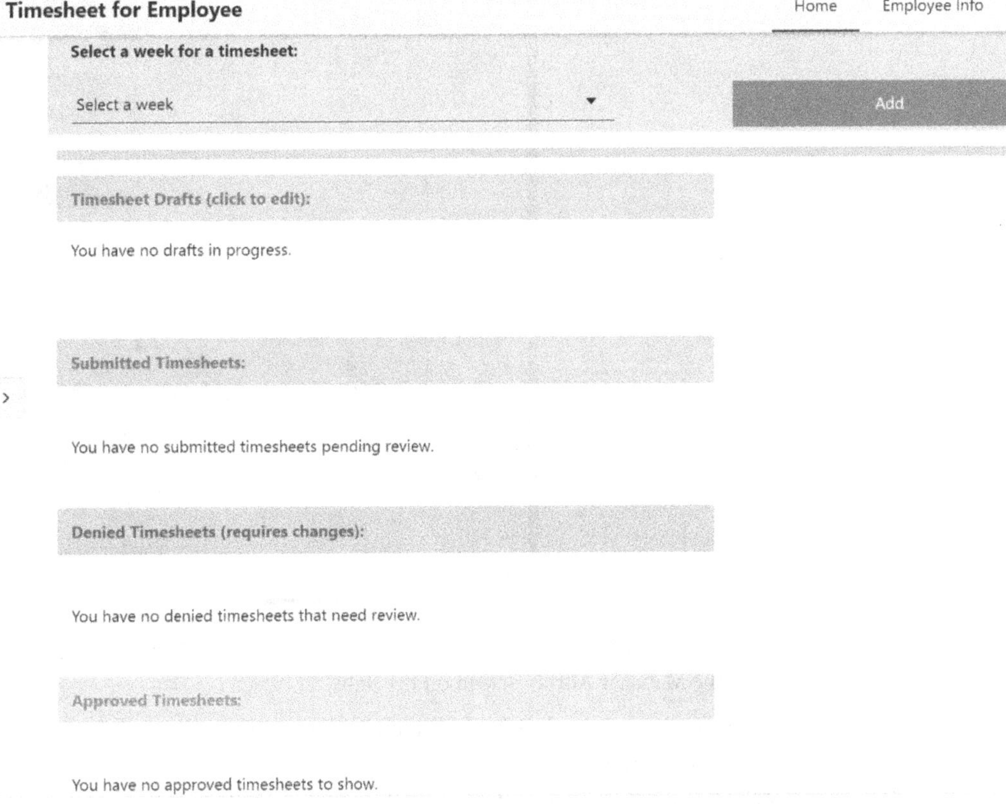

Figure 6.26 – A screenshot of the Timesheet for Employee app

The **Timesheet for Employee** app helps in creating, updating, and tracking the status of timesheets, as well as having a section for listing the approved timesheets.

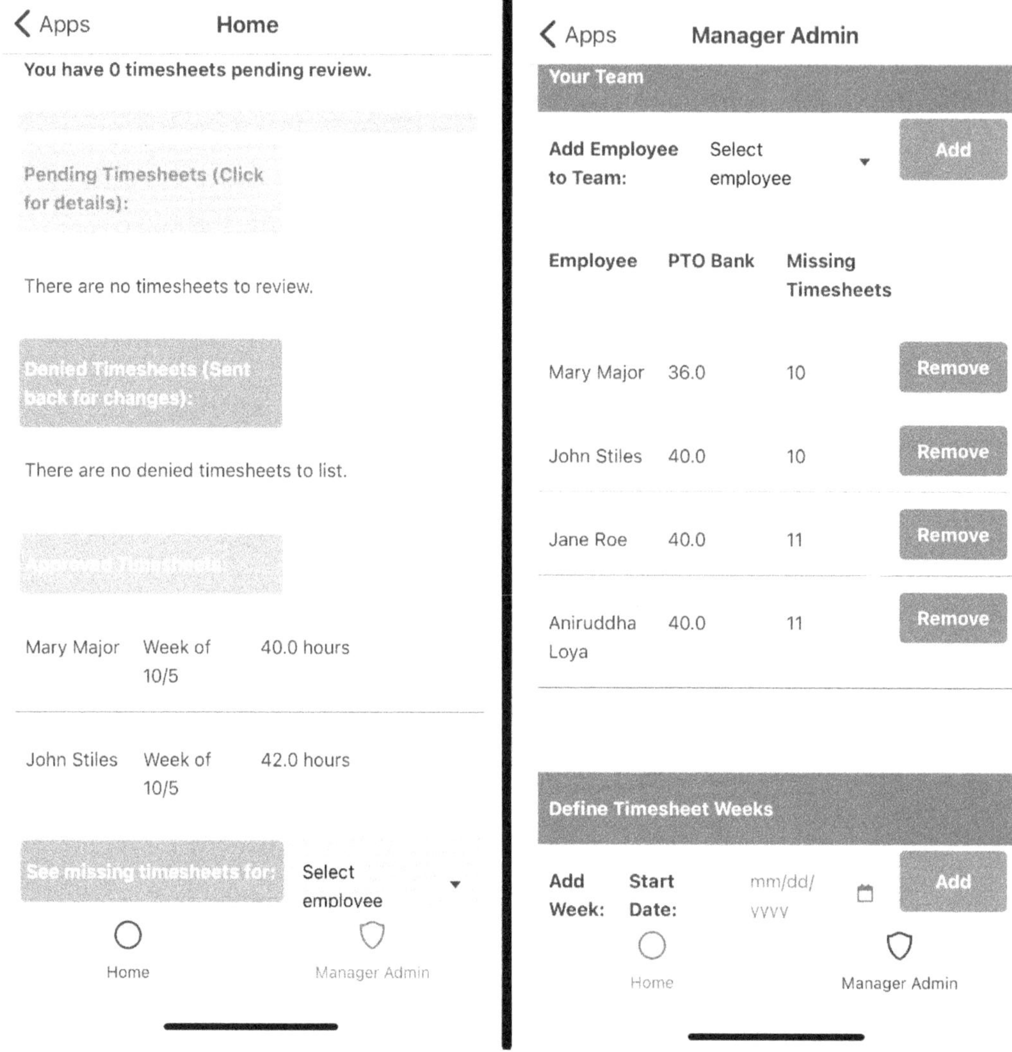

Figure 6.27 – Screenshots of the Timesheet for Manager app, showing the Home screen on the left and the Manager Admin screen on the right

The **Timesheet for Manager** app, on the other hand, helps in approving or rejecting a submission. It also provides functionality for a quick listing of weeks (or missing reports) for any selected employee. The app also contains an admin screen with functionality for adding new team members, as well as managing the weeks in which timesheets can be submitted.

While the template apps again do not provide the functionality of setting up the approval process, it does show how to maintain draft states of timesheets and submit them once ready.

Weekly Demo Schedule

This template provides a means to coordinate the presentation schedule within the team or organization, or even some open events. A common use case for this template would be sprint demos (which is also how the sample app is set up) or the Toastmasters meetings.

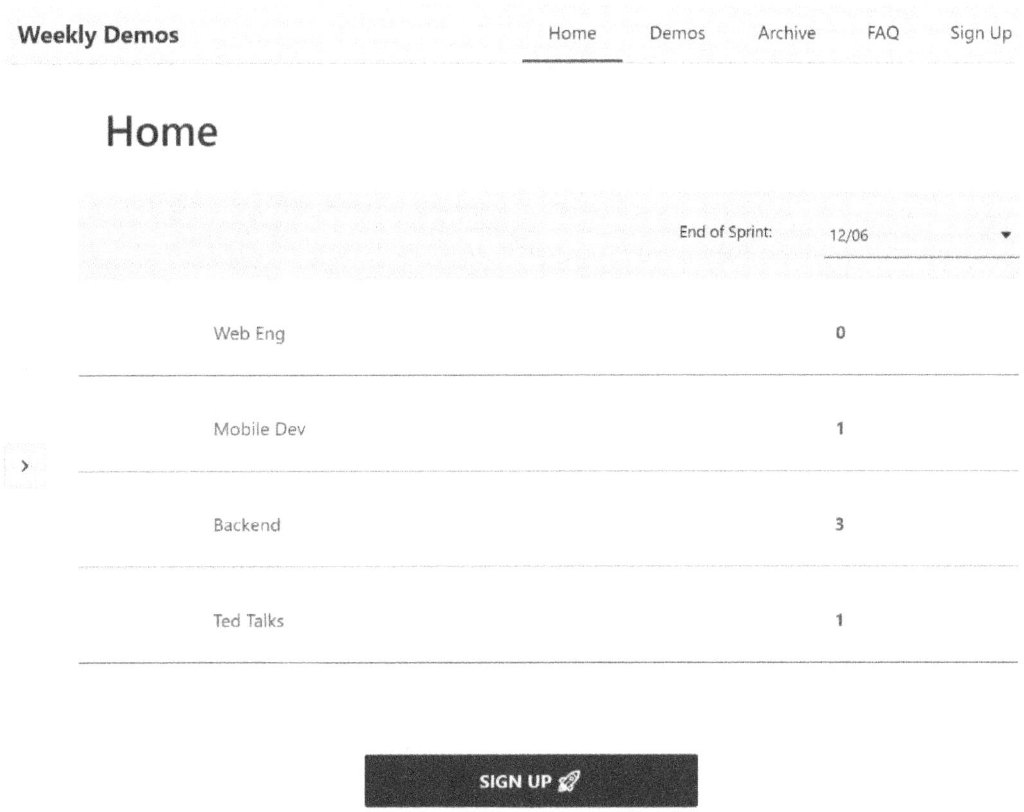

Figure 6.28 – A screenshot of the Weekly Demos app home screen

The template consists of a single **Weekly Demos** app that allows team members to sign up for their demos. It provides a dashboard view of the number of signups that are against each of the pre-defined groups and a screen for reviewing all the signups in a single list. It also provides an archive of the previously held demos.

This app, over a long period of use, can become a great knowledge source, especially for onboarding new members, who can review the recordings from the archive as they progress through their learning curve in the new team.

Summary

In this chapter, we reviewed all the templates currently offered by Amazon Honeycode. This will enable you to determine whether the problem you are trying to solve presents a pattern already solved in one of these templates and, therefore, can be used to jump-start new application development.

In the next chapter, we'll do a deep dive into learning about the **Instant Polls** template by reviewing the data model as well as the builder interface.

7
A Simple Survey Template

Surveys are a common occurrence in any organizational setup. They are effective for gathering information and opinions asynchronously and are sometimes also modeled as a learning tool. This template will help you get started with setting up surveys that can be accessed both on a mobile as well as desktops/laptops.

In this chapter, we'll learn how the app is set up and, thereby, empower you to customize the app as per your requirements.

In this chapter, we're going to cover the following main topics:

- Defining the app requirements
- Creating the Survey app
- Reviewing the data model
- Reviewing the app

Technical requirements

To follow this chapter, you'll need to have access to Amazon Honeycode, which requires a laptop with a web browser, preferably Google Chrome, and optionally a mobile device running either Honeycode's supported version of Android (currently Android 8.0 and up) or iOS (currently iOS 11 or later).

Furthermore, we'll use the Honeycode terminology and refer to the components that we covered in *Chapter 2, Introduction to Honeycode*, *Chapter 3, Building Your First Honeycode Application*, *Chapter 4, Advanced Builder Tools in Honeycode*, and *Chapter 5, Powering the Honeycode apps with Automations*, and therefore, we recommend you complete those first.

Defining the app requirements

We first introduced this section in *Chapter 3, Building Your First Honeycode Application*, where, before building our app, we listed down the requirements or the use cases that we wanted the app to fulfill. And we made use of that list throughout the chapters as a guide to define our data model, conceptualize the application interface, and visualize the interactions between various onscreen elements as well as the data displayed. Therefore, even though we are not going to be building this app here but only review what is built, I would encourage you to take 5 minutes to think about what your app should do and how it should look, and then make a list of the key points.

> To-Do
>
> Take 5 minutes and list how you would like your Survey app to work.

Here is what my list looks like:

1. App users must be able to create new surveys.
2. A survey should support different question types – single-selection, multiple-selection, and free-text.
3. App users must be able to participate in the existing surveys.
4. Questions in the survey can be marked as required or optional.
5. A survey cannot be submitted until all required questions are answered, and there should be visual feedback to participants for questions that require answers before submitting the survey.
6. There should be a setting to permit or block users to change their responses.
7. There should be a setting to permit or block navigation to the previous question.

8. The surveyor should be able to close the survey, or it should automatically close on a specified date.

Your list may have some or all of these requirements and may have even more, which is perfectly fine. You may find some of these missing in the template app and may want to extend that for yourself after you finish the chapter.

Creating the Survey app

Now that we know how we want our app to work, let's see what the template app can do. We begin creating the app and the associated workbook by following these steps:

1. On the dashboard, locate the **Create workbook** button at the top-right corner and click it.
2. You will see that a popup appears. Under the **USE A TEMPLATE** header (as shown in *Figure 7.1*), locate the tile labeled **Simple Survey** and click it:

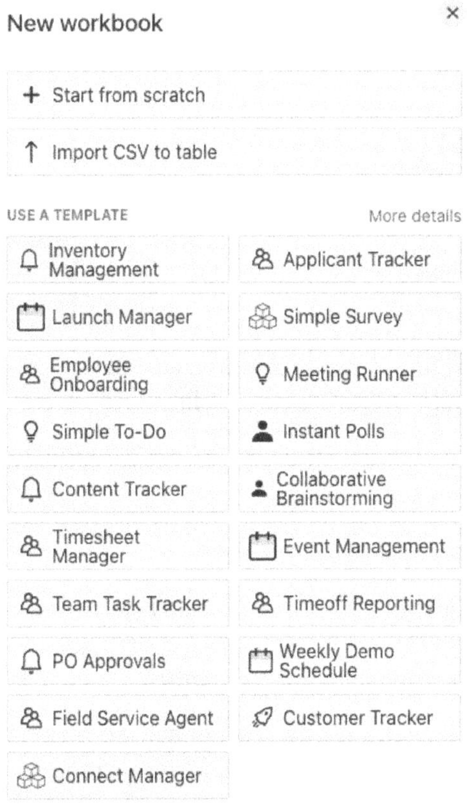

Figure 7.1 – Selecting the Simple Survey template

3. Next, you will see a pop-up box to name the workbook and choose a team. For now, let's leave the default values and click on the **Create** button.

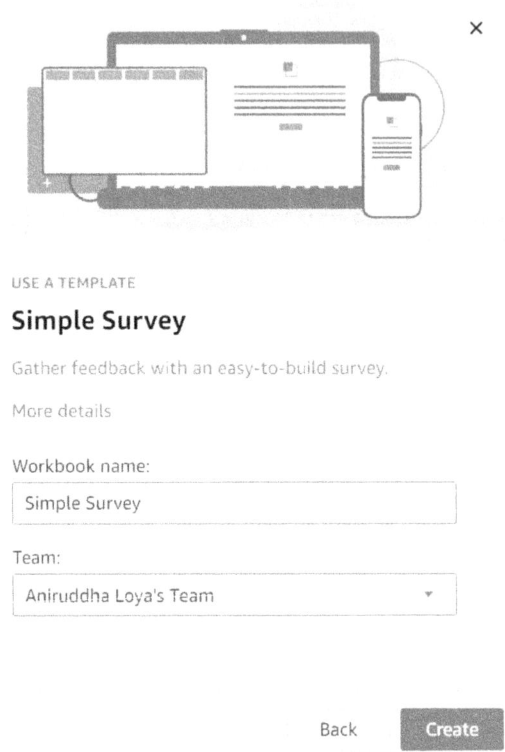

Figure 7.2 – Providing a workbook name and team details to create the workbook

4. This creates and loads our **Simple Survey** workbook and loads up the workbook in a table view, displaying the first table in the list.

And with that, we are now ready to review the Survey app and its data model.

Reviewing the data model

With the template workbook created, we are now ready to review how the Honeycode team set up the data model for the Survey app. However, before we dive into that, let's take a couple of minutes to think of the data that is required for such an app.

We are creating a survey, so it naturally has a set of questions and a set of provided answers for each question, or maybe an empty field for text input. We may need information about the survey itself, in terms of its creator, the creation date, and the end date. Furthermore, we'll need to store who our participants are and their responses.

Now that we have an idea of what data we need, let's explore the different tables in the workbook and the data they store, as well as the relationships between them, if any. This template comes with just four tables, as shown in *Figure 7.3*:

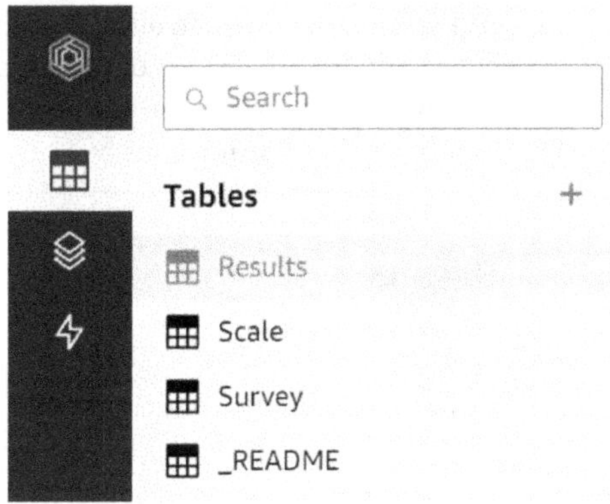

Figure 7.3 – A list of tables in a Simple Survey workbook

Let's understand each of these tables:

Results

This table aggregates the survey responses and stores the count, average, and mean values for each question:

Simple Survey > Results

Style Settings Wizards

fx On-Boarding

	A	B	C	D
1	Question	Responses	Average	Median
2	On-Boarding	1	5	5
3	Documentation	1	5	5
4	NPS	1	4	4

Figure 7.4 – The Results table

> **Note**
>
> The blue notches on the cells under the **Responses**, **Average**, and **Median** table columns are not errors. By default, Honeycode tries to nudge users to have the same formula and the format for the entire column, as it ensures that the data format for all cells in the column is consistent, which allows for the definition of consistent rendering as well as predictable behavior of the elements if rendered as a list in the app. However, this is, as we can see here, not a necessity for all use cases and ends up creating a visual annoyance.

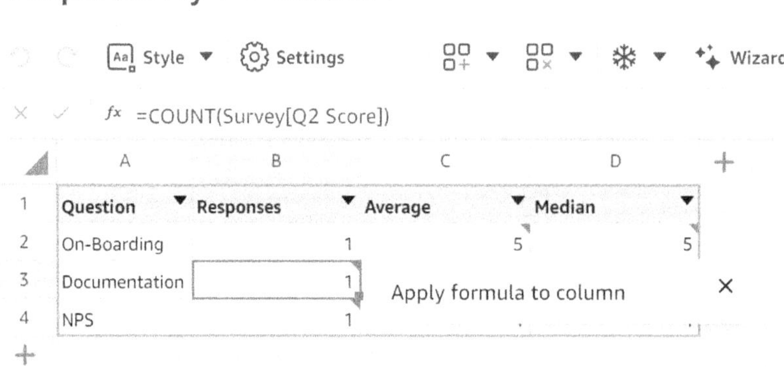

Figure 7.5 – A popup showing the Honeycode recommendation for a cell with the blue notch

Scale

Let's understand **Scale** as shown in *Figure 7.6*:

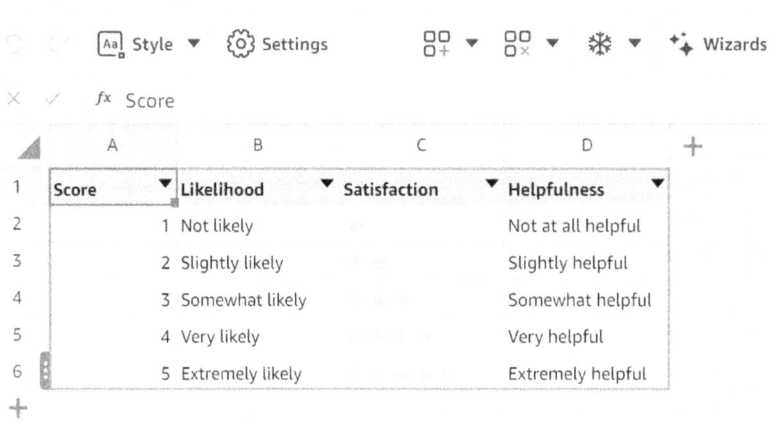

Figure 7.6 – The Scale table

Rating-type questions are a common occurrence in a survey, and they appear in various forms. Some examples are listed as follows:

- **As numbers**: On a scale of 1 to 5, rate your satisfaction with a particular item, with 5 being the highest.
- **As text**: How likely are you to recommend this book? Did the solution provided on this page address your problem?
- **As graphics such as stars or thumbs up or thumbs down**: How do you rate this book?

> **Note**
> The aforementioned examples are not the formal classification of rating-scale questions and are listed here for context and shared understanding. The formal classification of the question types with examples can be found in online resources such as *Rating scale – Wikipedia* or *Your Guide to Rating Scale Questions in 2021*or *qualtrics.com*, among many others.

This table is set to provide these alternatives for the different rating-type questions in the survey.

Survey

This table is the record of survey submissions. It records the name of the participant along with the selected response to every question.

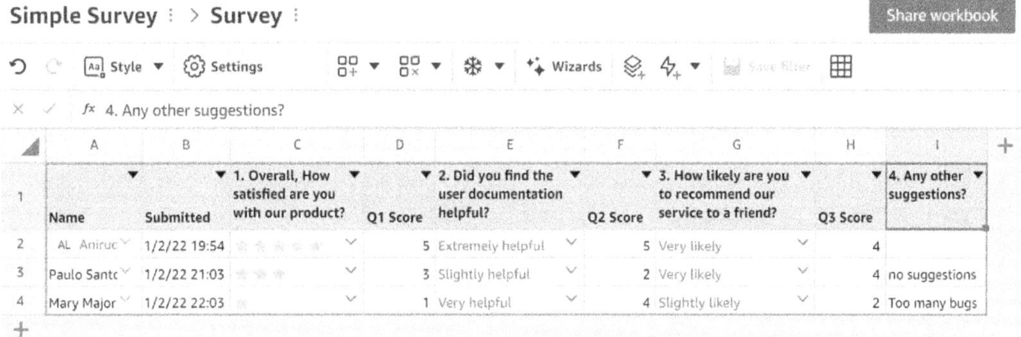

Figure 7.7 – The Survey table

The rating-scale questions in the **C**, **E**, and **G** columns are configured to use the values from the **Scale** table, and each uses a different representation. This is achieved using the **rowlink and picklist** format, with the **Customize display text** property used to select a column from the **Scale** table to use (see *Figure 7.8*). The last question is in the **I** column and is supposed to be a free-text question.

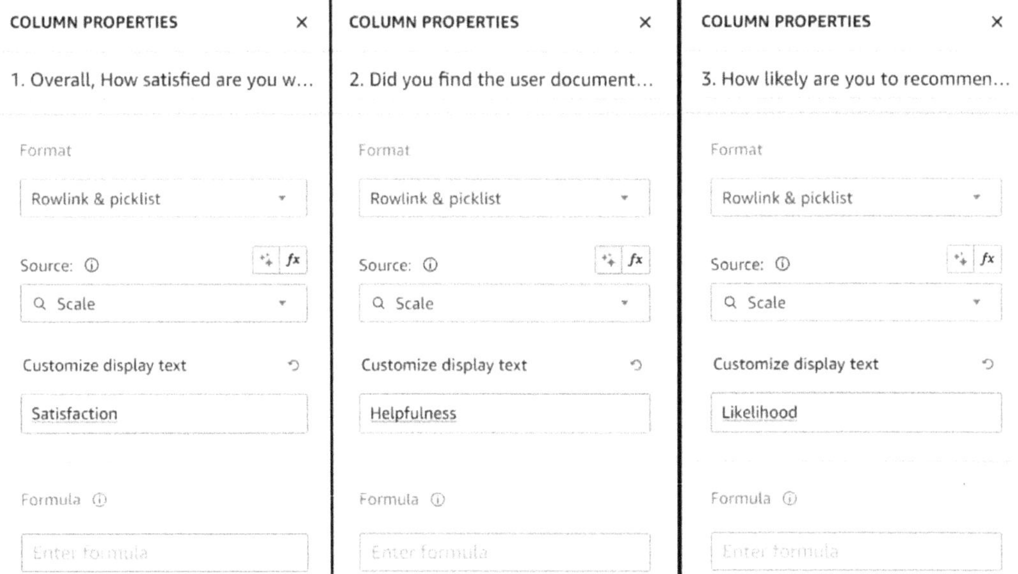

Figure 7.8 – The column properties for the C, E, and G columns, showing the format as well as the different customized display text settings on the same source table

However, the **D** (**Q1 Score**), **F** (**Q2 Score**), and **H** (**Q3 Score**) columns are then configured to translate that value to the number, allowing the computation of aggregate values in the **Results** table. This is achieved using the following formulas, which dereference the **Score** column for the rowlink created by the response to the question:

- The formula for **Q1 Score**: `=IFERROR([1. Overall, How satisfied are you with our product?][Score],"")`
- The formula for **Q2 Score**: `=IFERROR([2. Did you find the user documentation helpful?][Score],"")`
- The formula for **Q3 Score**: `=IFERROR([3. How likely are you to recommend our service to a friend?][Score],"")`

> **Rowlink**
>
> **Rowlink** is a term used for the value in a rowlink and picklist-formatted cell – that is, a reference to a table row. This is useful in building the relationship between tables. More details on rowlinks and picklists can be found here: `https://honeycodecommunity.aws/t/intro-to-rowlinks-picklists/88`.

> **Dereferencing**
>
> A rowlink is useful in building the relationship between tables. However, this is just a reference, and oftentimes, we need to be able to make use of the values from a specific column of that row. This value is retrieved by **dereferencing** the rowlink by specifying the column whose value is required. The value of the cell using the dereferenced value then becomes that of the column, and it no longer shows as a link to the table row. More details on dereferencing with examples can be found here: `https://honeycodecommunity.aws/t/dynamically-retrieve-a-rowlinks-column/4520`.

> **The IFERROR function**
>
> The `IFERROR` function returns the second argument in the event when the first argument evaluates to an error value. It is useful in providing a valid default value to a cell instead of displaying an error. In the preceding case, the score defaults to a blank value.

_README

In this template, this table is just a placeholder with a link to the Honeycode community's page for templates.

> **Note How Unlike Other Table Names, This Table Has _ as a Prefix**
>
> Honeycode lists tables in lexicographical order, and therefore, to ensure that this `Readme` table does not come in the middle of the list, it is prefixed with a _ to put it toward the rear end of that list.

Reviewing the app

In the previous sections, we have defined how we want our app to work, what data we will need for it, and how that data is being stored and linked in the template workbook. Now, let's review how the template created the Survey app and link it all together to work.

The entire app is comprised of only three screens – **Home**, **Thanks**, and **Survey** – as shown in *Figure 7.9*:

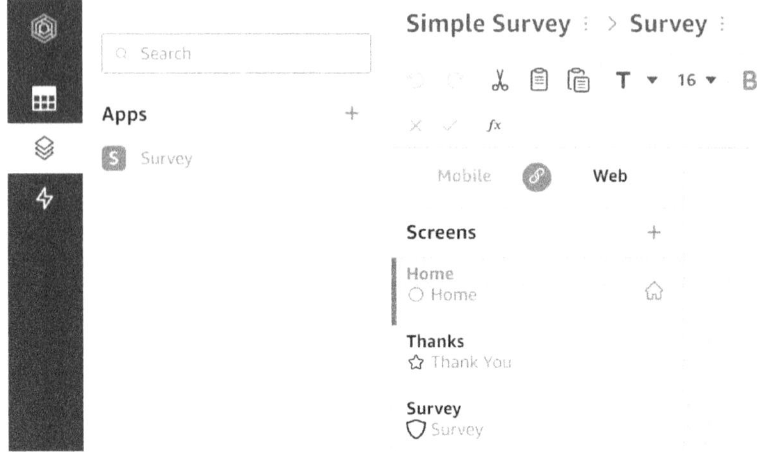

Figure 7.9 – The screens in the Survey app

Home

This is the home screen of the app. The screen consists of two main sections – a block at the top that contains a welcome note and a button to start the survey.

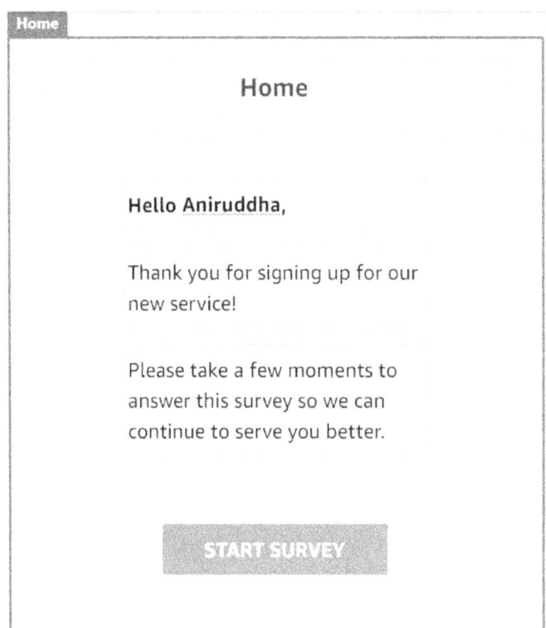

Figure 7.10 – The Home screen

The **START SURVEY** button has automation defined to first create or update a table row for the participant. It then navigates to the **Survey** screen while setting the `InputRow` variable to be the newly added or existing row for the participant (see *Figure 7.11*). This enables the **Survey** screen to display a blank survey if it wasn't previously taken or instead show the previous responses.

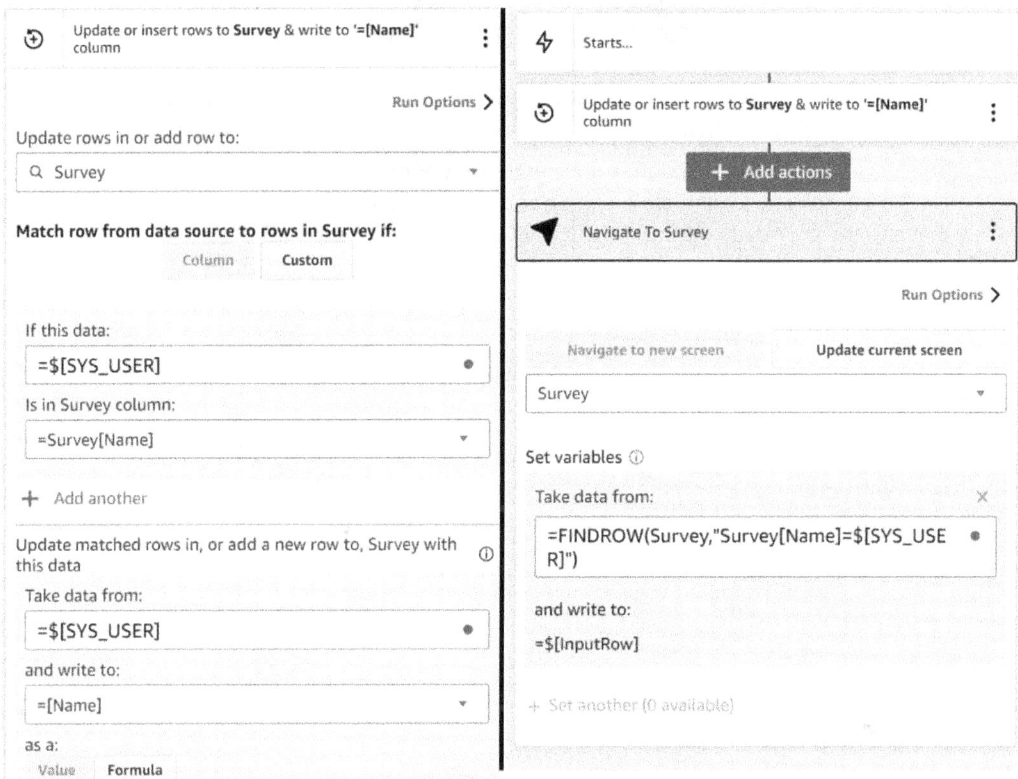

Figure 7.11 – Automation set on the START SURVEY button.
On the left is the verification and insertion/update of the table row for the participant, and on the right is navigation to the Survey screen along with a formula to set the InputRow variable

The value from the **Survey** table for the user is retrieved using the following `FindRow` function: `=FINDROW(Survey,"Survey[Name]=$[SYS_USER]")`

The FindRow Function

The **FindRow** function takes an ordered set of table rows as a source and returns the first row that matches the provided condition. Its syntax is similar to that of a `Filter` function, and in a sense, this can be viewed as a special case of it, wherein it only returns the first row instead of the entire matching set. Official documentation for the function can be reviewed at `https://honeycodecommunity.aws/t/findrow/889`.

The FindLastRow Function

The `FindLastRow` function is the counterpart of the `FindRow` function. It takes an ordered set of table rows as a source and returns the last row that matches the provided condition. Official documentation for the function can be reviewed at `https://honeycodecommunity.aws/t/findlastrow/896`.

Thanks

This screen is as basic as a screen can be in Honeycode. It contains a title, two static textboxes, and no other functionality, and simply says **Thank You** to the survey participant upon successful submission of the survey responses.

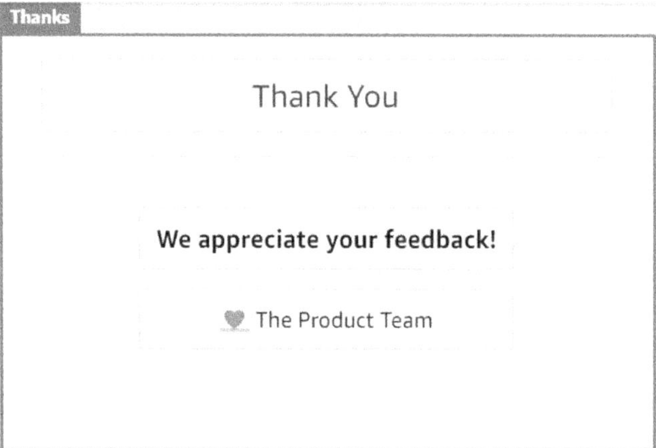

Figure 7.12 – The Thanks screen

Survey

This is the main functional screen of the app. The screen consists of the four questions that we saw in the **Survey** table, which are statically added one below the other, along with their corresponding input fields.

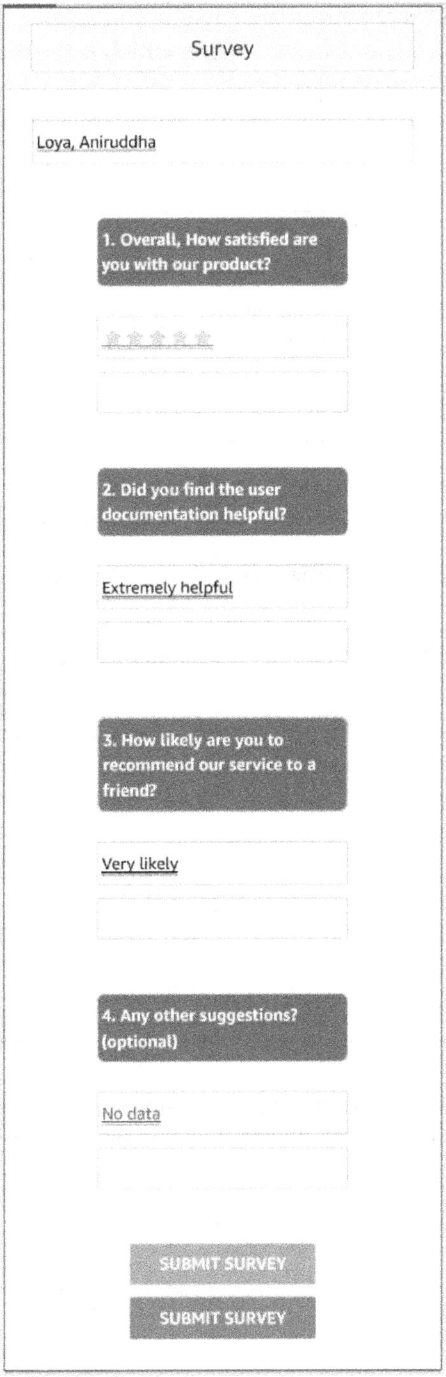

Figure 7.13 – The Survey screen

At the bottom of the screen, there appear to be two buttons, both labeled **SUBMIT SURVEY**. However, there is only one button there, which is the one with a green background. The one with a gray background is a content box with static text. However, when you open the app to take the survey, only one of the buttons is visible at any given time.

> **Can You Guess How?**
>
> Both the controls have a visibility condition set that is the opposite of each other. The visibility condition on the content box is `=OR($[Input1]="",$[Input2]="",$[Input3]="")` and that on the button is `=AND($[Input1]<>"",$[Input2]<>"",$[Input3]<>"")`.

The visibility condition here is used to enforce a response to required questions, but there is no visual cue or feedback to the participant about the same.

The button also has an automation defined to update the survey's **Submitted** date.

> **Can you think of why the automation is not updating the responses on submission?**
>
> This is the case because the fields for capturing responses are set as **shared**-type variables and are, therefore, tied directly to a cell's value on a particular row. See the *Understanding variables in Honeycode* section in *Chapter 5, Powering the Honeycode apps with Automations*, to review the details of variable types in Honeycode.
>
> While this is the setup provided by the Honeycode team in a template, I would not recommend this approach to any of you as it creates a risk of unintended updates.

> **Exercise 1**
>
> Can you think of how there can be unintended updates to survey responses?

Summary

In this chapter, we reviewed the **Simple Survey** template and learned its data model and how the app is built. We learned some new functions (`IFERROR`, `FindRow`, and `FindLastRow`) and terms (*rowlink* and *dereferencing*), and we also learned how visibility conditions can be used in an app as a means of input validation.

While there were some new concepts learned, the app and the template were very basic and lacked many features we typically see in a survey-creating or administering service. In this app, there is no means to create a new survey without updating the data model and the corresponding update to the survey screen. The template, therefore, has limited extensibility.

However, this is not to say we cannot have a more powerful survey app in Honeycode. In the next chapter, we will continue our template deep-dives and review the **Instant Polls** template, which will give you some ideas on how to improve this **Survey** app.

8
Instant Polls Template

Polls are a common feature in any organizational setup—they set up a democratic process as well as create an inclusive environment. We can get started with this template by setting up polls that can be accessed on mobile devices as well as desktops/laptops. In this chapter, we'll learn how the application is set up and thereby enable you with the power to customize the app as per your requirements.

In this chapter, we're going to cover the following main topics:

- Defining app requirements
- Creating an Instant Polls app
- Reviewing the data model
- Reviewing the app

Technical requirements

In order to follow this chapter, you'll need to have access to Amazon Honeycode, which requires a laptop with a web browser, preferably Google Chrome, and—optionally—a mobile device running either a Honeycode-supported version of Android (this currently requires *Android 8.0* or upward) or iOS (currently requires *iOS 11* or later).

Furthermore, we'll use the Honeycode terminology and refer to the components we covered in *Chapter 2, Introduction to Honeycode*, *Chapter 3, Building Your First Honeycode Application*, *Chapter 4, Advanced Builder Tools in Honeycode*, and *Chapter 5, Powering the Honeycode apps with Automations*, and therefore recommend you complete those first.

Defining app requirements

We first introduced this section in *Chapter 3, Building Your First Honeycode Application*, where, before building our app, we listed down the requirements or the use cases that we wanted the app to fulfill. We also made use of that list throughout the chapters as a guide to define our data model, conceptualize the application interface, and visualize the interactions between various onscreen elements as well as the data displayed. Therefore, even though we are not going to be building this app here but only review what is built, I'd encourage you to take 5 minutes to think about what your app should do and how it should look, and then make a list of this.

> **To-Do List**
>
> Take 5 minutes and list down how you would like your polling app to work.

Here is what my list looks like:

1. App users must be able to create new polls.
2. App users must be able to vote in existing polls.
3. The app should provide a visual clue of which polls I've already voted in and which I have not.
4. The app should provide an indication of which polls are new and which are closing soon.
5. Users should not see results until they have cast their vote.
6. The poll creator should be able to choose whether voters are allowed to change votes.
7. The poll creator should be able to delete the poll.
8. The poll creator should be able to close polls, or polls should close automatically.

Your list may have some or all these requirements and may have even more, and that is perfectly fine. You may find some of these missing in the template app and may want to extend that for yourself after you finish the chapter.

Creating an Instant Polls app

Now that we know how we want our app to work, let's see what the template app can do. We can now begin with creating an app and the associated workbook with the help of the following steps:

1. On the **Dashboard**, locate the **Create workbook** button in the top-right corner and click it.
2. On the popup that shows up, under the **Use a template** header, locate a tile with the label **Instant Polls** and click it, as demonstrated in the following screenshot:

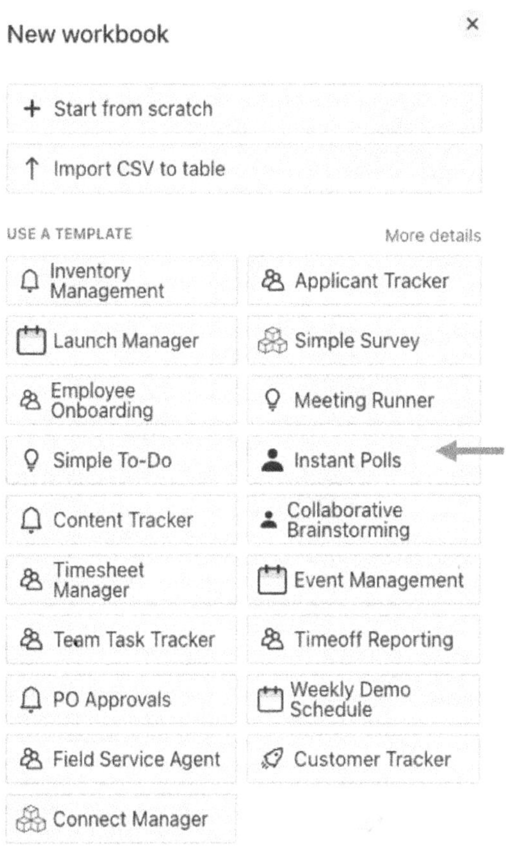

Figure 8.1 – Selecting the Instant Polls template

3. Next comes a popup to name the workbook and choose a team. For now, let's leave the default values and click on the **Create** button, as illustrated in the following screenshot:

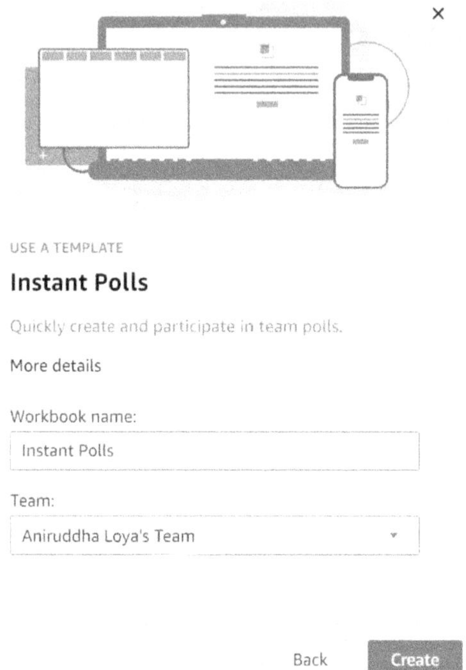

Figure 8.2 – Providing a workbook name and the team's details for creating a workbook

4. This creates and loads our **Instant Polls** workbook and loads up the workbook in the **Tables** view, displaying the first table in the list.

After completing the preceding steps, we are now ready to review the **Instant Polls** data model and the app.

Reviewing the data model

With the template workbook created, we are now ready to review how the Honeycode team set up the data model for the **Instant Polls** app. However, before we dive into that, let's take a couple of minutes to think of the data that is required for such an app.

We are creating a poll, so it naturally has a question and a set of options. We need information about the poll itself in terms of its creator, creation date, and due date. Furthermore, we need to store who our voters are and what they have voted for in individual polls.

Now that we have an idea of what sort of data we need, let's explore the different tables in the workbook and the data they store, as well as the relationship between them (if any). The template comes with six tables, as shown in the following screenshot:

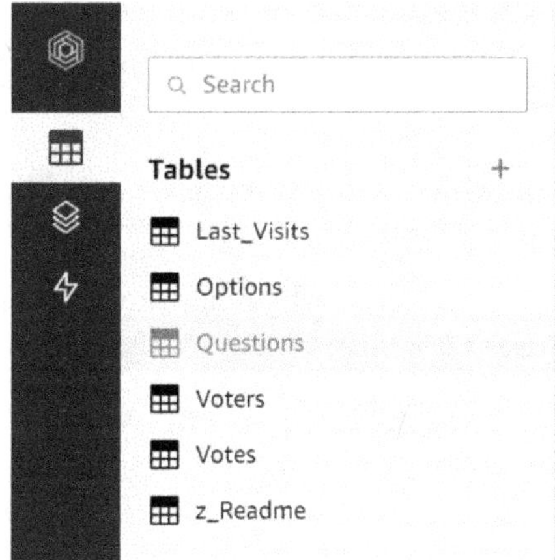

Figure 8.3 – List of tables in Instant Polls workbook

Let's understand the elements in *Figure 8.3*. We'll look at these next.

Last_Visits

A representation of the **Last_Visits** table is provided in the following screenshot:

Instant Polls > Last_Visits

Style | Settings | Wizards

fx Voter

	A	B	C	D
1	Voter	Questions Row	Timestamp	Votes_Since_Visit
2	AL Aniruddha I	Who will win the ga	12/26/21 2:52 AM	0
3	AL Aniruddha I	What topic would Y	12/26/21 2:52 AM	0
4	AL Aniruddha I	Which Avenger mo	12/26/21 2:58 AM	1
5	AL Aniruddha I	Which Avenger mo	12/26/21 3:06 AM	0

Figure 8.4 – Last_Visits table

This table keeps a log of polls that an app user has visited, what was the time of the last visit, and whether there have been any new votes on the poll since the last visit. It contains the following columns:

- The `Voter` column, which is formatted as a **contact**
- The `Questions Row` column, which is linked to the `Questions` table through a **picklist**
- The `Timestamp` column, which is formatted as date and time and stores the timestamp when the user last visited that poll
- The `Votes_Since_Visit` column, which has the following formula to calculate the votes cast since the user's last visit:

```
=FILTER(Votes,"Votes[Voter]<>% AND Votes[Timestamp]>% AND
 Votes[Questions Row]=%",[Voter],[Timestamp],[Questions
 Row])
```

This formula uses a `FILTER` function on the `Votes` table with multiple conditions combined using the AND operator. The first condition excludes rows where the `Voter` in the `Votes` table shown in *Figure 8.7* is the same as the `Voter` in this row, and then ensures that the timestamp on the vote must be greater than the timestamp of the last visit recorded for this voter, and finally (this one being the most obvious), that the questions must match.

> **Multiple Conditions inside the FILTER Function**
>
> We have seen some simple uses of the `FILTER` function in our previous chapters. In the preceding example, we see multiple conditions clubbed with the AND operator. You can also use other Boolean operations with `NOT` and `OR` operators as well as chain them.

> **Dynamic Arguments in the Filter Condition**
>
> Take a note of the use of the `%` symbol as a placeholder for arguments that are then provided after the condition and separated by a comma. This enables values to be supplied to the condition dynamically, directly as another formula, or as a value of one of the cells in the row using the **column reference** with **[Column_name]** notation.

Options

A representation of the **Options** table is provided in the following screenshot:

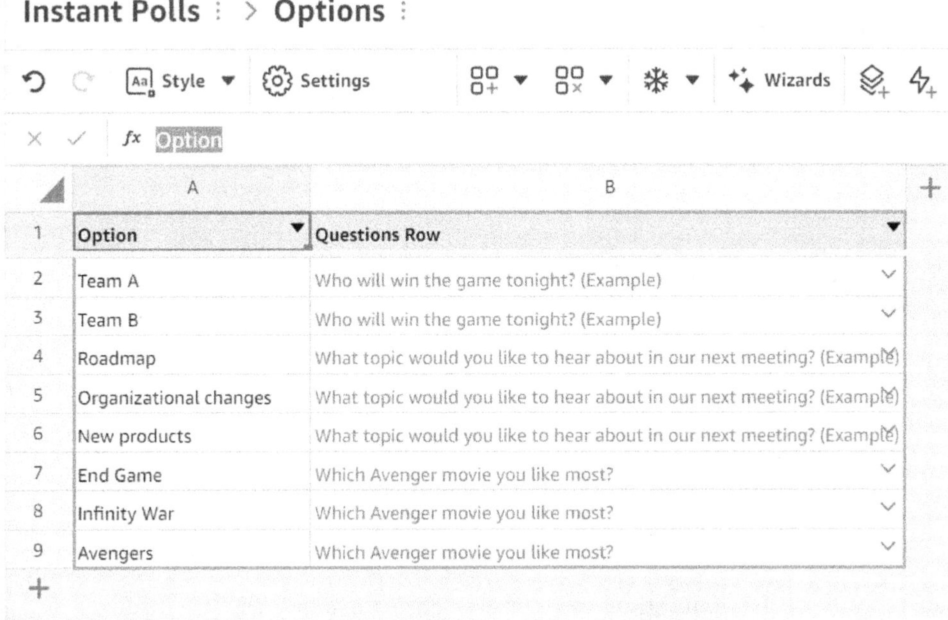

	A	B
1	Option	Questions Row
2	Team A	Who will win the game tonight? (Example)
3	Team B	Who will win the game tonight? (Example)
4	Roadmap	What topic would you like to hear about in our next meeting? (Example)
5	Organizational changes	What topic would you like to hear about in our next meeting? (Example)
6	New products	What topic would you like to hear about in our next meeting? (Example)
7	End Game	Which Avenger movie you like most?
8	Infinity War	Which Avenger movie you like most?
9	Avengers	Which Avenger movie you like most?

Figure 8.5 – Options table

This table is a list of options for all polls available in the app, and these options are mapped to their questions using the **picklist** in the **Questions Row** column.

Questions

A representation of the **Questions** table is provided in the following screenshot:

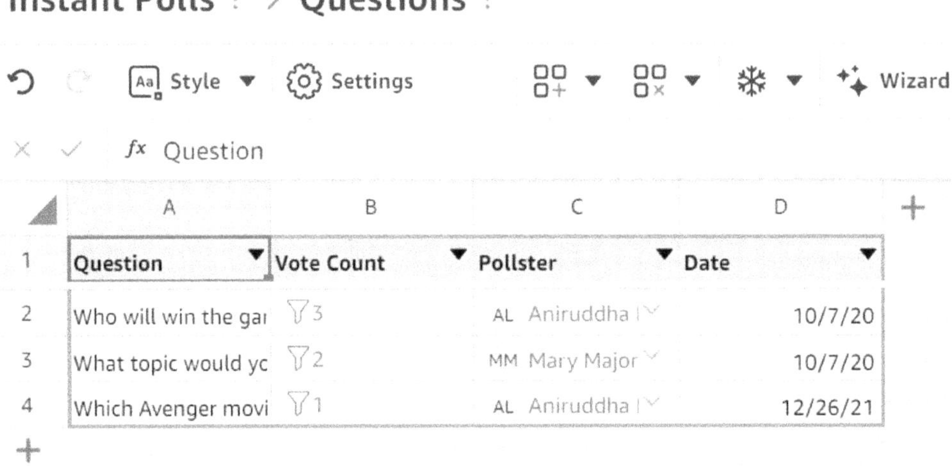

Figure 8.6 – Questions table

This table contains a list of questions or polls created so far, along with their related information, such as the following:

- Number of votes submitted for the poll
- Creator of the poll
- The date on which it was created

The value for the number of votes for each question is computed using the following filter formula:

```
=FILTER(Votes,"Votes[Questions Row]=%",THISROW())
```

While this formula is simpler than what we saw earlier in the **Last_Visits** table, it is of interest to review because it makes use of the `THISROW` function. The **Votes** table contains a reference to the question in this table using the picklist, and therefore, to filter the **Votes** table to retrieve the votes for a question, we have the condition to perform a comparison on the row reference for that question.

A comparison of the reference is important because it makes the condition invariant of the text in the cell and, therefore, any change in the referred row values—for example, a question is edited—will not break the condition.

> **ThisRow Function**
>
> The `ThisRow` function returns a self-reference (**rowlink**) to the table row containing the cell where this function is used.

Votes

This table is a simple list of all team members who have accessed the app at least once and went past the initial screen to see a list of questions available for voting. You can see a representation of the **Votes** table in the following screenshot:

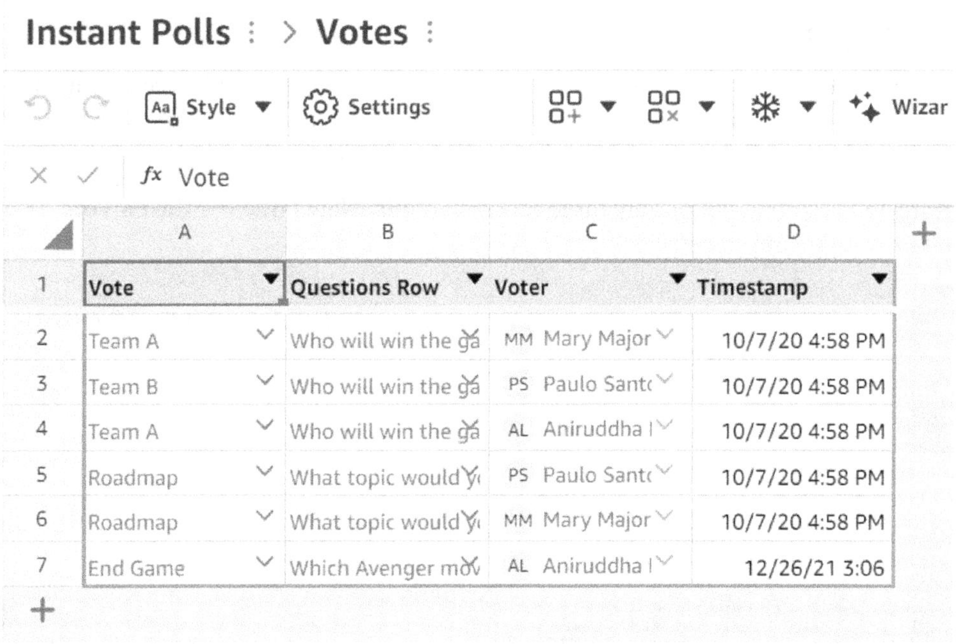

Instant Polls ⋮ > Votes ⋮

Style ▾ Settings Wizar

fx Vote

	A	B	C	D
1	Vote	Questions Row	Voter	Timestamp
2	Team A	Who will win the ga	MM Mary Major	10/7/20 4:58 PM
3	Team B	Who will win the ga	PS Paulo Santc	10/7/20 4:58 PM
4	Team A	Who will win the ga	AL Aniruddha I	10/7/20 4:58 PM
5	Roadmap	What topic would y	PS Paulo Santc	10/7/20 4:58 PM
6	Roadmap	What topic would y	MM Mary Major	10/7/20 4:58 PM
7	End Game	Which Avenger mo	AL Aniruddha I	12/26/21 3:06

Figure 8.7 – Votes table

This table is a collection of all votes across all polls, along with information about who cast votes and when. This table makes extensive use of picklists and stores data in the form of references to other tables, except for timestamps. References created using picklist selections allow us to provide a **single source of truth** (**SSOT**) for data and the creation of filter functions, as we observed in the case of the **Questions** table.

z_Readme

In this template, this table is just a placeholder with a link to the Honeycode community's page for templates.

> **Note (unlike Other Table Names, This Table Has z_ as a Prefix)**
>
> Honeycode lists tables in lexicographical order, and therefore, to ensure that this **Readme** table does not come in the middle of the list, it is prefixed with `z_` to put it toward the rear end of that list.

Reviewing the app

Now we know how we want our app to work, what sort of data we would need for it, and how that data is stored and linked in the template workbook, let's review how the template created in the **Instant Polls** app links it all together and works.

The entire app is comprised of only three screens—**Polls**, **New Poll**, and **Cast a Vote**, as shown in the following screenshot:

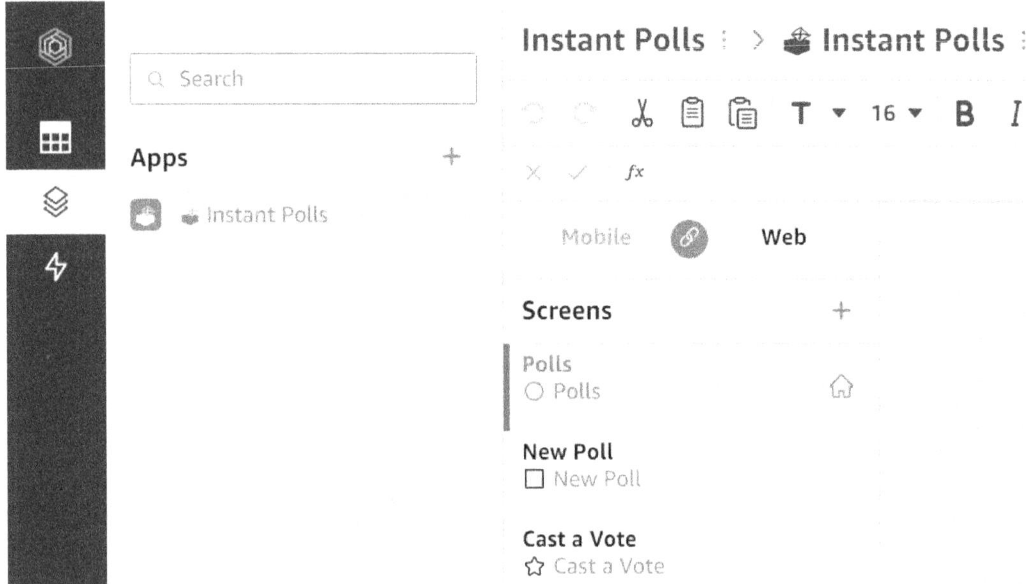

Figure 8.8 – Screens in the Instant Polls app

Let's look at these screens in more detail.

Polls

This is the home screen of the app. The screen consists of two main sections—**Block**, at the top, which contains a welcome note and general instructions about the app, and **Questions List**, as shown in the following screenshot:

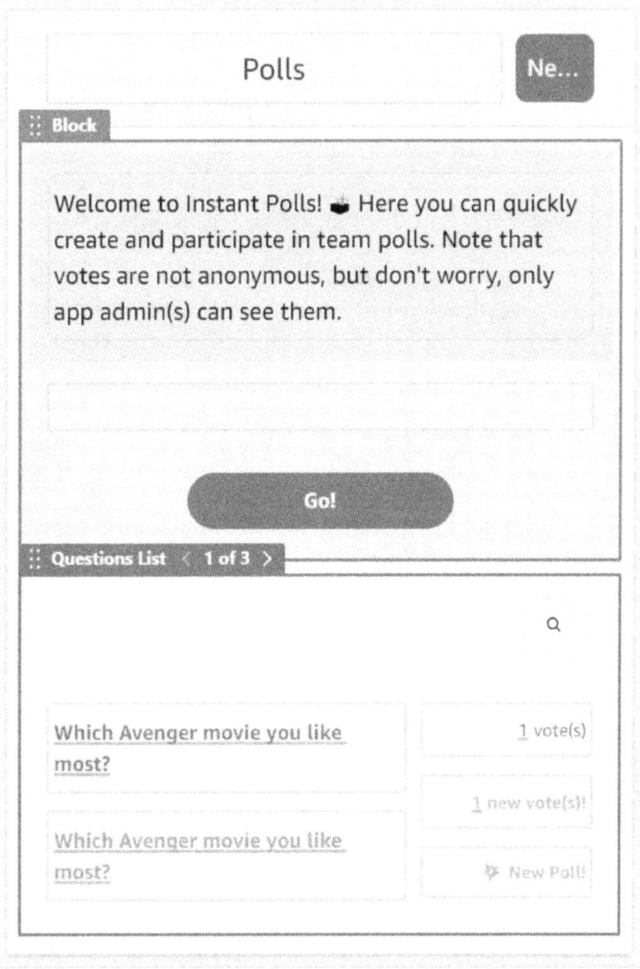

Figure 8.9 – Polls screen in builder view

Let's dive into the details of these two sections next.

Block

This section has the following two components:

- A textbox with a welcome message and a disclaimer about the anonymity of votes
- A button underneath

The interesting part of this section, though, lies in its properties. Let's take a look at the **Visibility** condition for the top **Block** section. It is set to show only when the app user is either opening the app for the first time or has never gone past this screen by hitting the **Go!** button. It uses the following formula to achieve that:

```
=FILTER(Voters,"Voters[Voter]=%",$[SYS_USER])=0
```

The **Go!** button has automation defined that adds a new entry to the **Votes** table for the app user, ensuring that the preceding condition will never again hold true.

Questions list

The **Questions List** section consists of a search control, and a set of content boxes for displaying questions, the number of votes, and updates (if any) since the user's last visit.

The list has its visibility set to be opposite of the top **Block** section, implying it will be the default view from the second visit onward but not on the first load of the app. The **Visibility** condition for this list is configured with the following formula:

```
=FILTER(Voters,"Voters[Voter]=%",$[SYS_USER])>0
```

Furthermore, even though this list displays all questions in the table, it is not sourced with `Table` as a source and instead uses the following `FILTER` function, which also introduces us to the concept of ordering:

```
=FILTER(Questions,"ORDER BY Questions[Date] DESC,
 Questions[Vote Count] DESC")
```

In this case, the condition inside the `FILTER` function is not for including/excluding table rows but simply defines the order in which rows should be displayed and also showcases how multiple ordering criteria can be pipelined such that they apply one after the other. The list is first ordered in descending order by creation date so that the oldest question is right at the top, and followed by ordering by the number of votes to order among questions created on the same date.

Finally, the list has an action defined on it that navigates to the **Cast a Vote** screen and also updates the **Last_Visits** table either by adding a new row if it is the app user's first view of the poll question or by updating the timestamp against the matching row.

The content boxes also heavily use the **Visibility** property. There are two content boxes for displaying the question, with the color of the text being the only difference to differentiate between questions you have already voted on from those that you have not. In the following screenshot, the third question in the list has already been voted on by the app user and therefore is in a lighter shade of green compared to the first two.

> **Exercise 1**
>
> Can you think of another way of achieving the same result of separate text color based on a specified condition?

< Apps	Polls	New +
Which Avenger movie do you like most?		0 vote(s) New Poll!
What topic would you like to hear about in our next meeting? (Example)		2 vote(s)
Who will win the game tonight? (Example)		3 vote(s)

< Apps	Polls	New +
Which Avenger movie do you like most?		1 vote(s) 1 new vote(s)!
What topic would you like to hear about in our next meeting? (Example)		2 vote(s)
Who will win the game tonight? (Example)		3 vote(s)

Figure 8.10 – Screenshot of app showing the New poll content box on the left and the new votes content box on the right

In *Figure 8.10*, there is another set of content boxes configured to display the number of new votes when there is a new vote on a question and **New Poll!** text if a question has been added since your last visit.

New Poll

This screen is for creating a new poll, similar to the **Add New Task** screen we created in *Chapter 3, Building YourFirst Honeycode Application*, but is packed with more functionality than that. It has a duplication check built on the question as well as the options. Also, unlike the **Add New Task** screen where we had a fixed set of fields, for polls we cannot know in advance how many options will be needed for a question, so the screen allows for dynamically adding options to questions. Lastly, it also includes a functionality to send a custom message to team members about a newly created poll, as shown in the following screenshot:

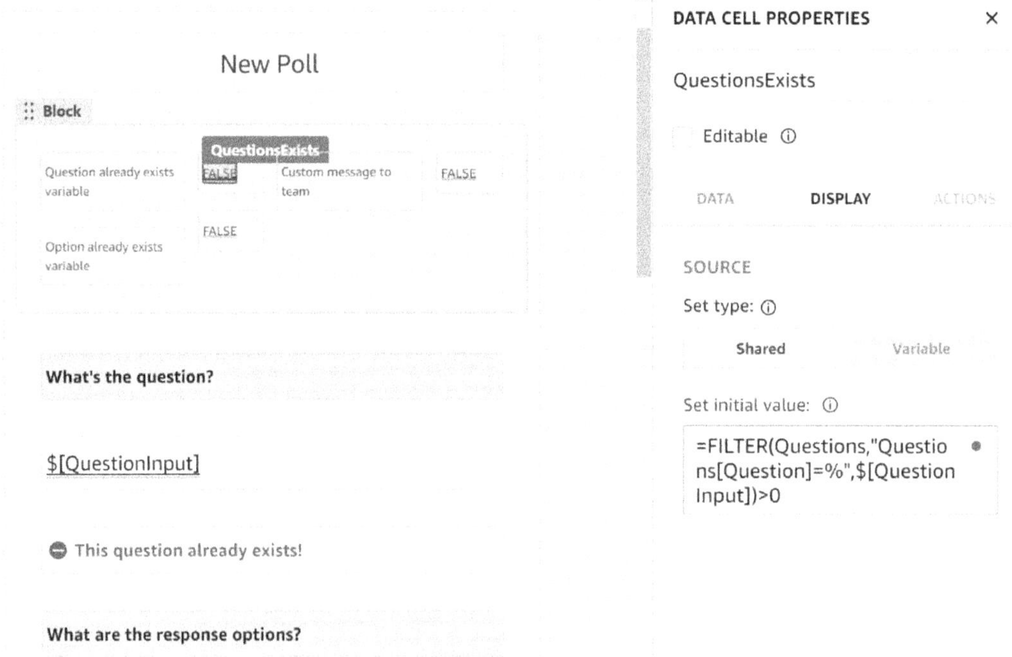

Figure 8.11 – New Poll screen with the QuestionsExists variable selected

The topmost block in this screen is a hidden block that serves two purposes, as outlined here:

1. It allows us to declare and collect all intermediate variables and states in a single place that can simply be referenced using the data cell name across this screen.
2. While building the app, the visibility of the block—and thereby all the variables it contains—can be set to `TRUE` as you test the app. This enables you to verify whether the formulas are working as expected and to identify incorrect values (if any) and fix them.

For instance, the **QuestionsExists** data cell shown in *Figure 8.11* is used to control the visibility of various controls on the screen, including the content box that displays **This question already exists!** error text. Furthermore, during development, you can make the block visible and see the value of this variable as you type in the question, and validate that the value is being correctly updated and thus also correctly applied to visible objects on the screen. This is extremely helpful in fixing any errors in the formulas.

In the following screenshot, we have controls that are displayed in the app as per the visibility conditions wherever set:

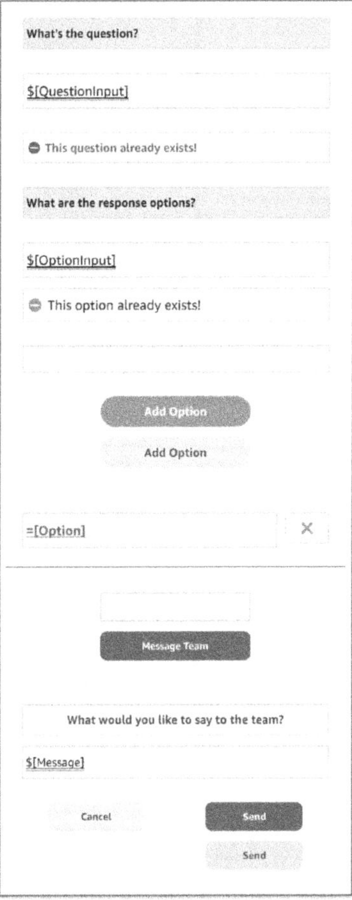

Figure 8.12 – New Poll screen with controls that are visible to the app user depending on the conditions

While duplicate buttons are obvious candidates for conditional visibility, as you go through each object on this screen, you'll notice an extensive use of visibility conditions to control screen elements. Compare *Figure 8.12*, which is the builder view of the screen, with *Figure 8.13*, this being a screenshot of the app.

The **Add Option** button with a green background has the automation set to add a question to the **Questions** table as well as add an option to the **Options** table. Once the first option is added, every subsequent click of this button will simply update the timestamp for this question in the **Questions** table while adding the new option in the **Options** table.

Once an option is added to the **Options** table, the content box below it becomes visible along with the **Message Team** button (check out its visibility condition), as illustrated in the following screenshot:

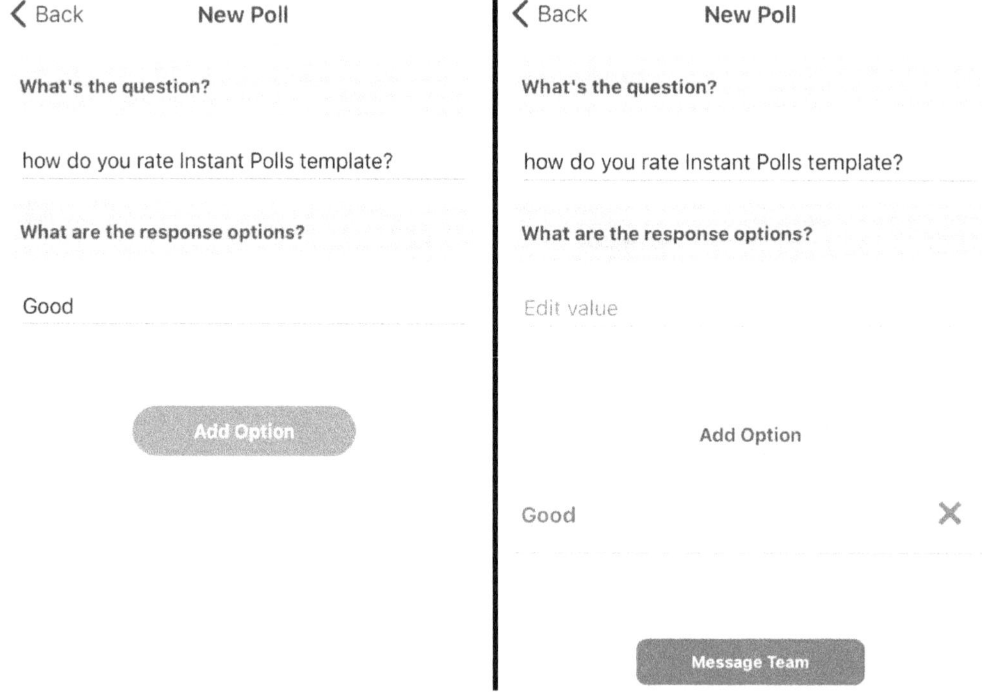

Figure 8.13 – New Poll screen as visible in the app; on the left is the screen before a question is added, and on the right is the screen after adding a question using the Add Option button

> **Note**
>
> A poll with a single option does not make much sense. Can you update the app so that at least two options are added before the poll is valid? Are you able to think of more than one way to achieve this?

Clicking on the **Message Team** button displays elements to input text that you'd like to add to the app notification and an email that will be sent to every voter from the **Voters** table, as illustrated in the following screenshot:

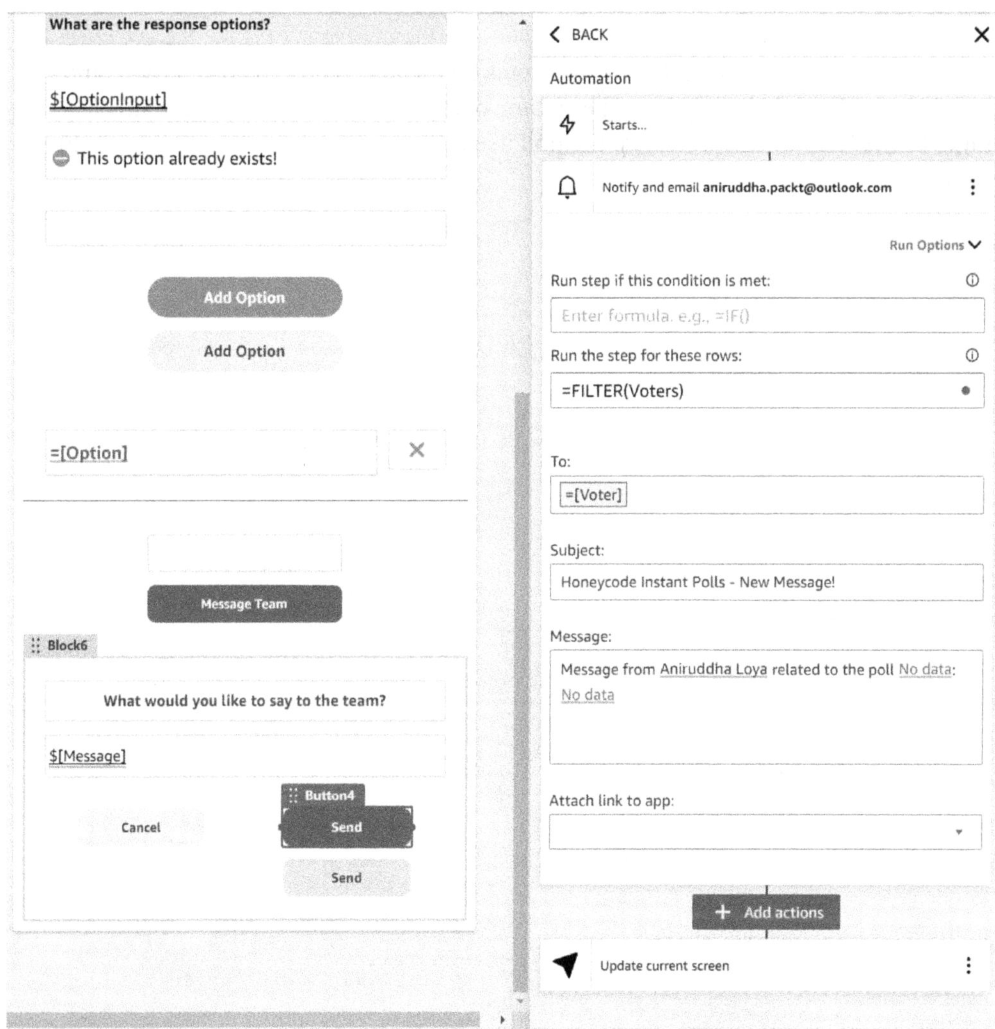

Figure 8.14 – Automation to send a message to all voters

Figure 8.14 shows the automation on the **Send** button. Notice the difference between how the **To:** field is set here compared to how we configured notifications in *Chapter 5, Powering the Honeycode apps with Automations*. Here, the **To:** field is neither a static list nor a specified user, but instead, it's a list of users that can keep changing as more and more users start using the app. Therefore, the **To:** field here is populated using the values from the **Voter** column of rows returned by the formula defined for the **Run the step for these rows** property.

Cast a Vote

This is a screen primarily meant for casting votes by the app users. However, there are hidden sections that are displayed to the creator of the poll to probably send reminder messages to the team or maybe delete the poll itself. You can see a representation of this screen here:

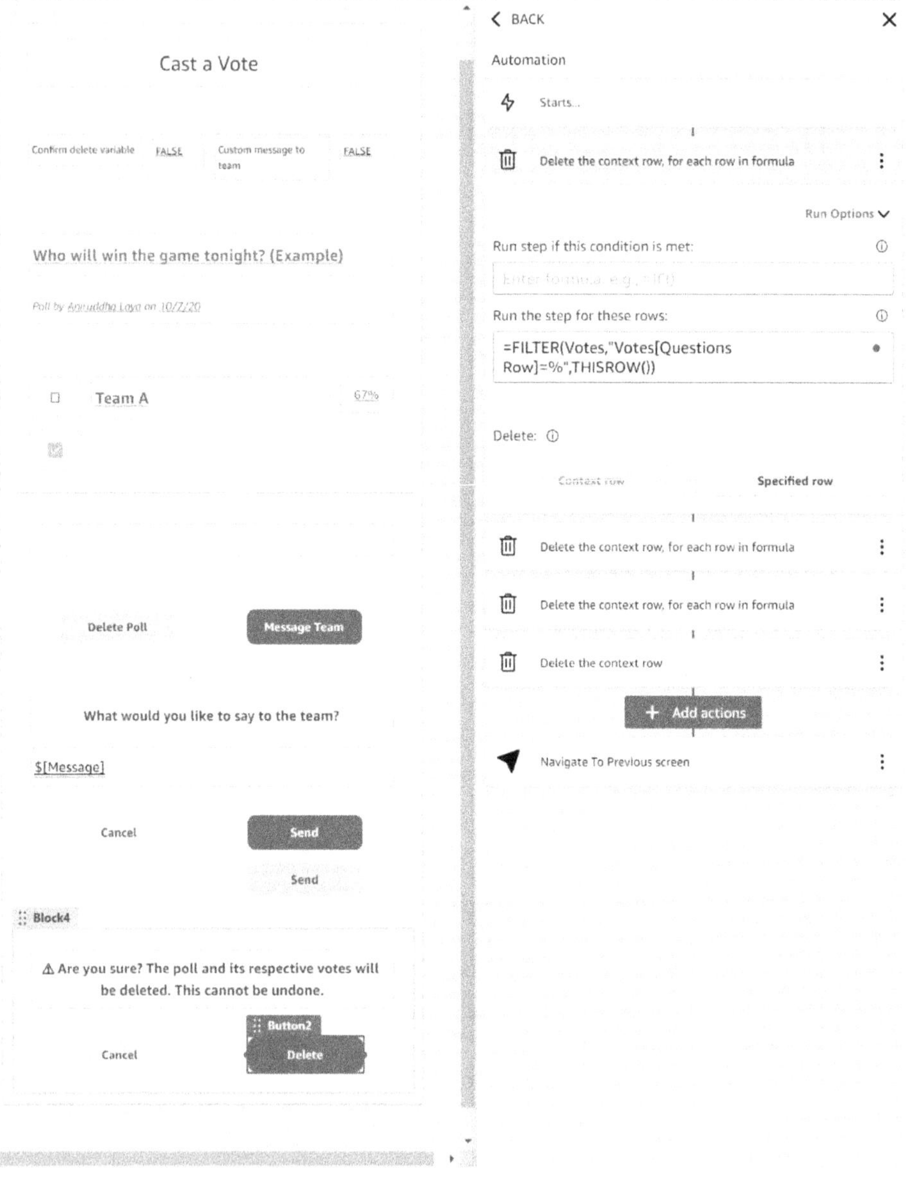

Figure 8.15 – Cast a Vote screen, along with automation for the Delete button

Similar to the **New Poll** screen, this screen also has a hidden block at the top, and the **Message Team** functionality is also a copy of it.

> **Note**
> The mention of the word *copy* might cause some frowning or raised eyebrows among readers with some background in programming and or familiarity with the **Don't Repeat Yourself** (**DRY**) principle. But Honeycode is not meant to be a programming paradigm, so the only way of duplicating a functionality is to copy it.

However, not everything here is a copy. The automation on the **Delete** button is a useful study. Now, you may be thinking: What's so special about it? We already know how to delete a row in a table from an app; we even implemented it in *Chapter 5, Powering the Honeycode apps with Automations,* to delete a task.

Well, the difference here is the cleanup of all artifacts of a poll from our data store (tables). In our earlier sections, we learned how different tables are interrelated, with references being created using **rowlinks** created with **picklists** or set with automations using the app. If we simply delete **Polled Question** from the **Questions** table, while the operation will succeed, it will leave the rows in the **Votes**, **Options**, and **Last_Visits** tables with a non-existing reference. Now, given that those values are not visible in the app without the question, it does not affect the functionality, but it does leave a lot of dead references and unused table rows in our tables.

The automation on the **Delete** button thus cleans up the tables in four steps, as outlined here:

1. It deletes all votes cast for the question being deleted. This is achieved by selecting all rows in the **Votes** table associated with the question using the following filter formula: `=FILTER(Votes,"Votes[Questions Row]=%",THISROW())`, as shown in *Figure 8.15*.
2. It deletes all options for this question. This is achieved by selecting all rows in the **Options** table associated with the question using the following filter formula: `=FILTER(Options,"Options[Questions Row]=%",THISROW())`.
3. It deletes all the last-visit records for this question. This is achieved by selecting all rows in the **Last_Visits** table associated with the question using the following filter formula: `=FILTER(Last_Visits,"Last_Visits[Questions Row]=%",THISROW())`.
4. It deletes the question itself.

> **Note**
>
> Such a manual cleanup is typically not required for commercial databases that usually allow cascading deletes, as defined by relations using **foreign keys** (**FKs**), and may be perceived as a shortcoming of Honeycode. As a platform for citizen developers, it is asking for more diligence and setup than what most professional developers would get out of the box.

Summary

In this chapter, we reviewed the **Instant Polls** template and learned about its data model and how the app is built. We learned about a new function (`THISROW`) and more complex use of the previously known `FILTER` function. We learned how we can build support for multiple polls in a single app and how to support a dynamic number of options. We also discovered how extensively visibility conditions can be used in an app to build multiple functions and features within a single screen.

Recall that in the previous chapter, we saw a static survey with fixed questions and response types. With the concepts learned in this chapter, you can now improve the **Survey** app from a simple to a professional version that, among other features, allows for the creation of new surveys on the fly with a varying number of questions and other response types.

In the next chapter, we will continue our template deep-dives and review the **Event Management** template.

9
Event Management Template

Organizing events is challenging on many levels. So many things need to fall into place logistically and operationally and across various departments and personnel to host an event successfully. You might find yourself thinking about the marketing, websites, flyers, guest registration desk, help desk, and more. In today's world, you might also need to decide whether or not to spend on building a custom mobile app for your guests to easily access the schedules and details of various ongoing events, register for them, and more.

On the other hand, participating in a large-sized event has its own challenges. Have you ever been to an event and found yourself constantly opening the website to find details of the next session to attend, check the whereabouts of your next registered session, or maybe stand/wait in a queue to register yourself when a registered participant does not show up?

If you have been in any of those categories, this template will give you a tool to take control of some of the moving parts and that too fairly quickly. In this chapter, we'll learn how the app is set up and, thereby, enable you with the power to customize the app as per your requirements.

In this chapter, we're going to cover the following main topics:

- Defining the app requirements
- Creating an Event Management app
- Reviewing the data model
- Reviewing the app

Technical requirements

To follow along with this chapter, you'll need to have access to Amazon Honeycode. This will require a laptop with a web browser, preferably Google Chrome, and optionally a mobile device running either a Honeycode-supported version of Android (currently, this requires Android 8.0 or later) or iOS (currently, this requires iOS 11 or later).

Furthermore, we'll use the Honeycode terminology and refer to the components that we covered in *Chapter 2, Introduction to Honeycode*, *Chapter 3, Building Your First Honeycode Application*, *Chapter 4, Advanced Builder Tools in Honeycode*, and *Chapter 5, Powering the Honeycode apps with Automations*. Therefore, we recommend that you complete those first.

Defining the app requirements

We first introduced this section in *Chapter 3, Building Your First Honeycode Application*, where, before building our app, we listed the requirements or use cases that we wanted the app to fulfill. Additionally, we made use of that list throughout the chapters as a guide to define our data model, conceptualize the application interface, and visualize the interactions between various onscreen elements and the data displayed. Therefore, even though we will not build this app here, we will review what is built. I would encourage you to take 5 minutes to think about what your app should do and what it should look like, and then make a list of your requirements.

> **To-Do**
>
> Take 5 minutes and make a list of how you would like your Event Management app to work.

Here is what my list looks like:

- App users must be able to view the event details along with the details of any sessions within the event.
- App users must be able to register for an event and/or various sessions within an event as a participant.
- Event participants must be able to provide feedback for the events/sessions they attended.
- Event participants must be able to de-register themselves from an event/session.
- Event participants should be able to put themselves on a waitlist for a full event/session.
- Event participants should be able to seek help from the organizers through the app.
- Event participants should be able to check in to an event through the app.
- The app should remind participants of their upcoming events.
- The app should prevent a participant from registering for two sessions that are held at the same time.
- Event organizers must be able to add or update details of the sessions or events.
- Event organizers must be able to view all participants of an event.
- Event organizers must be able to broadcast notifications for an event or a specific session to all of its participants.

Your list might have some or all of these requirements or even have more, and that is perfectly fine. You might find some of these missing in the template app and want to extend that for yourself after you finish the chapter.

Creating an Event Management app

Now that we know how we want our app to work, let's see what the template app can do. We will begin by creating the app and the associated workbook. Perform the following steps:

1. On the **Dashboard** screen, locate the **Create workbook** button from the upper-right corner and click on it.
2. You will see that a popup appears. Under the **USE A TEMPLATE** header (as shown in *Figure 9.1*), locate a tile with the label **Event Management** and click on it:

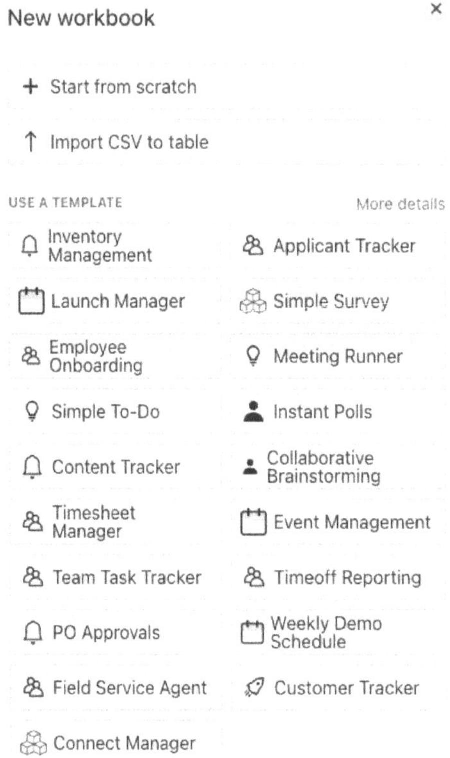

Figure 9.1 – Selecting the Event Management template

3. Next, you will see a popup asking you to name the workbook and choose a team. For now, let's leave the default values as they are. Click on the **Create** button.

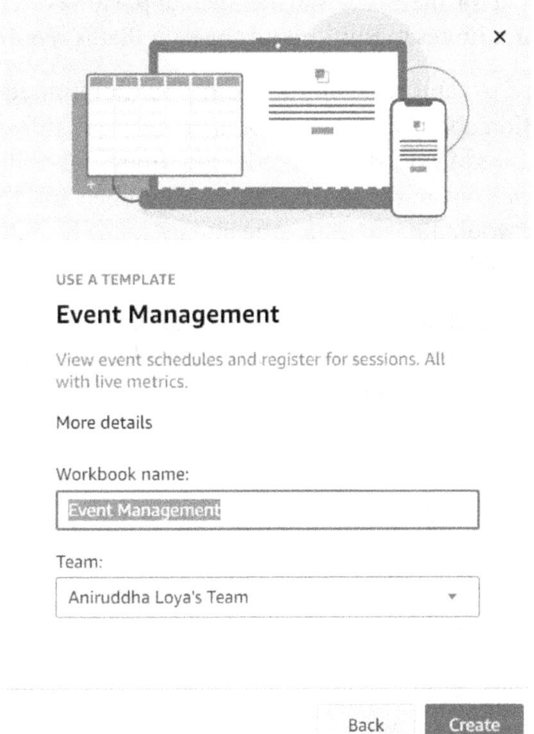

Figure 9.2 – Providing the workbook name and team details for creating the workbook

4. This creates and loads our **Event Management** workbook. Additionally, it loads up the workbook in the **Tables** view, displaying the first table in the list.

And with that, we are now ready to review the **Event Management** data model and the app.

> **Note**
>
> Similar to how the survey template has a single survey and no means to create a new survey using the app, the **Event management** template is also only created for a single event and lacks the feature to easily create and manage multiple events. Additionally, it is designed for a multiday conference-style event with multiple sessions throughout the day; therefore, it does not address all classes of events.

Reviewing the data model

With the template workbook created, we are now ready to review how the Honeycode team set up the data model for the Event Management app. However, before we dive into that, let's take a couple of minutes to think about the data that is required for such an app.

For an event, we will need to capture details including, but not limited to, the date, the time, the place, information about various activities in the event, information about the speakers or hosts in the case of workshops or talks, and more. We will need information about the participants, such as their name, their phone number, and their registration details. Furthermore, we would like to reduce the queues at the help desk by providing basic FAQs and self-help tools for participants.

Now that we have an idea of what data we need, let's explore the different tables in the workbook and the data they store, along with the relationship between them, if any. The template comes with nine tables, as shown in *Figure 9.3*:

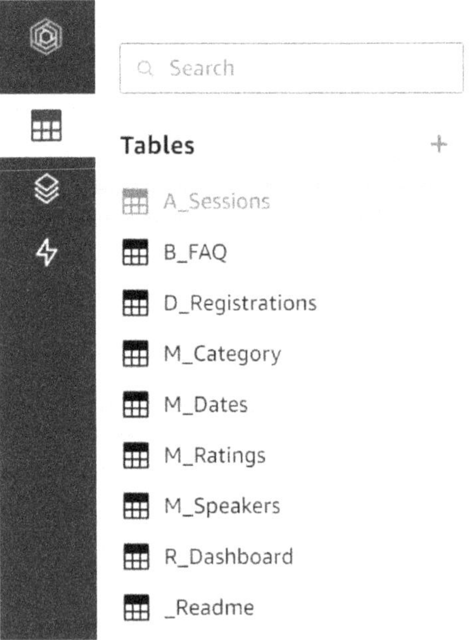

Figure 9.3 – The list of tables in the Event Management workbook

Let's understand each of these tables next.

A_Sessions

This table contains details about the sessions taking place at the event and all its related information, such as the date of the session, the start and end times, the category, the speaker, the location, and more.

Event Management > A_Sessions

Style Settings Wizards Save filter

fx Session

	Session	Category	Date	Date (Text)	Start Time	End Time	Speaker	Capacity	Registrations	IsFull	Percent Full	Location	Rating
2	Innovating with High Standards	#Keynote	3-Jan	9-Feb	9:00 AM	9:50 AM	Paulo Santos	5	4	FALSE	80.00%	Room 03-101	4.5
3	Getting Started with Machine Learning	#Advanced	3-Jan	9-Feb	10:00 AM	10:50 AM	Aniruddha Loy	3	3	TRUE	40.00%	Room 03-101	3.7
4	Integrating Customer Insights Into Your Product	#Foundatio	3-Jan	9-Feb	11:00 AM	11:50 AM	Jane Roe	10	3	FALSE	30.00%	Room 03-101	5.0
5	Lunch		3-Jan	9-Feb	11:50 AM	1:00 PM		10	4	FALSE	40.00%	Room 04-201	5.0
6	Future of AI	#Keynote	3-Jan	9-Feb	1:00 PM	1:50 PM	John Stiles	5	3	FALSE	60.00%	Room 04-201	3.3
7	Drive Your Product Strategy	#Foundatio	3-Jan	9-Feb	3:00 PM	3:50 PM	Aniruddha Loy	10	3	FALSE	40.00%	Room 04-201	3.3
8	Panel: Biggest Mistakes in My Career	#Foundatio	3-Jan	9-Feb	4:00 PM	4:50 PM	Mary Major	10	0	FALSE	0.00%	Room 04-201	Not Rated
9	Happy Hour!		3-Jan	9-Feb	5:00 PM	6:00 PM		25	4	FALSE	16.00%	Room 05-101	2.3
10	How Machine Learning is Transforming SaaS	#Keynote	4-Jan	10-Feb	9:00 AM	9:50 AM	Jane Roe	10	4	FALSE	40.00%	Room 05-103	5.0
11	Rapid Innovation at Amazon Scale	#Keynote	4-Jan	10-Feb	10:00 AM	10:50 AM	Mary Major	10	3	FALSE	30.00%	Room 05-101	5.0
12	Escalating Effectively and Responding to Escalations	#Advanced	4-Jan	10-Feb	11:00 AM	11:50 AM	Mary Major	10	3	FALSE	40.00%	Room 05-103	2.7
13	How to Build Roadmaps and Make Prioritization Decisions	#Foundatio	4-Jan	10-Feb	11:50 AM	1:00 PM	Mary Major	10	3	FALSE	30.00%	Room 05-101	4.0

Figure 9.4 – The A_Sessions table

So, if you are setting up your own event, here, you will be making changes for adding or updating the details of different talks, sessions, or workshops that will be running during the event. At the time of writing, there is an error in this template wherein the values of the **Percent Full** column are hardcoded. This means they will not change as you clear out the template and set it up for your own use and/or when participants register for your events. To fix that issue, select the **Percent Full** column and click on the **Setting** control from the toolbar, which will bring up the **COLUMN PROPERTIES** panel on the right-hand side. In the field labeled **Formula**, set the following formula: `=ROWS([Registrations])/[Capacity]`. Then, set the **Format** setting as **Percentage**, as shown in *Figure 9.5*:

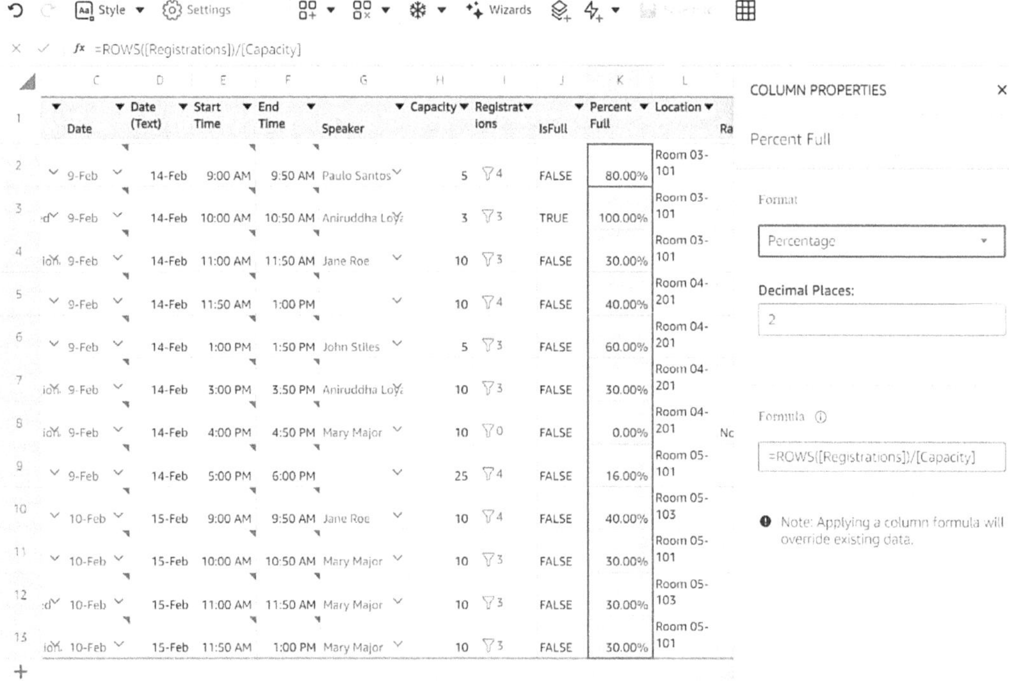

Figure 9.5 – Fixing the Percent Full column by setting the formula and its format

Another error you might notice is that the same room has been set at a different capacity. Now that is a very possible error that can take place. Fortunately, the fix for this is also simple. Can you think of how to do it?

> **Exercise 1**
>
> Fix the location-capacity mapping error to keep the information consistent.

> **The Rows Function**
>
> The Rows function returns the number of rows in the given reference. For example, in the case of a formula used for the **Percent Full** column, the reference provided is the value of the cell underneath the **Registrations** column of the same row. This cell contains the Filter function; therefore, the Rows function returns the number of rows that satisfy the filter condition.

> **Note**
>
> As mentioned earlier, in terms of a template being designed for only a single event, you will notice that the mapping between the session and the event does not exist in this table, and there is no separate table for that either.

B_FAQ

This table is very straightforward. It has two columns, namely, **Question** and **Answer**. They list the different questions that we might want to provide as FAQs and enables participants to help themselves through the app.

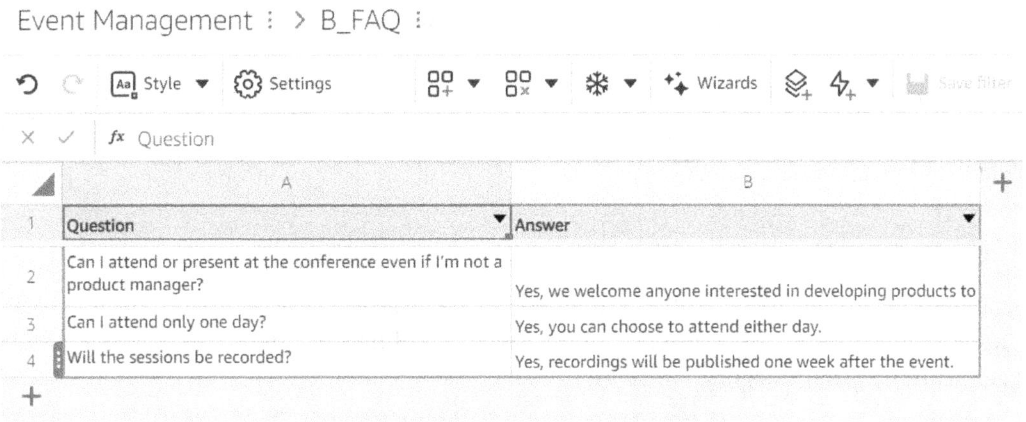

Event Management > B_FAQ

	A	B
1	Question	Answer
2	Can I attend or present at the conference even if I'm not a product manager?	Yes, we welcome anyone interested in developing products to
3	Can I attend only one day?	Yes, you can choose to attend either day.
4	Will the sessions be recorded?	Yes, recordings will be published one week after the event.

Figure 9.6 – The B_FAQ table

Try to have more than three questions that you know would need to be on the FAQ page for your event. You can do this by just adding another row and typing the question in.

D_Registrations

This table contains the information about all of the registrations, for example, who is attending which session, when they registered for it, and how they rated the session afterward.

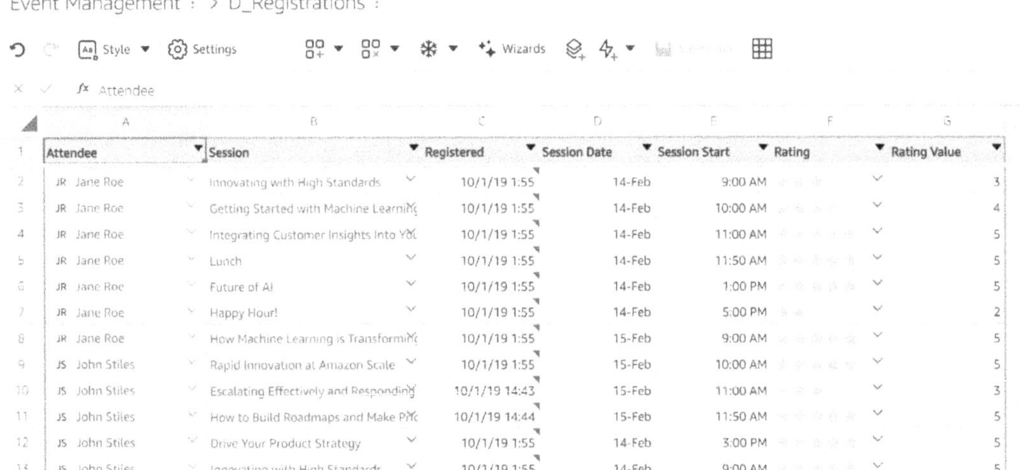

	Attendee	Session	Registered	Session Date	Session Start	Rating	Rating Value
2	JR Jane Roe	Innovating with High Standards	10/1/19 1:55	14-Feb	9:00 AM		3
3	JR Jane Roe	Getting Started with Machine Learnin	10/1/19 1:55	14-Feb	10:00 AM		4
4	JR Jane Roe	Integrating Customer Insights Into Yo	10/1/19 1:55	14-Feb	11:00 AM		5
5	JR Jane Roe	Lunch	10/1/19 1:55	14-Feb	11:50 AM		5
6	JR Jane Roe	Future of AI	10/1/19 1:55	14-Feb	1:00 PM		5
7	JR Jane Roe	Happy Hour!	10/1/19 1:55	14-Feb	5:00 PM		2
8	JR Jane Roe	How Machine Learning is Transformin	10/1/19 1:55	15-Feb	9:00 AM		5
9	JS John Stiles	Rapid Innovation at Amazon Scale	10/1/19 1:55	15-Feb	10:00 AM		5
10	JS John Stiles	Escalating Effectively and Respondin	10/1/19 14:43	15-Feb	11:00 AM		3
11	JS John Stiles	How to Build Roadmaps and Make Pr	10/1/19 14:44	15-Feb	11:50 AM		5
12	JS John Stiles	Drive Your Product Strategy	10/1/19 1:55	14-Feb	3:00 PM		5
13	JS John Stiles	Innovating with High Standards	10/1/19 1:55	14-Feb	9:00 AM		5

Figure 9.7 – The D_Registrations table

Similar to *Chapter 7*, *Simple Survey Template*, the values in the **Session Date**, **Session Start**, and **Rating Value** columns are derived using **Dereferencing**, while the **Rating Value** column uses the **IFERROR** function to get the rating's numerical value or provide a default value in the absence of one.

M_Category

This is another simple table that lists the different categories for classifying the various sessions along with the **Filter** function in the **Sessions** column to get a list of all sessions tagged with the given category.

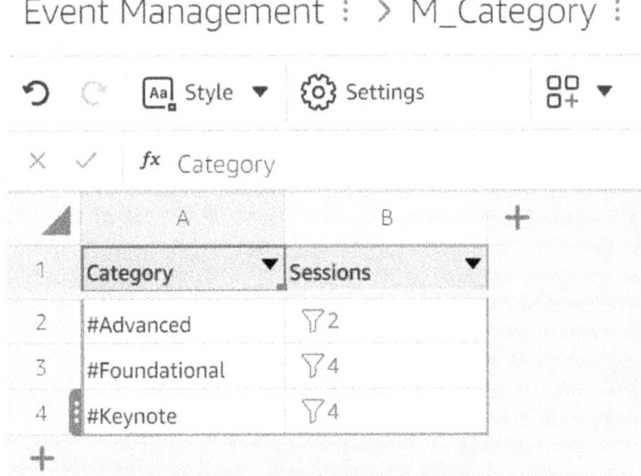

Figure 9.8 – The M_Category table

If you need to add a new category for your events, this is the table on which you can add a new row.

M_Dates

Similar to the **M_Category** table, this table lists the dates of the event and a similar `Filter` function to get a list of all the sessions scheduled for that day.

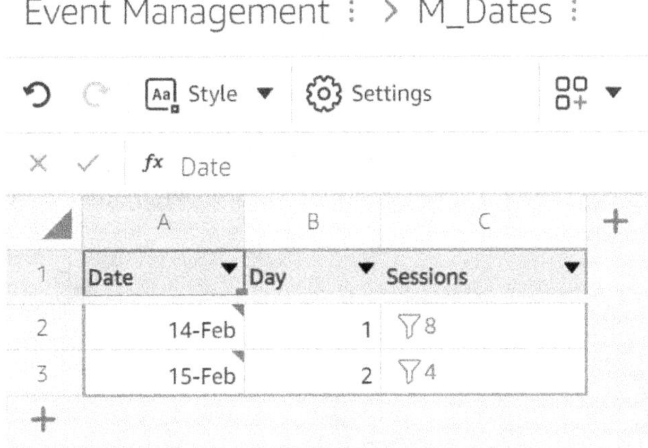

Figure 9.9 – The M_Dates table

If you need to change the number of days for your event, this is the table where you can add or remove a row.

M_Ratings

This table is used for standardizing the feedback from the sessions by defining a five-point rating scale, which is represented using stars. The numerical value associated with each rating is set in the **Value** column, and the `Filter` function in the **Related D_Registration** column collects all of the feedback for the given rating. This allows for a quick audit for each rating along with a general review in terms of the number of counts.

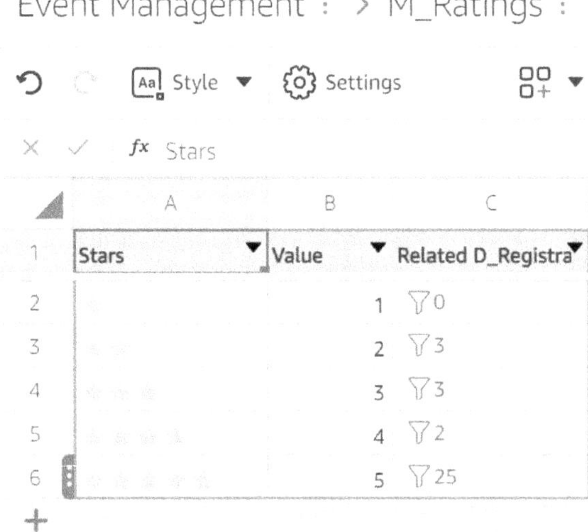

Figure 9.10 – The M_Ratings table

M_Speakers

This table is used for setting up the details of the speakers of the sessions. Similar to the earlier tables, a `Filter` function is used to collect and provide quick access to all the sessions they are speaking in.

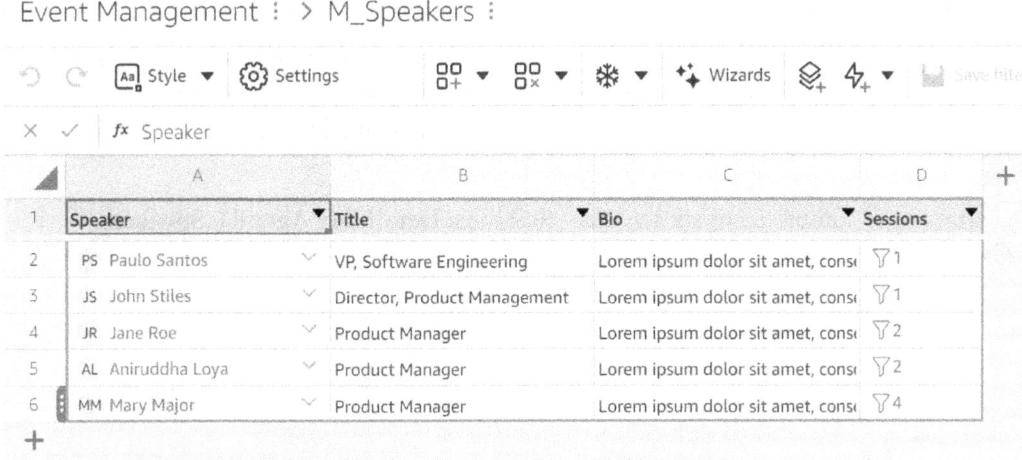

Speaker	Title	Bio	Sessions
PS Paulo Santos	VP, Software Engineering	Lorem ipsum dolor sit amet, cons	1
JS John Stiles	Director, Product Management	Lorem ipsum dolor sit amet, cons	1
JR Jane Roe	Product Manager	Lorem ipsum dolor sit amet, cons	2
AL Aniruddha Loya	Product Manager	Lorem ipsum dolor sit amet, cons	2
MM Mary Major	Product Manager	Lorem ipsum dolor sit amet, cons	4

Figure 9.11 – The M_Speakers table

R_Dashboard

This table is set to provide a collection of metrics, but it appears to have been left only half done. As we review the app in the next section, you'll notice that it is not used at all.

_Readme

In this template, this table is just a placeholder with a link to the Honeycode community page for templates.

> **Note the Table Naming Convention**
>
> While reviewing the templates previously, we noticed the workaround that was used to put the **Readme** table at the end of the list. However, often, there are some tables that you want to appear at the top because they are being used most frequently. In this chapter, we achieved this by appending alphabets combined with the underscore (_) character at the start of the table name to make the lexicographical ordering work as we desire.

Reviewing the app

Now we know how we want our app to work, what data we would need for it, and how that data is being stored and linked in the template workbook. Let's review how the template that created the **Product Conf Attendee** app links it all together and how it works.

The entire app is comprised of six screens – **Sessions**, **Detail**, **My Agenda**, **Speakers**, **By Category**, and **FAQ**, as shown in *Figure 9.12*:

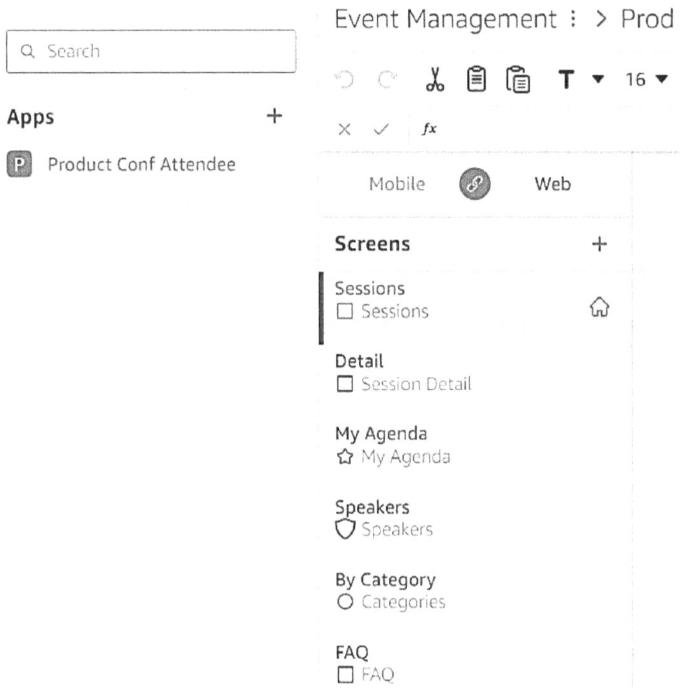

Figure 9.12 – Screens in the Product Conf Attendee app

Looking at the **APP NAVIGATION PROPERTIES** section, we can see that only four of these screens are directly navigatable. They are shown in the **Global Navigation** section of the app (please refer to *Chapter 2, Introduction to Honeycode*, for more details). In comparison, the other two screens are meant to be navigated based on actions.

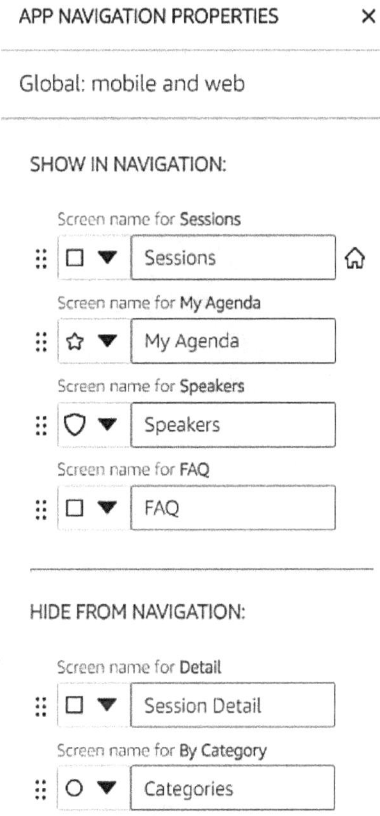

Figure 9.13 – APP NAVIGATION PROPERTIES

Sessions

This is the home screen of the app. The screen consists of three main sections:

- A **Block** section, as shown in *Figure 9.14 (a)*, at the top, which contains event date toggles
- A hidden block underneath, as shown in *Figure 9.14 (b)*, containing the **SelectedDate** variable for storing the value of the date selected in the upper block

- A **List** section at the bottom, as shown in *Figure 9.14 (c)*, containing the sessions scheduled for the selected date:

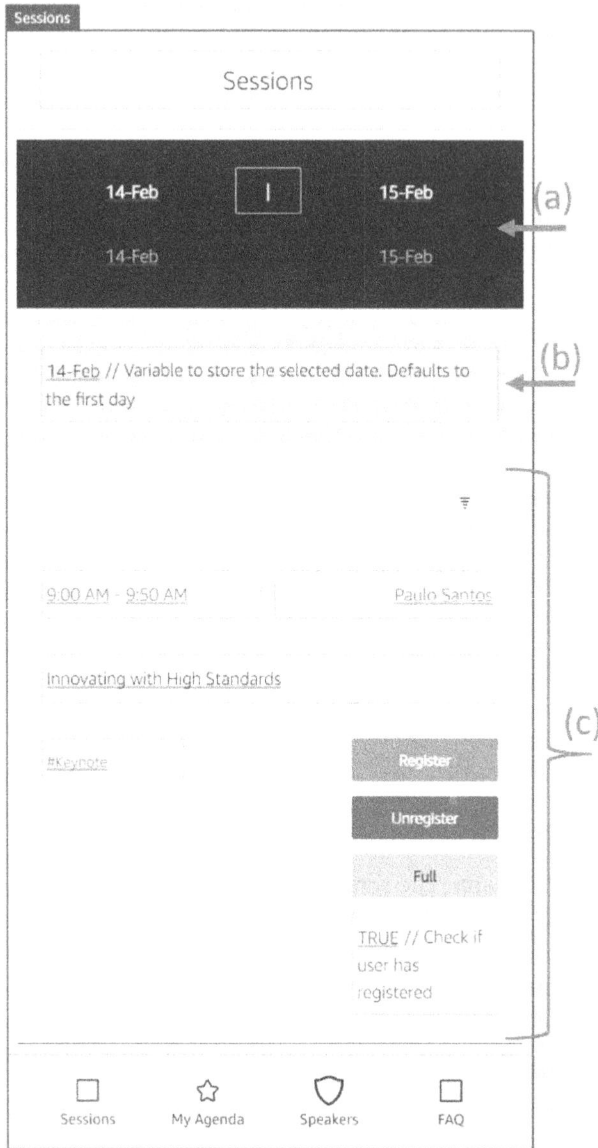

Figure 9.14 – The Sessions screen in builder view

Block

This block contains two pairs of buttons with a separator in the middle. Each set is linked to the day of the event. Recall that the **M_Dates** table has two dates and has a visibility condition that is set to display only one of the two buttons from each set.

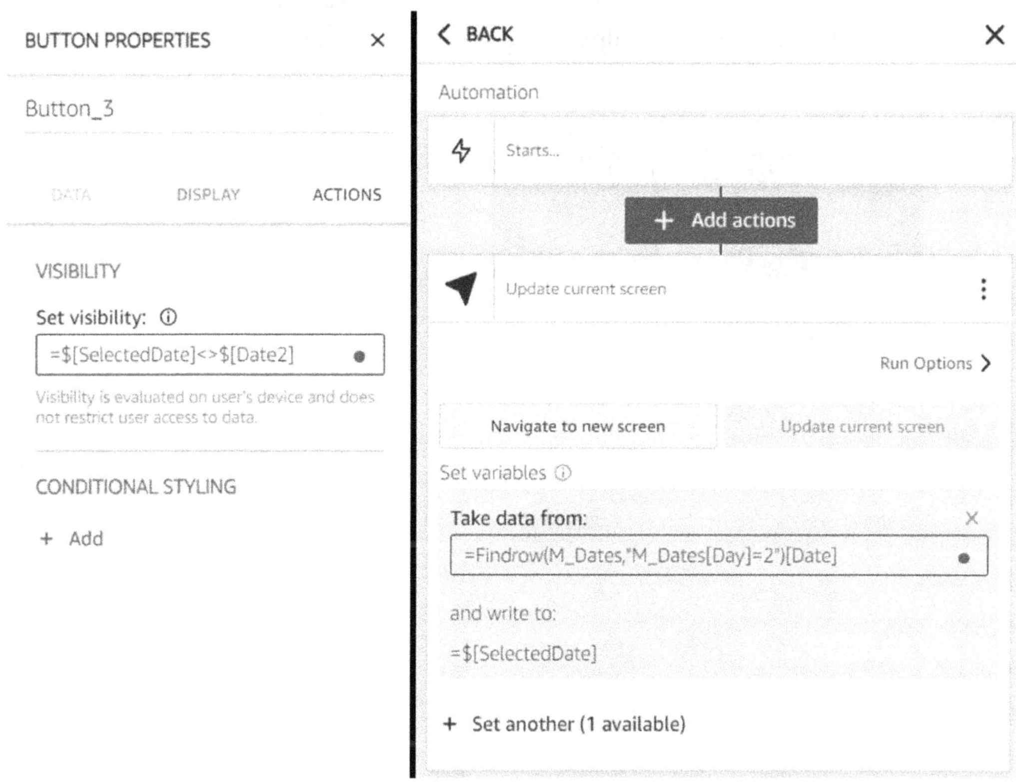

Figure 9.15 – Setting the visibility condition and automation in Button_3

If you review the **Visibility** condition for the buttons, you'll notice that the **SelectedDate** variable from the hidden block is being used to achieve the toggle behavior. Also, review the automation setting on the buttons representing the unselected case, with text shown in the color white. The automation is simply setting the value of this variable with that of the date linked to the button, as shown in *Figure 9.15*. You might recall that a similar pattern was used in the *New Poll* screen of the *Instant Polls* app, which we reviewed in *Chapter 8*, *Instant Polls Template*.

List

For the most part, the list is quite straightforward. It has controls added and linked to display various details about the various sessions on the day that was selected in the first block. It does this by making use of the **SelectedDate** variable in the filter function, which is set as the data source of the list. Additionally, it has three buttons that are conditionally visible based on whether the user has already registered and whether the event is full.

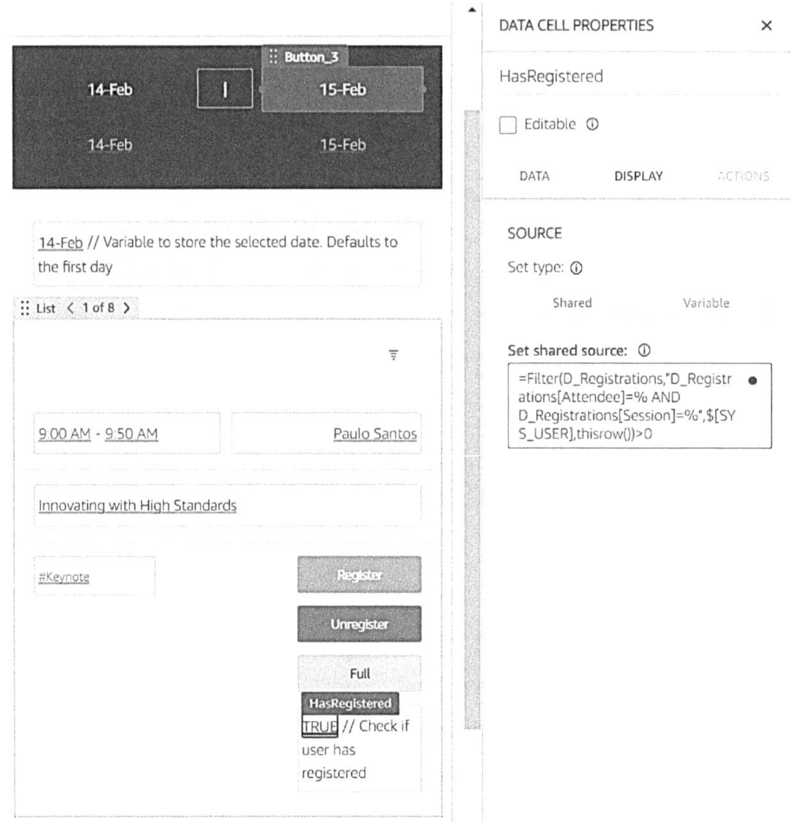

Figure 9.16 – The hidden HasRegistered variable to simplify and reuse the formula

However, there is an interesting pattern here to review. Take a look at *Figure 9.16* and make a note of the **HasRegistered Data Cell**. This variable is used for the visibility conditions of the **Register** and **Full** buttons. This variable is not visible in the app itself and is used as a local variable, which is similar to **SelectedDate** and the other examples we looked at earlier. But why is this variable declared inside the list and not outside of the list alongside **SelectedDate**? That is because the value of this variable is dependent upon the context of each row. Did you notice the use of the **ThisRow** function within the **Filter** function?

> **Note**
>
> A variable declared within a **List** control will be evaluated for every data row that is represented in that list. This implies that it might lead to an impact on performance if there are a large number of rows returned in the list data source. Therefore, it should only be used to reduce the duplication of the computation within the list and when the value is dependent on the row context.

You'll find the automations for the **Register** and **Unregister** buttons, as shown in *Figure 9.17* and *Figure 9.18*, simple to follow:

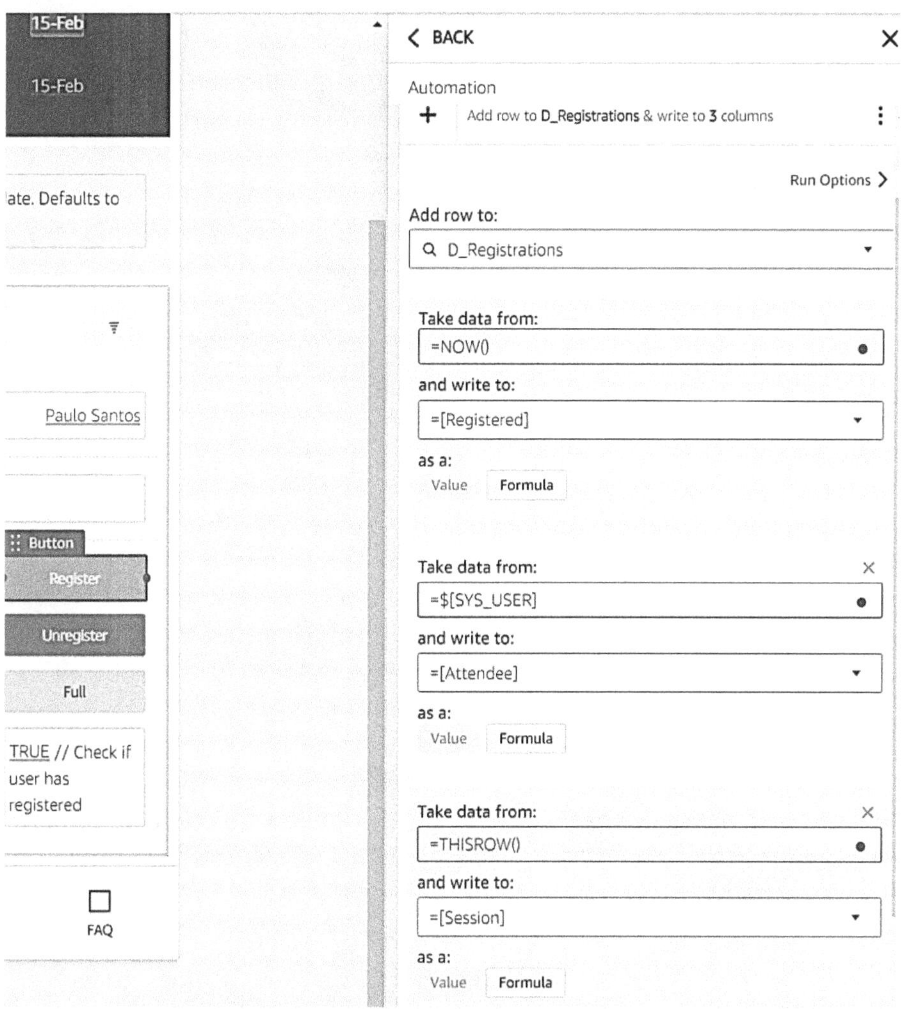

Figure 9.17 – The automations set on the Register button

The **Register** button automation is adding a new row in the **D_Registration** table along with the relevant column values. In comparison, the **Unregister** button is deleting the specified row from the same table.

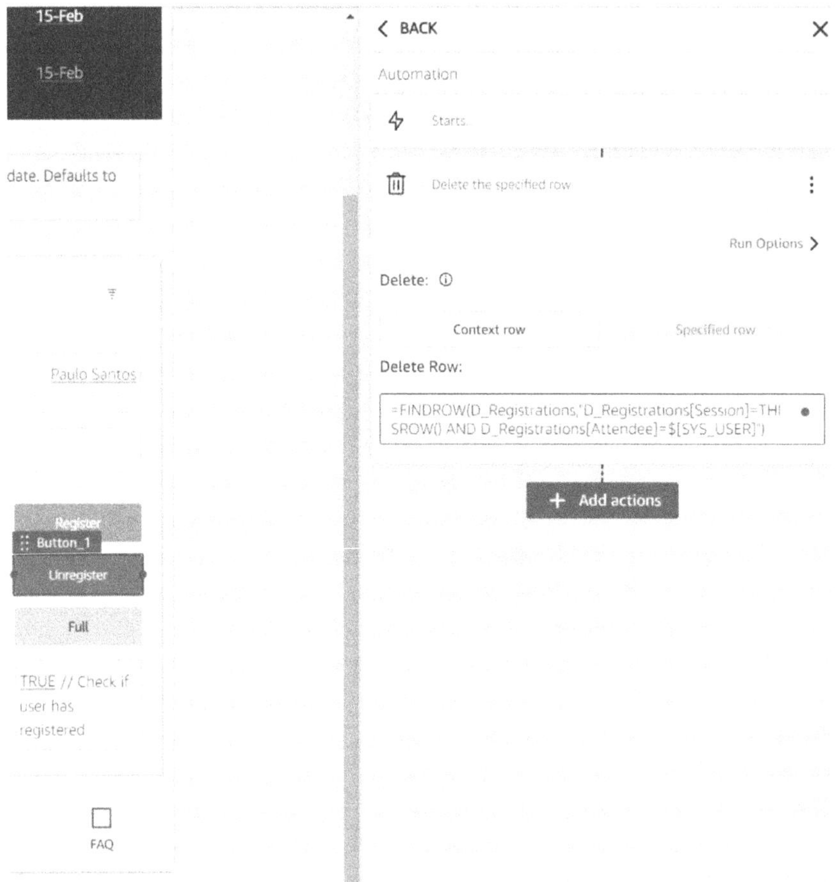

Figure 9.18 – The automations set on the Unregister button

The content box with the session's category has a **Quick Action** option that is defined, which navigates to the **By Category** screen and passes the category as the value of the **InputRow** variable of that screen.

Additionally, the list itself has a **Quick Action** option that is set to navigate to the **Detail** screen when clicked on and provide the current row as context by setting it to the **InputRow** variable of the **Detail** screen. It is also configured to allow sorting and filtering on a couple of selected columns.

Detail

The **Detail** screen displays all of the values from the row it was navigated from along with additional fields that display where the session will be taking place and the option for registered participants to rate the session. The screen contains the same set of three buttons and the hidden **HasRegistered** variable.

> **Note the Minor Difference in the Formula**
>
> In this case, it uses the value of `InputRow` rather than `ThisRow`, as the variable is no longer within the context of a **List** control, and the screen displays data from a single table row that was passed using the `InputRow` variable.

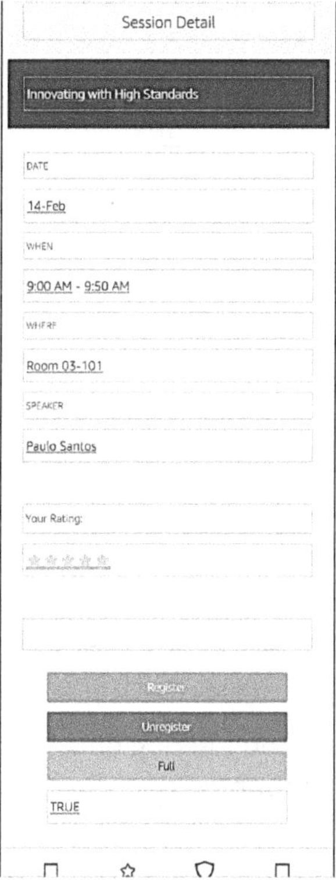

Figure 9.19 – The Session Detail screen

The visibility of the block containing the rating functionality is also determined by the **HasRegistered** variable, so the option is only shown to registered participants.

My Agenda

This is a simple screen containing a **List** control that has been configured to show the sessions that the current app user is registered for. It makes use of the `Filter` function to use the registration data from the **D_Registrations** table and present it in ascending order of the session's data and time.

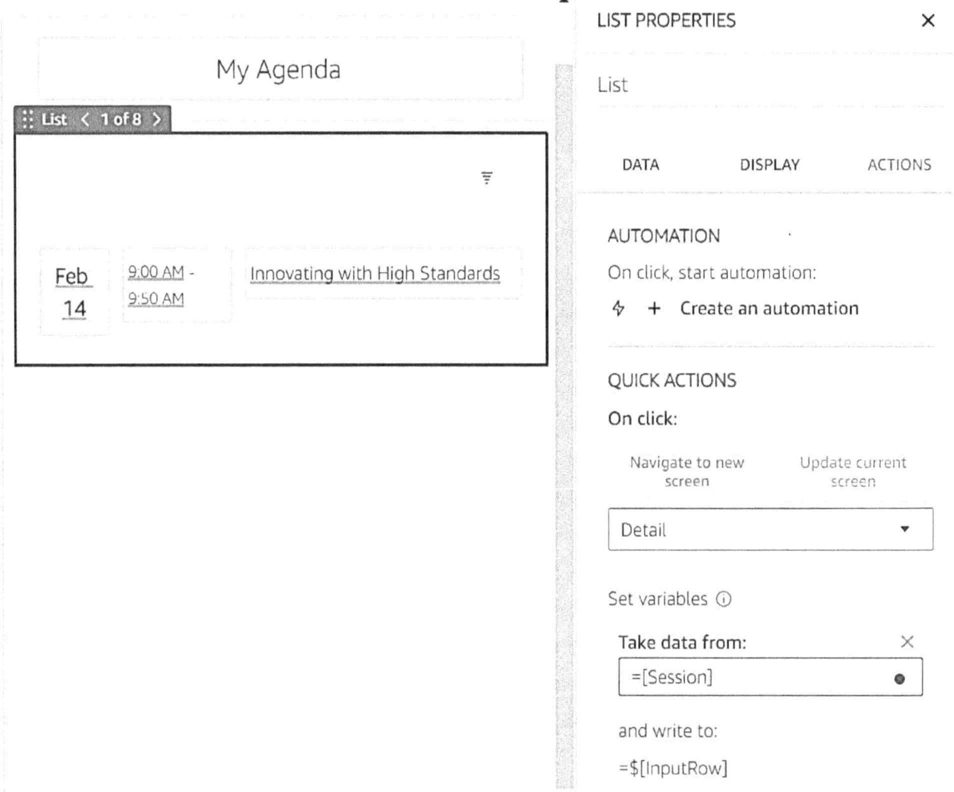

Figure 9.20 – The My Agenda screen showing QUICK ACTIONS defined on the list

The list is configured to allow sorting and filtering on a couple of selected columns. Additionally, it has a **Quick Action** option set that allows you to navigate to the **Detail** screen when clicked on and provide a value from the **Session** column of the selected row of the **D_Registrations** table as context. It does this by setting it to the **InputRow** variable of the **Detail** screen.

Why Do You Not Use the ThisRow Function in the Same Way as the List on the Sessions Screen?

That is because, for the **Sessions** screen, the table row was from the **A_Sessions** table. Therefore, it could be provided directly. However, in the **My_Agenda** screen, the table row was from the **D_Registrations** table, and this row would have provided that table row. The controls in the **Detail** screen are set to expect a row from the **A_Sessions** table and would not have resulted in an error displaying the data.

Speakers

This is another simple screen that is set to display the values from the **M_Speakers** table in the **List** control. The list is configured to allow you to sort by the speaker's name and filter using their name and title.

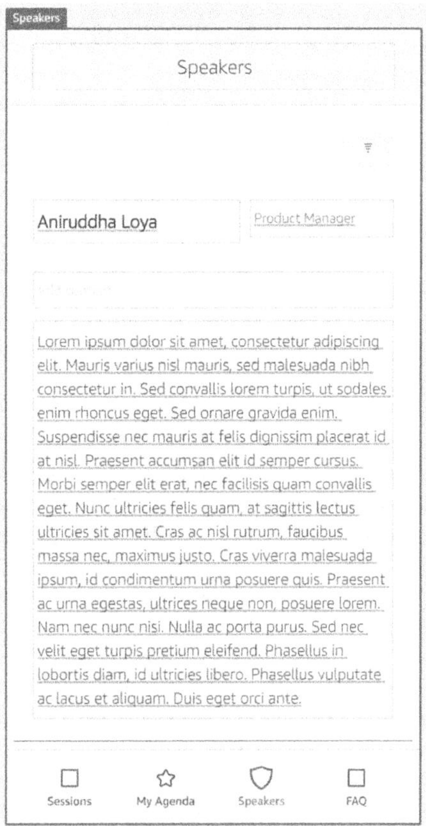

Figure 9.21 – The Speakers screen

By Category

The screen is designed for simply listing all of the sessions under a given category. It displays the category at the top and has a **List** control displaying all of the sessions tagged against the provided category. This allows participants to have quick access to a group of sessions that might be aligned with their interests.

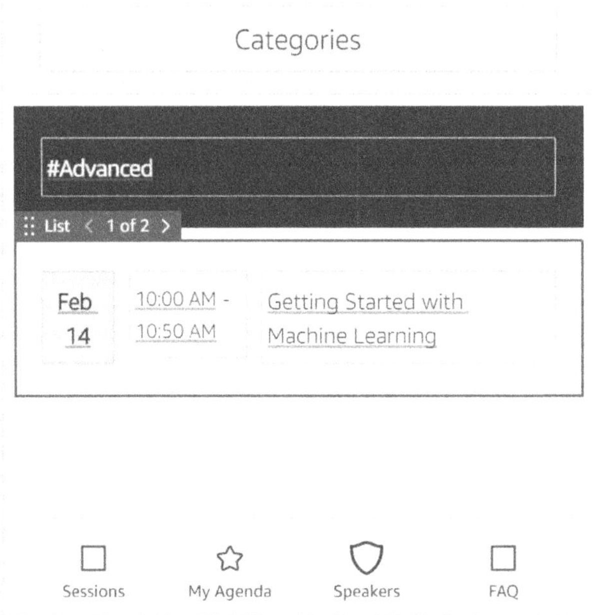

Figure 9.22 – The Categories screen

FAQ

This is another simple screen with a **Block** section at the top containing a bunch of content boxes to provide static information. This is directly added to the content boxes. Additionally, there is a **List** control at the bottom that displays the content directly from the **B_FAQ** table.

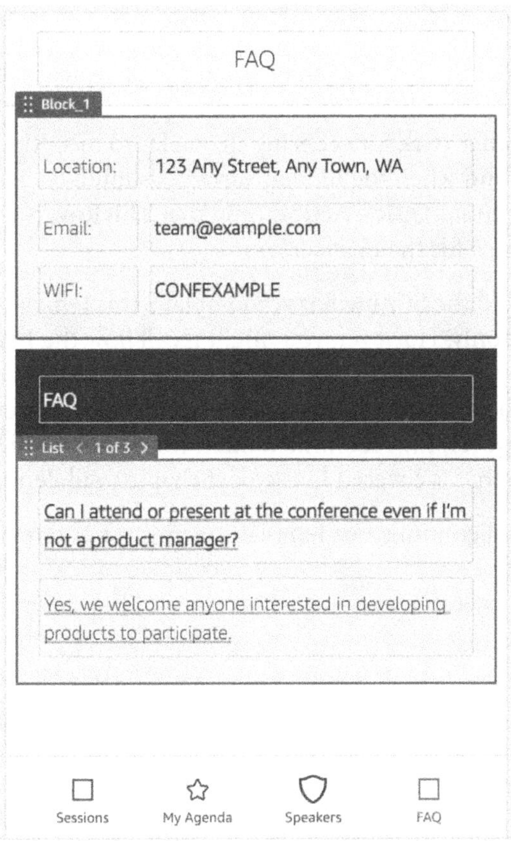

Figure 9.23 – The FAQ screen

This concludes our app review. Here, we saw the visibility condition being used heavily to achieve various results such as data filtering, data validation, and more. Additionally, we noticed that given that the app is designed mostly for read-only purposes, the screens built are fairly simple, and often, they directly use the tables as the source.

Summary

In this chapter, we reviewed the **Event Management** template, learned about its data model, and learned how the **Product Conf Attendee** app is built. We learned a new function (`Rows`) and another trick for ordering the tables to meet the needs of our usage pattern. We learned how and when the hidden variables should be used within a **List** control. Also, we learned about **Quick Actions** and that **ThisRow** is not always the row to be passed as the context row for the next screen.

We learned that similarly to the **Simple Survey** template, this template is also provided to be used for a single event only. However, you might recall how the **Instant Polls** template allows you to create new polls with the app. Additionally, you might recall the pointers left throughout the chapter to help convert this to also support multiple events if your use case calls for such an app. However, my recommendation would be to use two separate apps instead, and the template that is covered in the next chapter will demonstrate just that.

In the next chapter, we will continue our template deep-dives and review the **Inventory Management** template.

10
Inventory Management Template

Almost every business requires the management of an inventory. Some require it for internal use, while others might need it for the goods they sell, and then some might need it for both internal operations and sales. This template is aimed at internal use only, catering to the concept of keeping a record of company assets and their allocations. This could include the bookkeeping of only high-value assets, such as laptops, printers, TVs, and software licenses, or might also include consumable supplies such as printing paper, stationery, and more. Depending on the size of the business and what you are managing, a simple document or a spreadsheet might suffice for some, while others might need a full-fledged piece of software. However, you might not want to make a big jump from a spreadsheet to professional software but still need something more configurable and customizable than a simple spreadsheet. That is where this template fits in.

However, this template review is not just about being able to use Honeycode for inventory management. With this template, we will also introduce the construct of how to build multiple apps in the same workbook that derives data from the same underlying data model. You will learn how this concept can be applied to separate concerns and/or separate users through privileges, which you could find handy in various scenarios.

In this chapter, we're going to cover the following main topics:

- Defining the app requirements
- Creating an Inventory Management app
- Reviewing the data model
- Reviewing the apps

Technical requirements

To follow along with this chapter, you'll need to have access to Amazon Honeycode. This requires a laptop with a web browser, preferably Google Chrome, and optionally a mobile device running either a Honeycode-supported version of Android (currently, this requires Android 8.0 or later) or iOS (currently, this requires iOS 11 or later).

Furthermore, we'll use the Honeycode terminology and refer to the components that were covered in *Chapter 2, Introduction to Honeycode*, *Chapter 3, Building Your First Honeycode Application*, *Chapter 4, Advanced Builder Tools in Honeycode*, and *Chapter 5, Powering the Honeycode apps with Automations*. Therefore, we recommend that you complete those first.

Defining the app requirements

We first introduced this section in *Chapter 3, Building Your First Honeycode Application*, where, before building our app, we made a list of all the requirements or use cases that we wanted the app to fulfill. Additionally, we made use of that list throughout the chapters as a guide to define our data model, conceptualize the application interface, and visualize the interactions between the various onscreen elements and the data being displayed. Therefore, even though we are not going to build this app here, we will review what is built. I would encourage you to take 5 minutes to think about what your app should do and what it should look like, and then make a list of the requirements.

> **To-Do**
>
> Take 5 minutes and make a list of how you would like your Inventory Management app to work.

Here is what my list looks like:

1. App users must be able to view the inventory, its allocation, and its availability.
2. App users must be able to record to whom the asset has been assigned.
3. App users must be able to remove the assignment when the asset is returned, thus making it available for others but still keeping a history of the assignment and its duration.
4. App users must be able to request an asset or consumable supplies.
5. App users must be able to add and remove assets from the inventory.
6. App users must be able to add and update the consumable inventory.
7. App users must be able to categorize assets and also look up assets by category.
8. App users must be able to search, filter, and sort the assets.
9. The app should allow you to classify an inventory as a consumable supply or an asset.

Note that your list might have some or all of these requirements or even have more, and that is perfectly fine. You might find some of these missing in the template app and want to extend that for yourself after you finish the chapter.

Creating an Inventory Management app

Now that we know how we want our app to work, let's see what the template app can do. Let's begin by creating the app and the associated workbook. Perform the following steps:

1. On the **Dashboard** screen, locate the **Create workbook** button from the upper-right corner and click on it.

2. You will see that a popup appears. Underneath the **USE A TEMPLATE** header (as shown in *Figure 10.1*), locate a tile with the label **Inventory Management** and click on it:

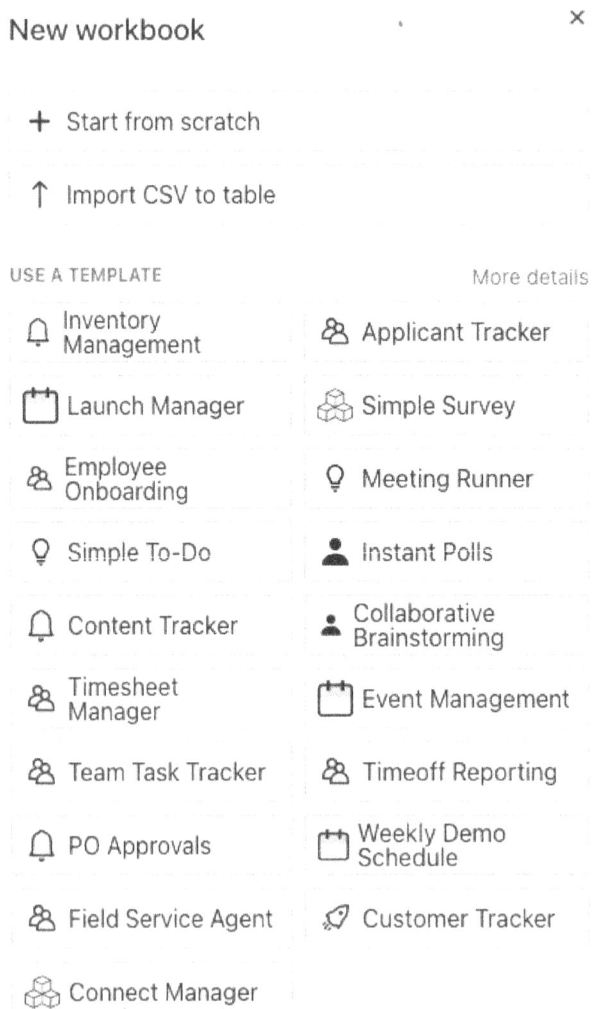

Figure 10.1 – Selecting the Inventory Management template

3. Next, you will see a popup allowing you to name the workbook and choose a team. For now, let's leave the default values as they are. Click on the **Create** button.

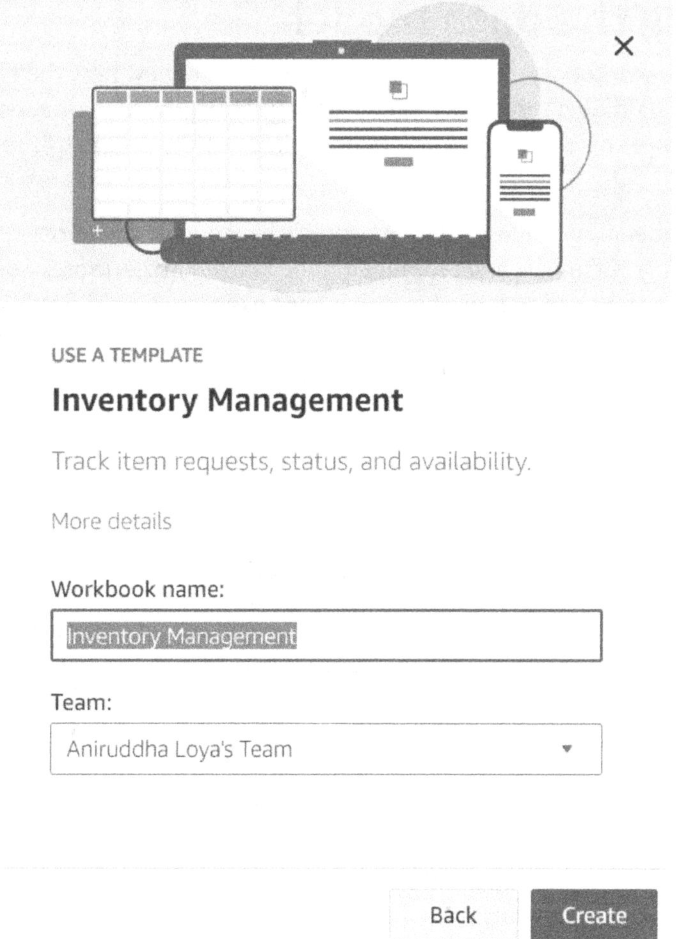

Figure 10.2 – Providing the workbook name and team details for creating the workbook

4. This creates and loads our **Inventory Management** workbook and loads up the workbook in the **Tables** view, displaying the first table in the list.

And with that, we are now ready to review the **Inventory Management** data model and the app.

> **Note**
>
> The template is only designed to enable the management of assets, but extending it to manage the inventory of consumable supplies should not be difficult.

Reviewing the data model

With the template workbook created, we are now ready to review how the Honeycode team set up the data model for the Inventory Management app. However, before we dive into that, let's take a couple of minutes to think about the data that is required for such an app.

From the requirements, as listed earlier, we would need to have categories defined for the consumables supply versus the asset for tagging the items. Furthermore, we would require sub-categorizations for each of those types. For each asset, we would need to maintain its category and sub-category, along with its status regarding whether it's available, has been requested (and who requested it), has been assigned (and to whom), and more. We would also like to maintain a record of the following:

- Who all the assets were assigned to
- When it was assigned
- When it was returned for auditing and usage purposes

Now that we have an idea of what data we need, let's explore the different tables in the workbook, the data they store, and the relationship between them, if any. The template comes with six tables, as shown in *Figure 10.3:*

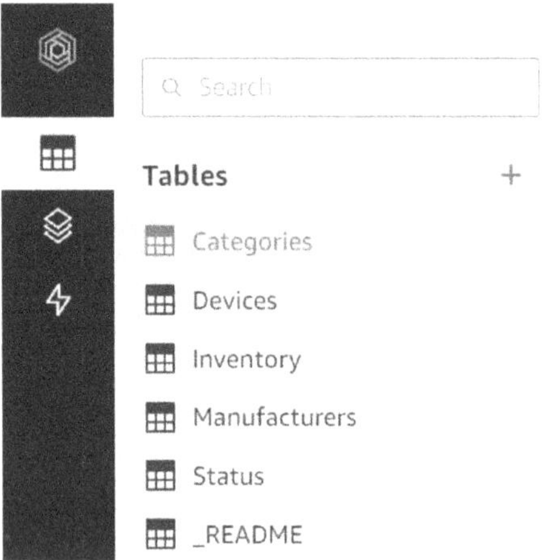

Figure 10.3 – The list of tables in the Inventory Management workbook

Let's understand each of these tables next.

Categories

We discussed the requirement of being able to categorize inventory items, and this table is exactly about that. We already noted that the template is only designed for asset types. Therefore, the table lists the different categories that can be used under the **Category** column and uses a filter function to list the number of devices for each of them in the **Devices** column.

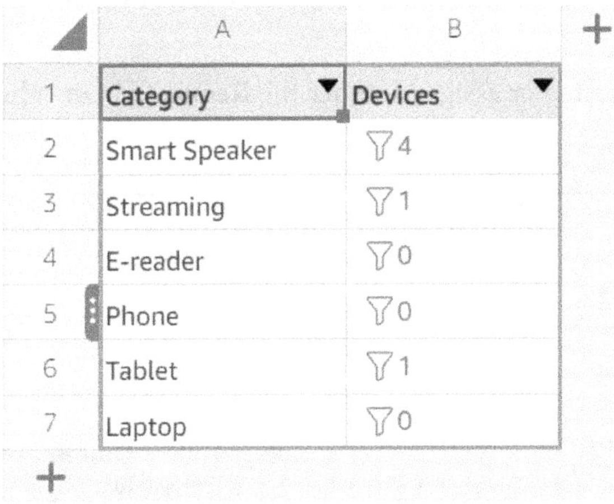

Category	Devices
Smart Speaker	4
Streaming	1
E-reader	0
Phone	0
Tablet	1
Laptop	0

Figure 10.4 – The Categories table

What happens if you need more categories? Well, just add rows to this table.

Devices

This table contains details about the different devices in the inventory, including their description, their category, the total number of devices in the inventory, the number of devices available or assigned or requested, and the device's manufacturer.

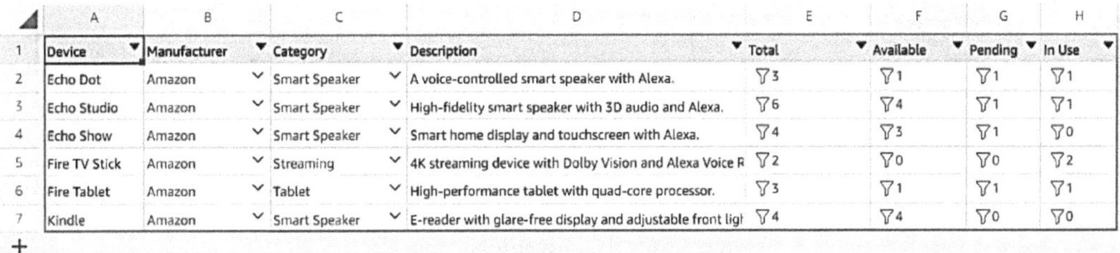

Device	Manufacturer	Category	Description	Total	Available	Pending	In Use
Echo Dot	Amazon	Smart Speaker	A voice-controlled smart speaker with Alexa.	3	1	1	1
Echo Studio	Amazon	Smart Speaker	High-fidelity smart speaker with 3D audio and Alexa.	6	4	1	1
Echo Show	Amazon	Smart Speaker	Smart home display and touchscreen with Alexa.	4	3	1	0
Fire TV Stick	Amazon	Streaming	4K streaming device with Dolby Vision and Alexa Voice R	2	0	0	2
Fire Tablet	Amazon	Tablet	High-performance tablet with quad-core processor.	3	1	1	1
Kindle	Amazon	Smart Speaker	E-reader with glare-free display and adjustable front ligl	4	4	0	0

Figure 10.5 – The Devices table

Similar to the tables that we have seen in other templates, the information is collected with the help of **Rowlinks**, which are created through **Picklists** and various **Filter** functions.

Inventory

This is the table that contains the most critical pieces of information. It lists every device individually with its assigned **Asset ID** field. Additionally, it captures information on the following topics:

- Whether the device has been requested or assigned under the **Status** column
- Who requested it or who it has been assigned to under the **Assigned To** column, and
- When it was requested or assigned under the **Requested On** column.

	A	B	C	D	E
1	Asset ID	Device	Assigned To	Status	Requested On
2	3613	Echo Show	JR Jane Roe	Requested	12/4/21
3	3869	Echo Show			
4	3613	Echo Show			
5	1443	Echo Show			
6	2415	Fire TV Stick	AL Aniruddha Loya	Approved	11/4/21
7	8452	Fire TV Stick	AL Aniruddha Loya	Approved	11/19/21
8	8888	Fire Tablet	AL Aniruddha Loya	Approved	11/14/21
9	5395	Fire Tablet	AL Aniruddha Loya	Requested	11/24/21
10	8839	Fire Tablet			
11	8432	Echo Studio	JS John Stiles	Approved	12/4/21
12	9968	Echo Studio	JR Jane Roe	Requested	11/19/21
13	8929	Echo Studio			
14	2269	Echo Studio			
15	5732	Echo Studio			
16	2406	Echo Studio			
17	7997	Echo Dot	JS John Stiles	Approved	11/19/21
18	2037	Echo Dot	JR Jane Roe	Requested	11/14/21
19	1955	Echo Dot			
20	1762	Kindle			
21	5979	Kindle			
22	7863	Kindle			
23	8466	Kindle			

Figure 10.6 – The Inventory table

Now, you might question why the requester's information is in the **Assigned To** column and why the assigned date has been captured in the **Requested On** column? Yes! You are absolutely right about questioning it. This limits the functionality and can cause errors.

Exercise 1

Can you think about the limitation(s) that this overloaded use of **Assigned To** and **Requested On** columns will create?

Manufacturers

This is a very simple table and is similar to the **Categories** table. It contains the list of manufacturers and devices that are present in the inventory.

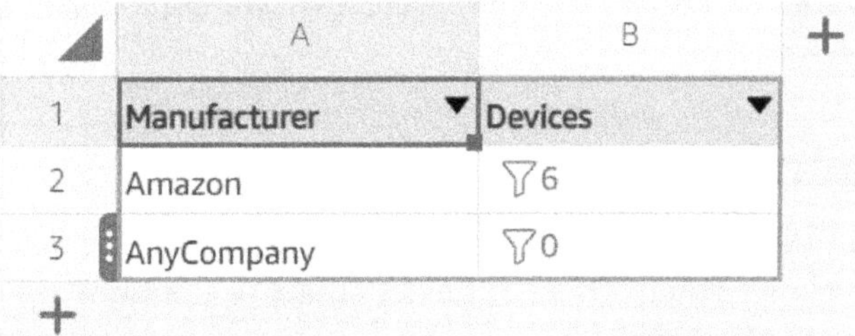

Figure 10.7 – The Manufacturers table

Given that there are only two rows in this table, this table will certainly be needing many additional rows to list the manufacturers of various devices being used in your business.

Status

Similar to the **Categories** and **Manufacturers** tables, this table enumerates two statuses – **Requested** and **Approved** – under the **Status** column and then provides the list of devices against each status in the **Devices** column.

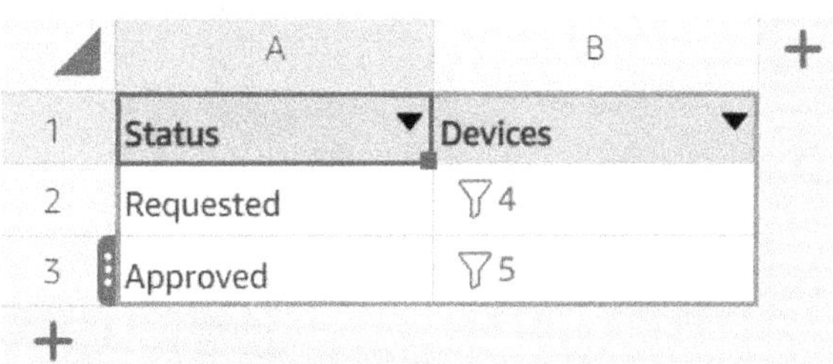

Figure 10.8 – The Status table

While the two statuses listed here should suffice in most use cases, you now understand where to add an entry if more statuses are required.

_Readme

In this template, this table is just a placeholder with a link to the Honeycode community page for templates.

Reviewing the apps

Now we know how we want our app to work, what data we would need for it, and how that data is being stored and linked in the template workbook. Let's review how the template that created the **My Devices** and **My Devices – Manager** apps linked it all together and how they work.

My Devices app

This app is for enabling users to review the inventory and request the devices that they need. Additionally, it allows you to view the devices that are already assigned or requested in the past or are awaiting approval.

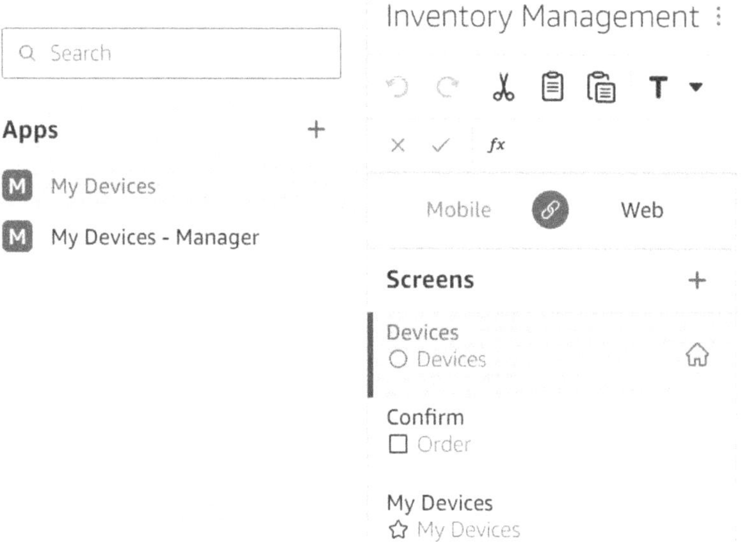

Figure 10.9 – Screens in the My Devices app

The entire app is comprised of three screens:

- **Devices**
- **Confirm**
- **My Devices**

This is shown in *Figure 10.9*.

Devices

This is the home screen of the app. The screen is one of the simplest that we have seen so far. It consists of only a list control that is configured to show the data from the **Devices** table. Also, it has been set up to allow filtering on the **Device** and **Category** columns and sorting on the **Device** and **Availability** columns from the table.

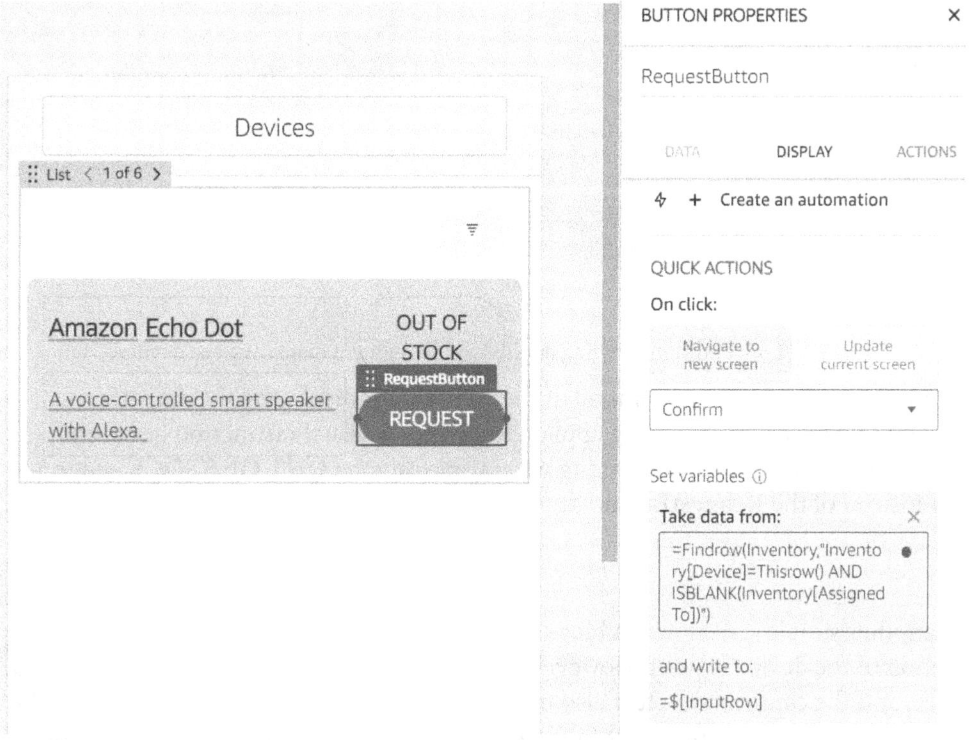

Figure 10.10 – The Devices screen in builder view along with the RequestButton actions set up

The **RequestButton** option, as shown in *Figure 10.10*, is configured to only display when the **Available** number of devices is greater than zero. It has a **Quick Action** option defined on the click that allows you to navigate to the **Confirm** screen and pass a row from the **Inventory** table, which contains an unassigned device of this type as an input parameter.

> **Note**
> The row that is passed as input to the next screen is not the context row on which the click is made. Instead, we are using the `FindRow` function to fetch and pass a specific row from the **Inventory** table.

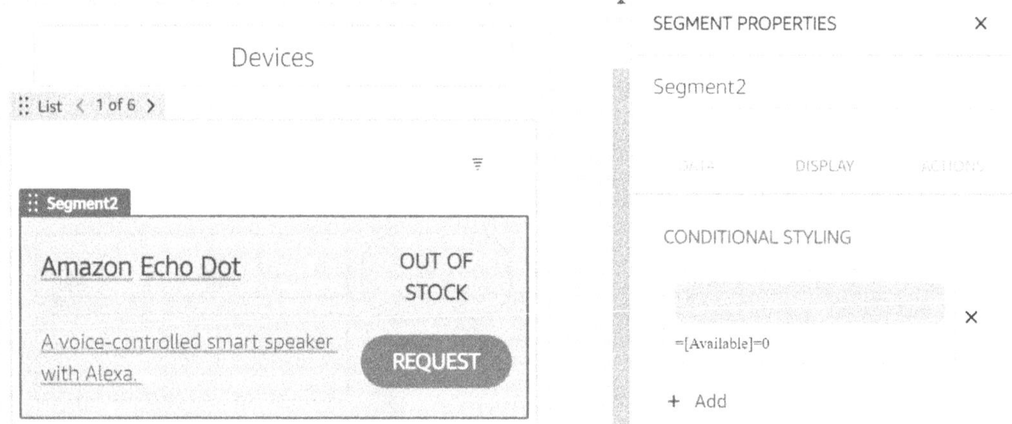

Figure 10.11 – Conditional styling applied to the segment displaying the devices

The segment in the list that contains all of the controls for displaying the data has a **CONDITIONAL STYLING** setting applied to provide a visual distinction between available and out-of-stock devices, along with displaying the **OUT OF STOCK** section as content instead of the **RequestButton** section. See *Figure 10.11*.

Confirm

This is another simple screen with a block that prompts the user to confirm the completion of the device request. However, given that the screen has its header displayed as **Order**, and the confirmation text also mentions orders instead of a request, it could be confusing to users. So, this is something worth changing before rolling out the app for your business.

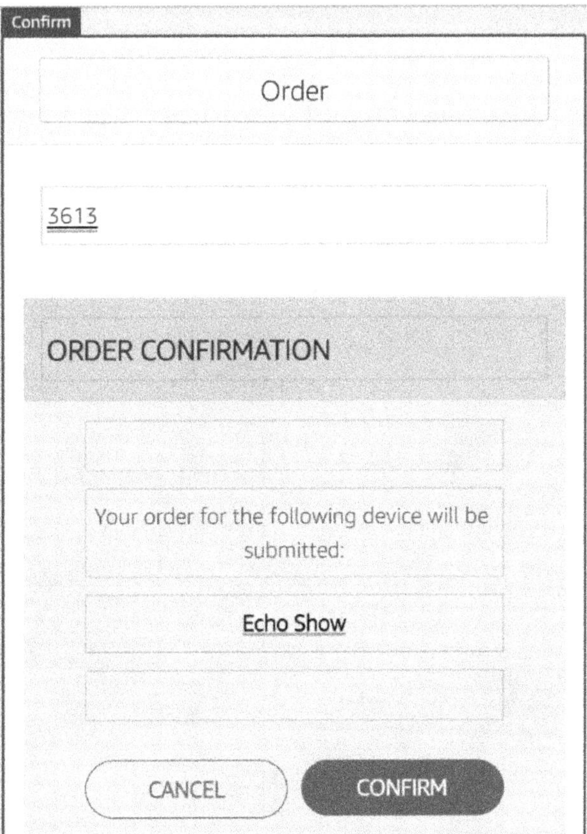

Figure 10.12 – The Confirm screen

There is automation on the **CONFIRM** button that updates the **Assigned To**, **Status**, and **Requested On** columns of the **Inventory** table row that was provided as context to this screen. Then, it navigates to take the user to the **My Devices** screen.

> **Exercise 2**
>
> Can you think of a scenario where this device request flow could result in an issue?

My Devices

The screen is similar to the **Devices** screen except that it starts by loading the inventory items assigned to the app user using only the `Filter` function and then derives the additional data about the device using the **dereferences** on the **Rowlinks** created in the data model. The data is displayed in descending order on the **Status** column values to show all of the requests first. The list control is configured to allow sorting on the **Device** and **Requested On** columns and also filtering on the **Device** and **Status** columns.

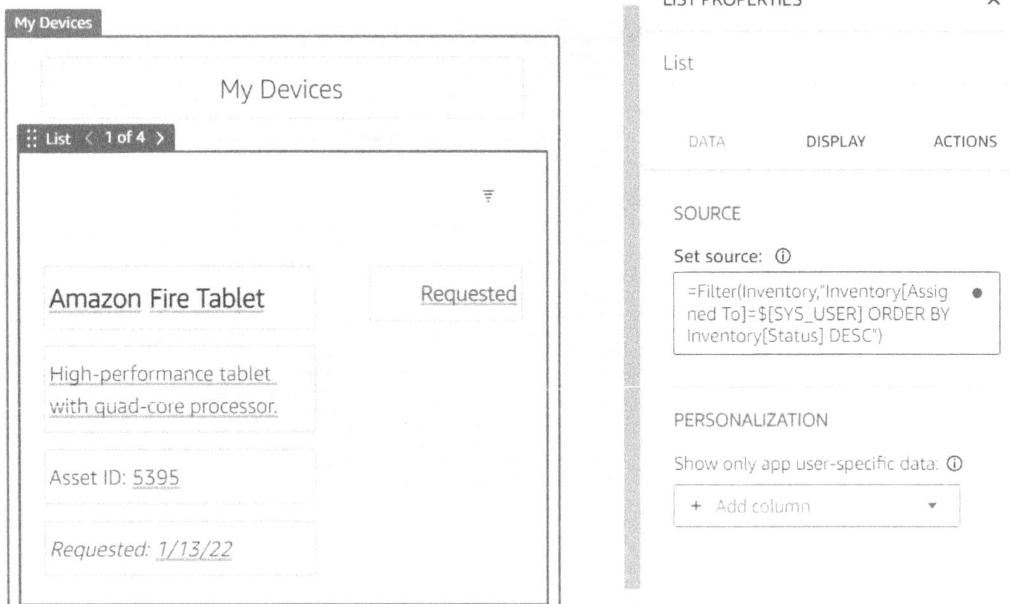

Figure 10.13 – The My Devices screen

In order to provide the visual distinction between the requested and assigned devices, **Conditional Styling** is configured on **Segment1** and **ContentBox3**, which displays the **Status** setting.

And that's all there is to this app. It is very basic and functional, and I'm sure you already have a few ideas to extend it for the missing use cases, such as withdrawing or canceling a request or requesting a device that is not available in the inventory.

My Devices - Manager app

This app is meant for the person or team managing the inventory. The app provides the feature to add an entirely new device or a new unit to an existing device. Similarly, it provides the functionality to delete the device and remove a single unit of the device. It also provides functionality to approve requests and assign devices, and it provides a view to track who currently has a given asset.

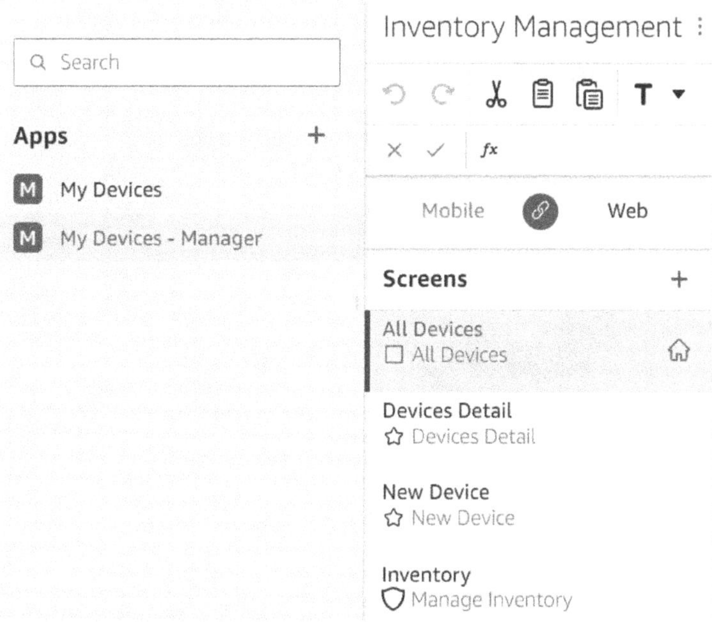

Figure 10.14 – Screens in the My Devices - Manager app

The entire app is comprised of four screens – **All Devices, Devices Detail**, **New Device**, and **Inventory** – as shown in *Figure 10.14*.

All Devices

This is the home screen of the app. The screen is fairly simple and lists all the devices from the **Devices** table. It consists of two containers – **Devices Block** and **Devices List** – as shown in *Figure 10.15*. Here, the former contains the column headers, and the latter contains the list of devices.

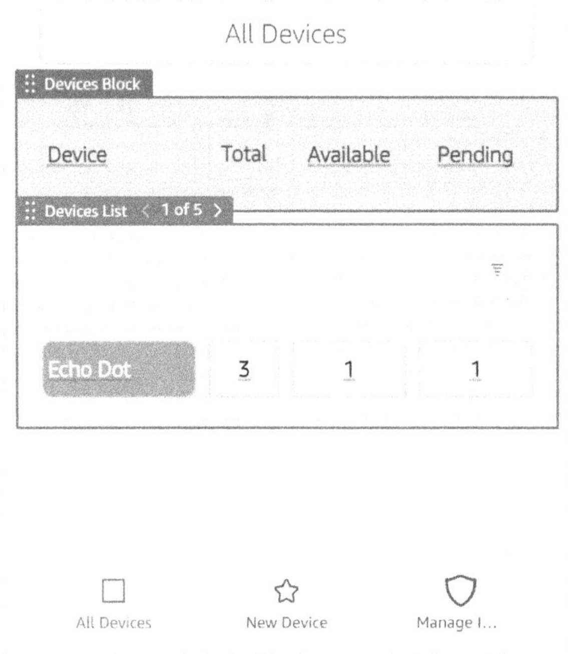

Figure 10.15 – The All Devices screen in the builder view

The preceding list has a **Quick Action** option configured to navigate to the **Devices Detail** screen. It passes the row clicked, as context, to the details screen.

Devices Detail

This is the details screen for each device and shows the metadata about the device such as its manufacturer and category, which can be updated here if needed. The screen also provides details about the status of the inventory of this device, including how many devices there are, who they are assigned to, and who has requested them. The app user can also manage their requests on this screen. Lastly, there is a **DELETE DEVICE** button at the bottom that allows you to delete the device.

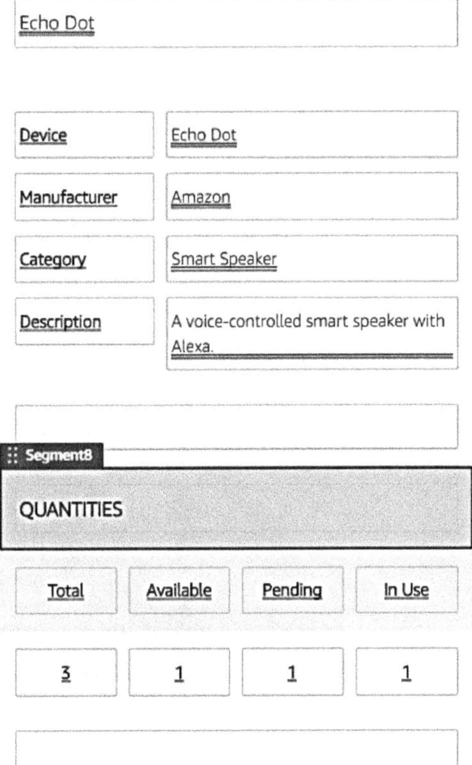

Figure 10.16 – The Devices Detail screen

> **Caution**
>
> The **DELETE DEVICE** button does not have a check for reconfirming the action. So, when clicked, accidentally or otherwise, it will go ahead and delete the device along with its asset details. Can you fix it?

New Device

This screen is a standard form type that gets users to provide inputs through various data fields and then adds that as a new record to the binding table. In this case, the binding table is the **Devices** table, and the action to add the record to the table is configured on the **Submit** button, as shown in *Figure 10.17*:

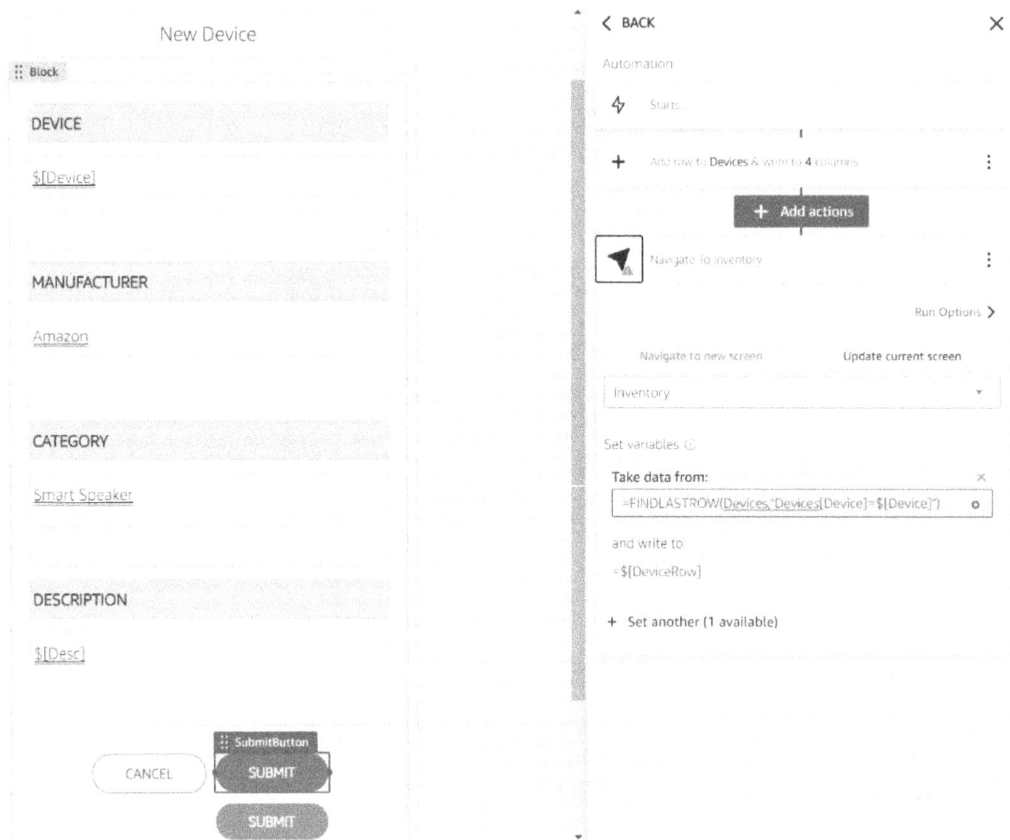

Figure 10.17 – The New Device screen in builder view with the SUBMIT button automation

The second block of the automation adds the rows to the **Devices** table and writes the four values to their corresponding columns. However, the interesting part of that automation is the last block defining the navigation. This makes use of the `FindLastRow` function to retrieve the just added row from the table and pass that as the context row while the app navigates to the inventory screen. This will result in the newly added device being displayed on the inventory screen.

Inventory

This screen displays the inventory for each device type and provides options to add and delete assets for the selected device type.

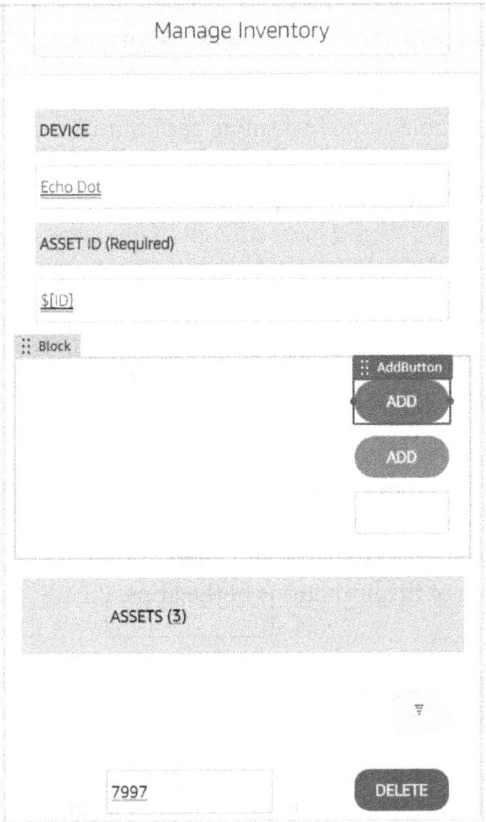

Figure 10.18 – The Manage Inventory screen

The **AddButton** option, as selected in *Figure 10.18*, has the following formula set as its visibility condition:

```
=AND((FILTER(Inventory,"Inventory[Asset ID]=$[ID] AND
Inventory[Device]=$[DeviceRow]")=0),$[ID]<>"")
```

> **Exercise 3**
>
> Explain why this formula is required? Can you identify any shortcomings of this formula?

In the last two sections, we deconstructed the two apps that, together, complete the process of managing an inventory. We learned that both apps are built on the same set of underlying tables, and the updates from one table flow to the other. More importantly, we learned how we can have different apps created for different use cases and be shared with only relevant groups. The **My Devices** app is meant to be shared and used by every member of the organization, while the **My Devices - Manager** app is only meant for the individual or team responsible for management.

So, could we have built the whole thing in a single app? Sure, we could have. But that would have either exposed the entire functionality to everyone, defeating the purpose of the request-approve flow, or it would have required complex logic to show or hide controls based on the roles. That would have still felt inadequate, as Honeycode does not allow visibility conditions on the screens. So, we would have either ended up with poorly named screens or a set of empty screens based on the user's group.

Summary

In this chapter, we reviewed the **Inventory Management** template, learned about its data model, and demonstrated how the **My Devices** and **My Devices - Manager** apps are built. Additionally, we learned how multiple apps can be created from a single data source, allowing us to separate the concerns and keep the apps both separate and simple. We also learned about some of the limitations of the apps. Finally, we saw the use of the `FindLastRow` function.

With this set of four template deep-dive chapters, you now have a framework for deconstructing and reviewing other templates yourself.

Therefore, in the next set of chapters, we will change gears and build some more apps. In the next chapter, we will start by learning how to build an app for managing **Shopping Lists**.

Part 3: Let's Build Some Apps

In this part, you will be able to walk through some specific use cases and learn how they can be solved using Honeycode, either from scratch or by modifying an existing template. The chapters in this section build on the lessons learned from the previous two sections and are presented in order of increasing complexity in terms of both use cases and content.

This section comprises the following chapters:

- *Chapter 11, Building a Shopping List in Honeycode*
- *Chapter 12, Building a Nominate & Vote App in Honeycode*
- *Chapter 13, Conducting Periodic Business Reviews Using Honeycode*
- *Chapter 14, Solving Complex Problems through Multiple Apps Within a Workbook*

11 Building a Shopping List App in Honeycode

Shopping is one of our recurring activities, particularly for groceries, with most of us making a list of items that we need to buy. This list could be on paper, a note-taking application, or simply in your mind. Quite often, the list builds up right after purchasing our groceries because we forget to buy something or forgot to take the paper list sitting on our desk or on our kitchen counter to the shop. Does this sound familiar?

Have you reached the store and felt distraught about forgetting the list again? If you go shopping alone, do you find yourself making multiple calls to your partner to ask them about items because you did not take the paper list or do not have access to the list that was made on their phone or laptop?

If the answer to any of these questions is yes, you may find yourself with a very handy application at the end of this chapter. How do I know it's handy? It's because I've been using this very application for the past 2 years along with my partner. If the answer is no, well, you'll still gain a lot from this chapter by practicing your app-building skills.

In this chapter, we'll build an app for an everyday use case of maintaining an updated shopping list. The app we'll build will support the creation of multiple lists tied to different stores you may be purchasing your groceries from, along with a single unified view of all the items to buy. We will also go through some exercises to further your understanding as well as extending the functionality.

In this chapter, we're going to cover the following main topics:

- Defining the app requirements
- Translating requirements to app interactions
- Defining the data model
- Building the app

Technical requirements

To follow this chapter, you'll need to have access to Amazon Honeycode, which requires a laptop with a web browser, preferably Google Chrome, and optionally a mobile device running either a Honeycode supported version of Android (currently requires Android 8.0 and up) or iOS (currently requires iOS 11 or later).

Furthermore, we'll use the Honeycode terminology and refer to the components that we covered in *Chapter 2, Introduction to Honeycode*, *Chapter 3, Building Your First Honeycode Application*, *Chapter 4, Advanced Builder Tools in Honeycode*, and *Chapter 5, Powering the Honeycode apps with Automations*, and therefore, recommend you complete those first.

Defining the app requirements

We first introduced this section in *Chapter 3, Building Your First Honeycode Application*, where before building our app, we listed down the requirements or the use cases that we wanted the app to fulfill. We made use of that list throughout the chapter as a guide to define our data model and conceptualize the application interface and visualize the interactions between various onscreen elements as well as the data displayed. Therefore, I'd encourage you to take 5 minutes to think about what your app should do and what it should look like, and then make a list.

> To Do
>
> Take 5 minutes and list down how you would like your **Shopping List** app to work.

Here is what my list looks like:

1. I must be able to view all the items I need to buy.
2. I must be able to view the items I need to buy from a specific store.
3. I must be able to review the items I've already bought.
4. I must be able to add items to my list and assign the store I want to buy them from.
5. I must be able to update or delete items from the list.
6. I must be able to mark items as bought.
7. I must be able to clear items bought from my shopping list.
8. I must be able to move items from one list to another.

Your list may have some or all of these requirements and may have even more and that is perfectly fine. You may find some of these missing in the app we will build and may want to extend it for yourself after you finish the chapter.

Translating requirements to app interactions

Based on the requirements we listed in the previous section, our app will have two primary views:

- List of all items to buy
- List of all stores to buy items from

> **Note**
>
> You may choose to only have a single view of all items and then enable the filtering capability to get the same result, but it will take more than one click to achieve that and, therefore, my preferred way is to build the two screens.

Next, we will need navigation from the list of stores to a page that displays all the items to be bought from those stores. Similarly, clicking on an item should provide a screen to edit or delete the item.

We will also need a form to add new items to our list. This form should allow adding the name of the item and selecting the store to buy it from.

Now that we have an understanding of the key interactions and screens of the app, let's define how our data model enables these screens and interactions.

Defining the data model

Our app has two clearly distinct entities:

- Stores
- Items to buy

Therefore, we will need two separate tables for them. The *items to buy* will need to be mapped to the *stores* as well as the status – *Bought* or *To buy*. Since we do not expect to have too many items in our list, we can maintain the status as simple text or even a **Boolean** field as there are only two values. However, in general, the recommended approach for fields that repeat is to set them using a **picklist** and, hence, define them in a separate table.

Will we need something else? No, that is all we need for now to enable the cases listed above.

Building the app

Alright, so now we know what our app should do, have a fair idea of what it should look like to enable the listed use cases, as well as how we will structure and store the data to power the app. Let's start building it up by creating our workbook.

Creating a new workbook

By now, you must be familiar with the process of creating a workbook. So, go ahead and create an empty workbook from the dashboard and name it **Shopping List**.

Creating tables

In the previous section, we noted that we will be needing two tables:

- One for listing stores
- One for listing items to buy

So, let's set them up.

The stores table

We set up this table by renaming our default table and updating it by following these steps:

1. Rename **Table1** to `Stores`.
2. Rename **Column1** to `StoreName`.
3. Add the list of stores you typically go to for shopping in the rows in the **StoreName** column. See *Figure 11.1* for reference:

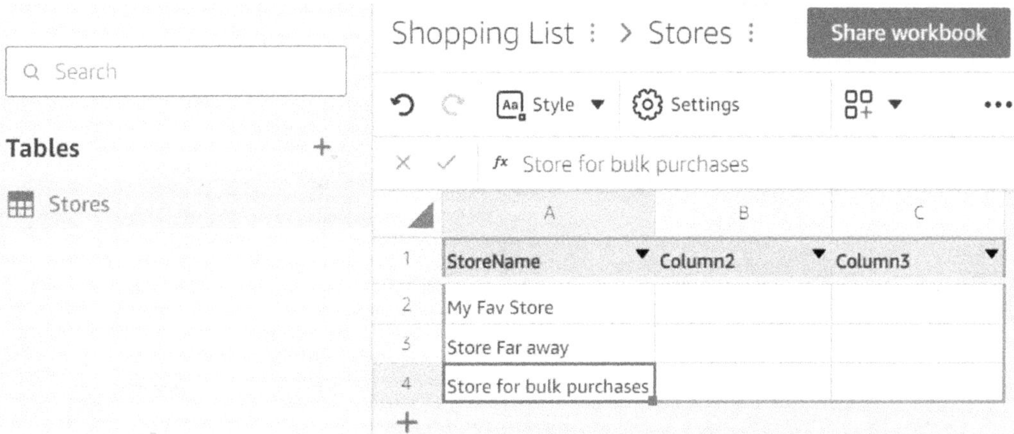

Figure 11.1 – Table1 renamed to Stores, Column1 renamed to StoreName, and a list of stores

4. Rename **Column2** to `ItemsToBuy`.

5. Click on the **Settings** control in the toolbar to open the **Column Properties** panel and type in the following formula:

```
=Filter(Items, "Items[Store] = THISROW() AND
Items[Bought] = FALSE")
```

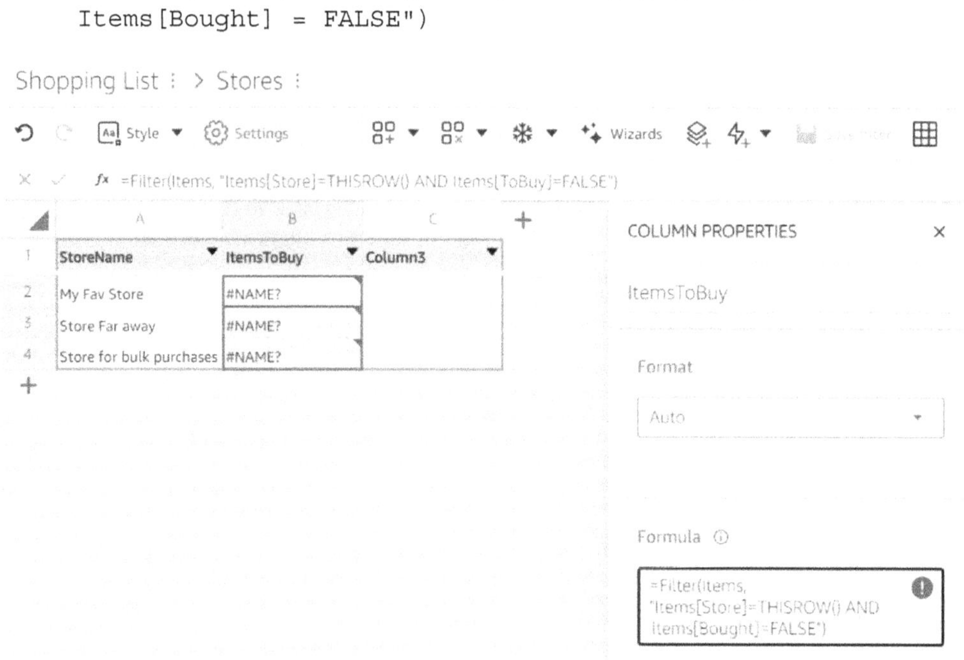

Figure 11.2 – The ItemsToBuy column filled with the filter formula returning the #NAME? error code

> **Note**
>
> The formula will apply to the column and all cells will show the `#NAME?` error code as shown in *Figure 11.2*. This is because we have used a name in the `formula (Items)` that does not exist in the workbook. This will be automatically fixed once we create the table and the columns.

> **Exercise 1**
>
> Can you explain what this formula returns? Why do we need the data returned by this formula?

6. And finally, delete **Column3** to complete the setup.

Now, let's create the **Items** table.

The Items table

We set up this table by creating a new table and updating it by following these steps:

1. Click on the + control to add a blank table to the workbook.
2. Rename the newly created **Table1** to `Items`. See *Figure 11.3*:

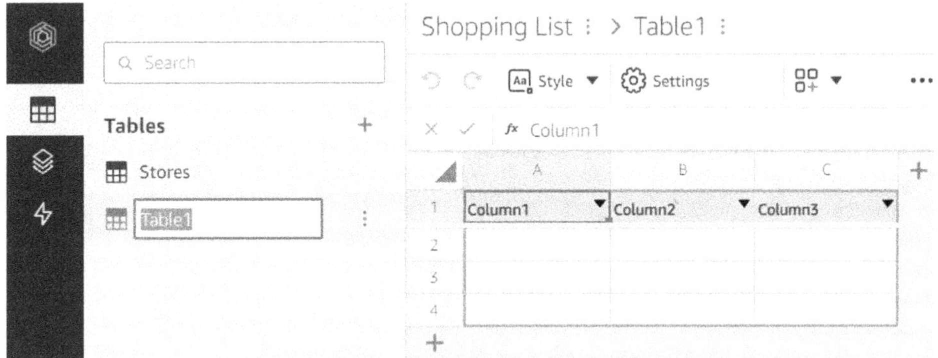

Figure 11.3 – Rename the newly added table to create the Items table

3. Rename **Column1** to `Item`.
4. Rename **Column2** to `Store`.
5. Click on the **Settings** control in the toolbar to open the **Column Properties** panel. Set **Format** as **Rowlink & picklist** by selecting it from the dropdown and set the **Stores** table as the **Source** as shown in *Figure 11.4*:

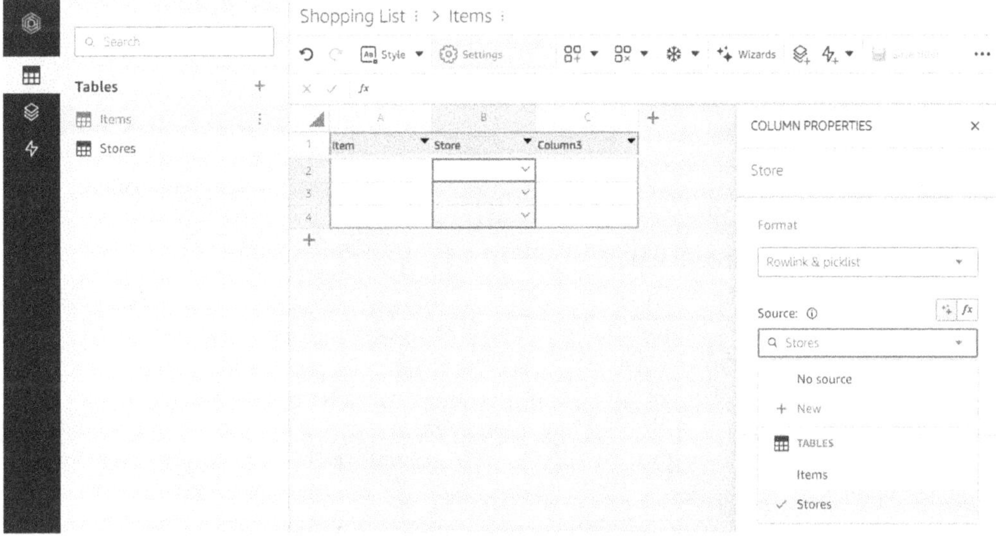

Figure 11.4 – Setting Picklist on the Store column

Why Set the Rowlink & picklist Format on the Store Column?

There are two reasons for this: 1. This will help you create the UI with a dropdown to select from a defined set of stores. 2. Recall the formula we set on the **ItemsToBuy** column.

We used the **ThisRow()** function there to reference all the rows with the given store name and the **Rowlink & Picklist** format enables us to create the reference when the value is set.

6. Rename **Column3** to `Bought`.
7. In the **Column Properties** panel, set **Format** as **Checkbox**, which will convert all the column cells to empty checkboxes as shown in *Figure 11.5*:

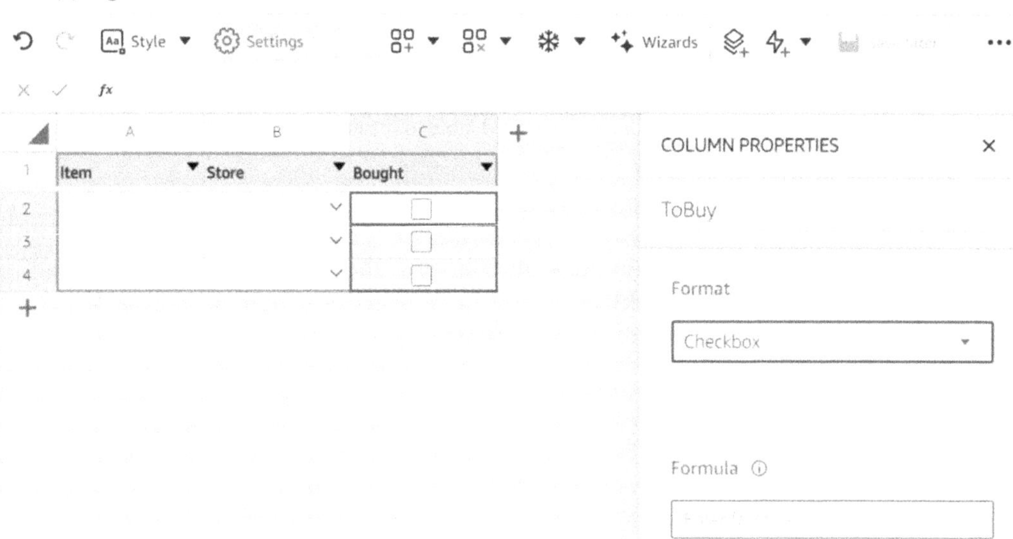

Figure 11.5 – Table1 renamed to Stores, Column1 renamed to StoreName, and a list of stores

Why Use the Checkbox Format?

In the *Defining the data model* section, we discussed different ways in which we can set this column, and given there are only two values for this field, we can work with the checkbox format, which has two states – *checked or TRUE* and *unchecked or FALSE.* For our case, an unchecked state will represent that we need to buy that item and checked that it has been bought. Furthermore, using the checkbox format will provide us with an intuitive checkbox display out of the box.

You may choose to add another column for the date of purchase if you want to check when you last bought the item, but for now, our table setup is complete.

But before we complete this section, let's verify a few things. Recall that while setting the **Stores** table, we made a note of the error code returned for the **ItemsToBuy** column and mentioned that it will automatically fix itself. Did it? Navigate to the **Stores** table. What do you see?

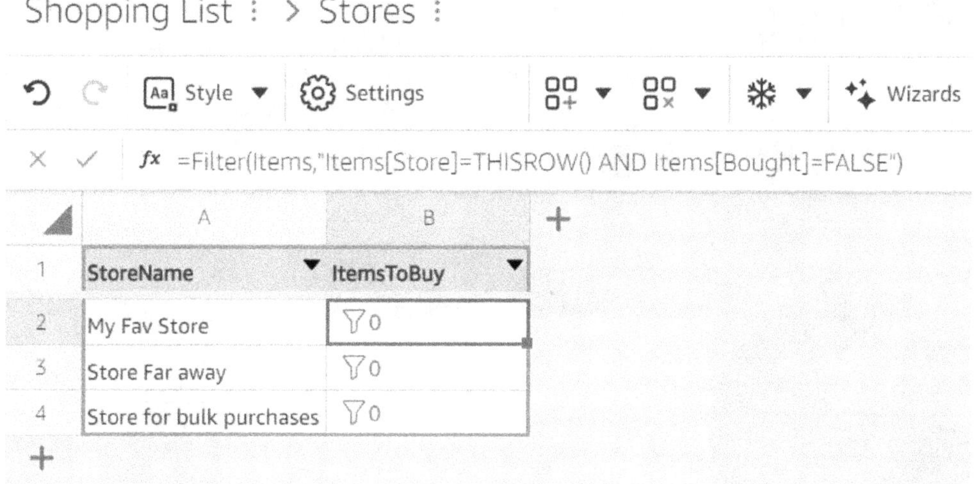

Figure 11.6 – Fixed ItemsToBuy column

The column values did get fixed and all of them now display the filter icon with `0` values representing no items to buy for that store.

So now that our formula is fixed, let's also verify that it is updating as expected:

1. Go to the **Items** table.
2. Fill the row(s) with some items to buy, select the store to buy from, and check the checkbox to set some of them as bought while leaving others as items to buy.
3. Navigate to the **Stores** table and verify that the value against each store is updated as expected.

You may leave the test values in the table as they will come in handy for previews on the app screens while building it.

Creating the app

In the previous sections of this chapter, we covered what use cases our app should support and then discussed what views and screens we will need to build for our app to enable them. So, let's start building them by first creating an empty app by following these steps:

1. Click on the + icon to add a new app.
2. On the popup, select the tile **Build from scratch**.

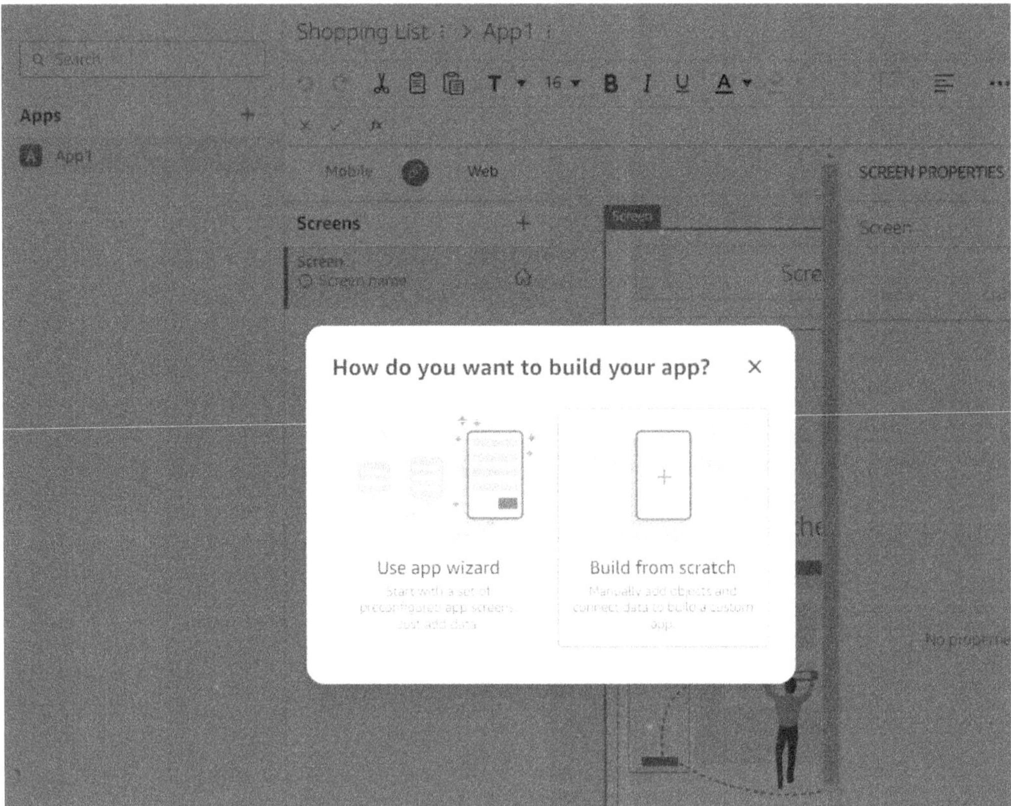

Figure 11.7 – Choose Build from scratch to create an empty app

> **Why Not Use the Wizard Tile?**
> We will show that option as well in the next section but let's first practice building with the individual components to practice and improve our skills.

3. Rename the newly created empty **App1** to `Shopping List`.

Now let's create the screens by following the steps in the next subsections.

Shopping List by Store

As mentioned earlier, my personal preference is to have single-click access to all items I want to buy from a store. Therefore, I prefer my first screen of the app to be a list of all stores I typically buy from. So, let's create it with the help of the following steps:

1. We start by renaming the default-created **Screen name** to `Shopping List by Store` and keep the local name consistent too by setting the same value for it.
2. Next, click the **Add objects** button and select **Column list** as the object to be added to the screen.
3. On the following pop-up screen, check the option to **Add a detail screen** as shown in *Figure 11.8* and then click **Create**.

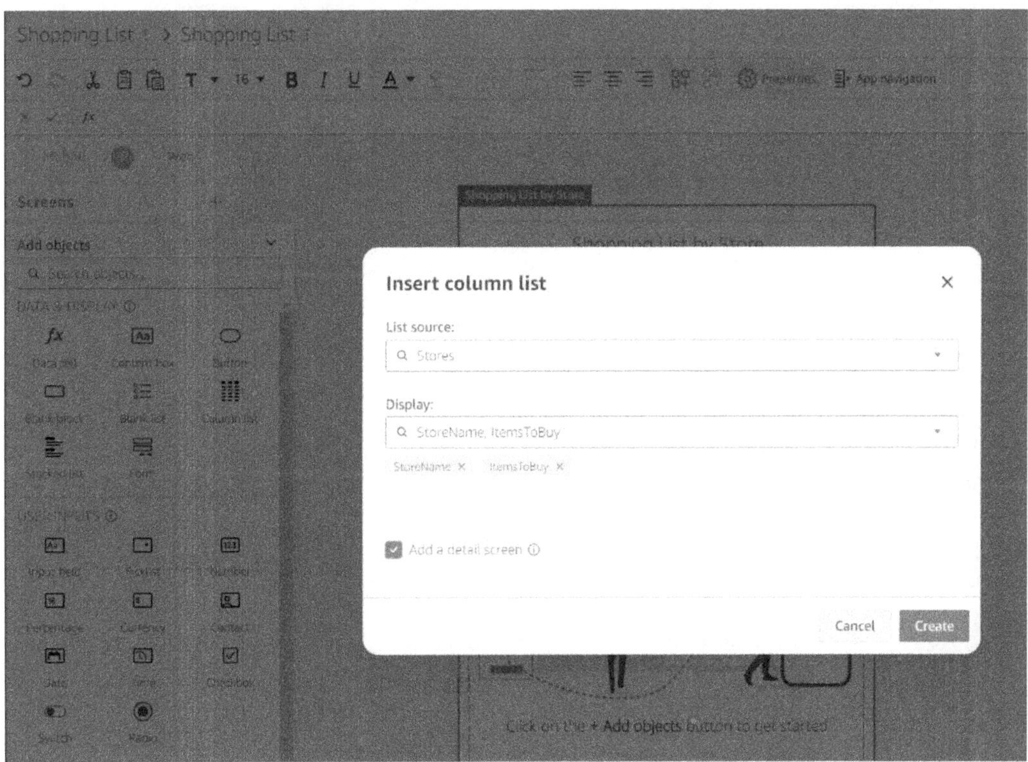

Figure 11.8 – Add a column list and select the option to add a detail screen

This will update the **Shopping List by Store** screen with a list control as well as a block to display the list headers. And in the screens list, you will also see a new screen named **Stores detail** has been added. See *Figure 11.9*:

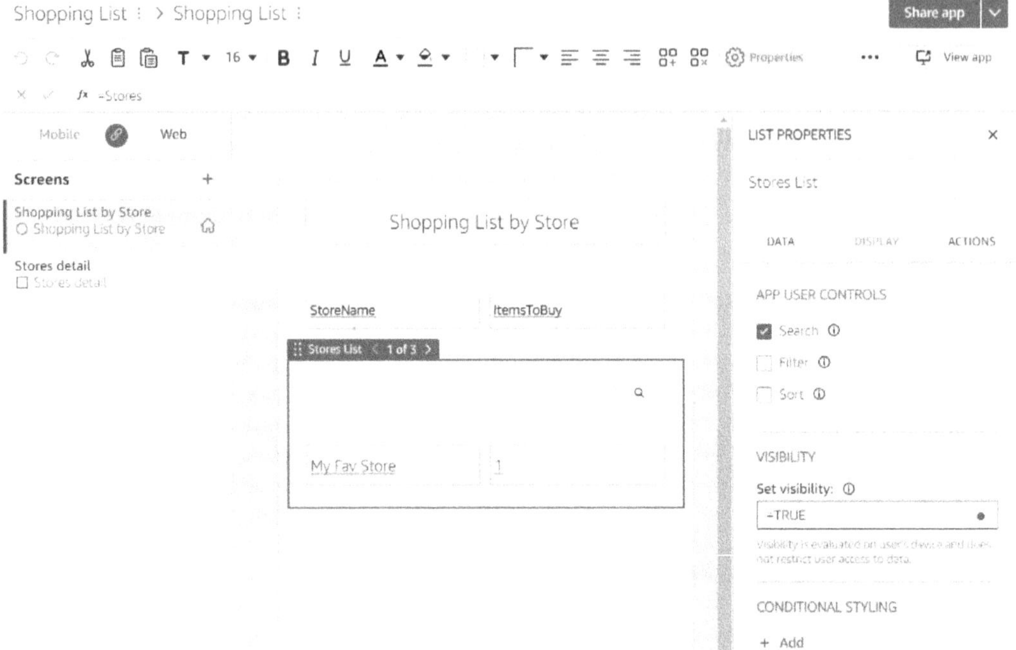

Figure 11.9 – The Shopping List by Store screen after adding the list of stores and the detail screen

Given that the number of stores we typically buy from is low, there is no real need for a search bar so you can remove it from the **Display** tab under the **List Properties** panel of the **Stores List**.

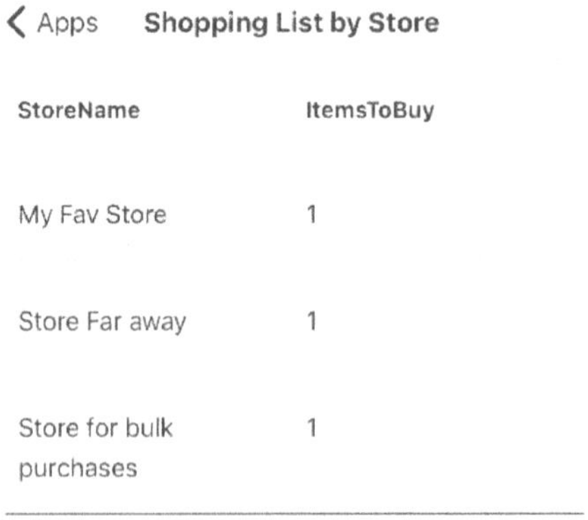

Apps
Shopping List by Store
StoreName
ItemsToBuy
My Fav Store
1
Store Far away
1
Store for bulk purchases
1

Figure 11.10 – Screenshot of the Shopping List by Store screen on an iPhone

With that, we have our screen ready. The mobile version of the screen is shown in *Figure 11.10*.

Stores detail

Next, let's customize our auto-created **Stores detail** screen shown in *Figure 11.11*. Looking at this screen, what we ideally want is to see a list of items to buy from this store and not just how many we need to buy. Also, the store name is shown twice, which is redundant, so let's fix it.

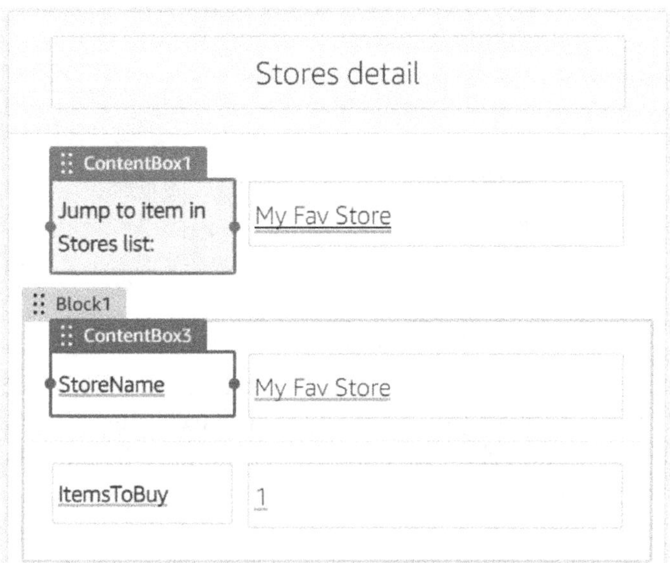

Figure 11.11 – The auto-created Stores detail screen

1. The first thing is to replace the control named **ContentBox1** with **ContentBox3**. And then delete the **segment** containing **ContentBox3**.

2. Next, let's add a **Blank list** object to display the items to buy from this store as shown in *Figure 11.15*. In the **Set source** field, type in the following formula to reference the filter function we have in the **ItemsToBuy** column cell for the displayed row of the **Stores** table: `=$[InputRow][ItemsToBuy]`.

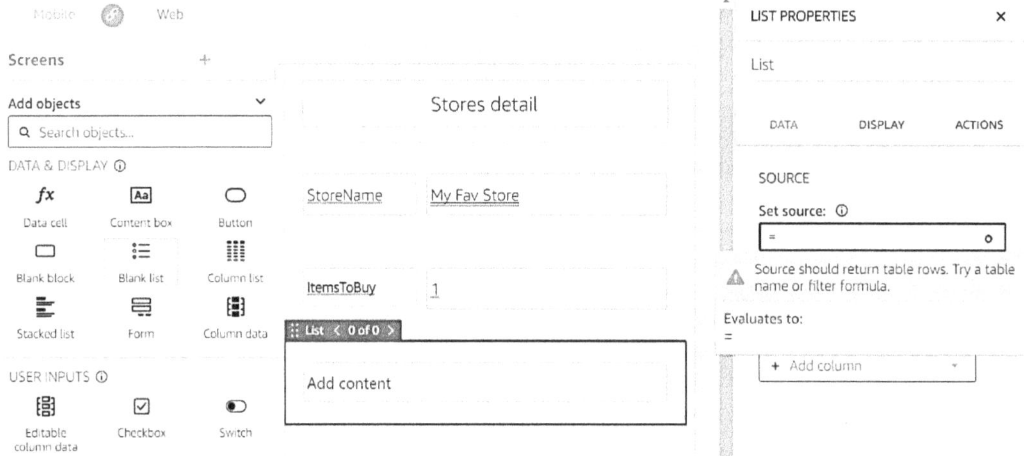

Figure 11.12 – Adding a blank list to display the items to buy

3. Next, resize the content box inside the list control to make some space on the left and then add a checkbox control. Update the **Set shared source** field with the formula `=[Bought]` and clear the `Label` text from the **Display data from** field as shown in *Figure 11.16*:

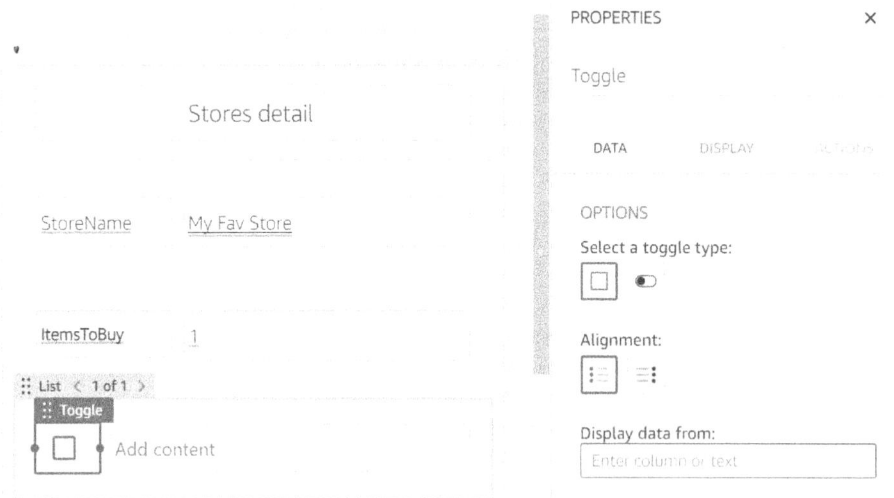

Figure 11.13 – A toggle control added and the display text for the control cleared

4. With **ContentBox1** selected, open the **peeking sheet** if not already visible and click on the + icon on the **Item** column header, as shown in *Figure 11.14*, to add a data cell mapped to the **Item** column and display the item to buy.

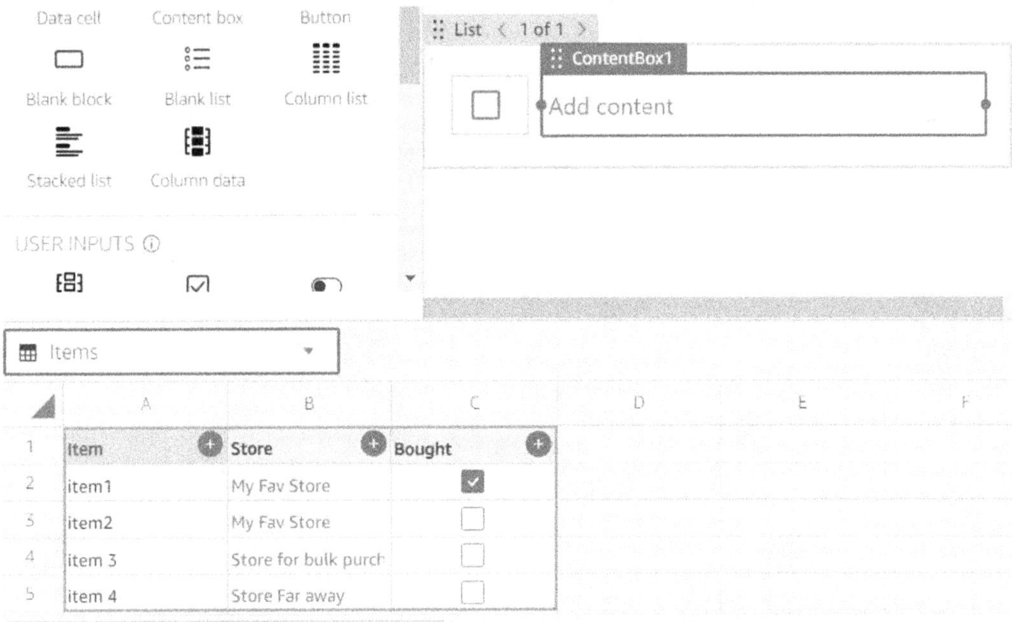

Figure 11.14 – Add the Item column to display in the list

5. Next, we want to add new items to buy from this store. So, let's add a form control that allows us to take user input. From the displayed popup, shown in *Figure 11.15*, choose the option to add a **Button + form screen**, select the **Items** table as the destination in the **Add form data to** field, and then remove the **Bought** column in the **Display** field as we know that we want to buy the item we are adding and therefore can default the value when adding the data.

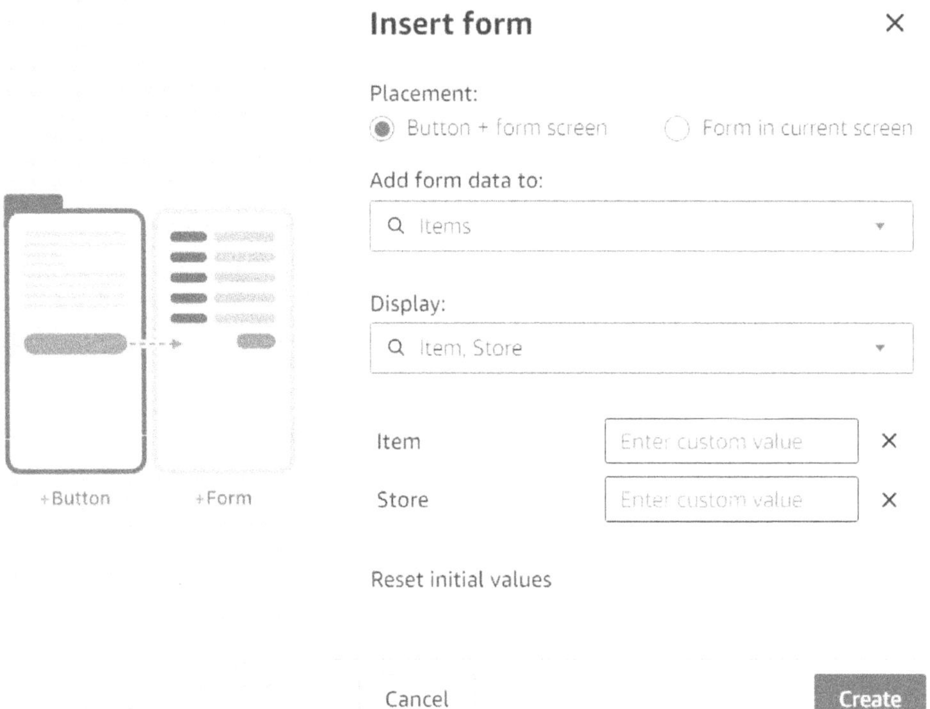

Figure 11.15 – Adding the form to be able to add new items to the list

6. Click on **Create**. This will add a new button on the screen as well as a new screen named **Items form**. Update the button text to `Add Item`.

> **Exercise 2**
> Set the store name on the form by default to the store we are adding the item to.

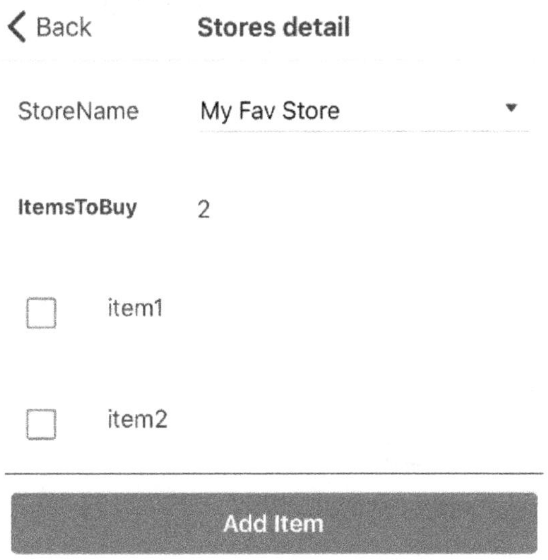

Figure 11.16 – Screenshot of the Shopping List by Store screen on an iPhone

And with that, we have our screen ready. The mobile version of the screen is shown in *Figure 11.16*.

The Item form

Next, let's look at the item form screen created in the previous section. Recall that while creating the screen, we removed the **Bought** field as we know the default value to be set. But that also resulted in this field not being set by the auto-create automation on the **Done** button, as shown in *Figure 11.17*. So, let's fix that.

1. Select the **Done** button and then navigate to the **Actions** tab in the properties panel to edit the automation.

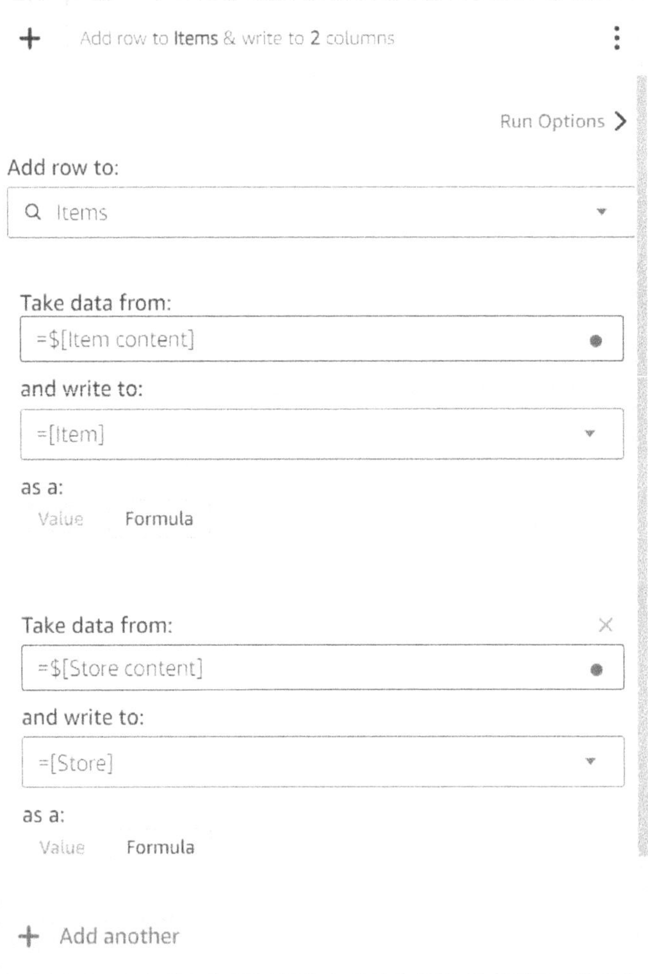

Figure 11.17 – Editing the automation on the Done button

2. Click on + **Add another** and set `=False` in the **Take data from** field, and select the **Bought** column in the **and write to** field.

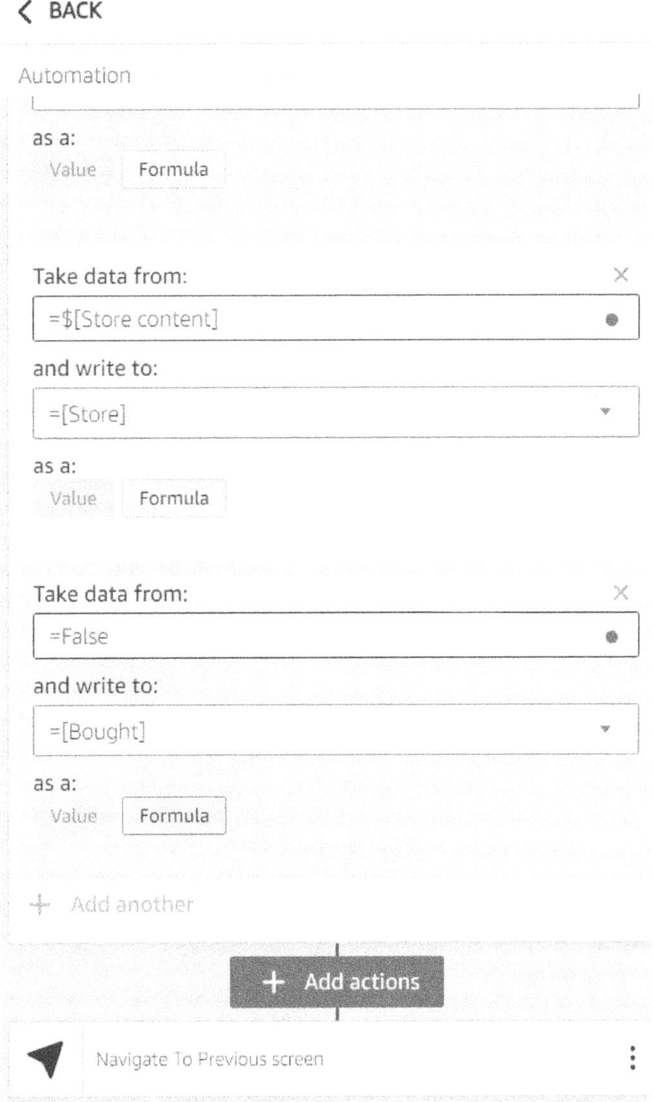

Figure 11.18 – Updated automation

3. Click **Back** to finish editing the automation.

Items to Buy

Now let's create the second view of the shopping list by items.

> **To Do**
>
> Create a similar set of list and detail screens for **Items to buy** by following the steps for creating the **Shopping List by Store** screen.

Upon completion, you should have a screen similar to *Figure 11.19*.

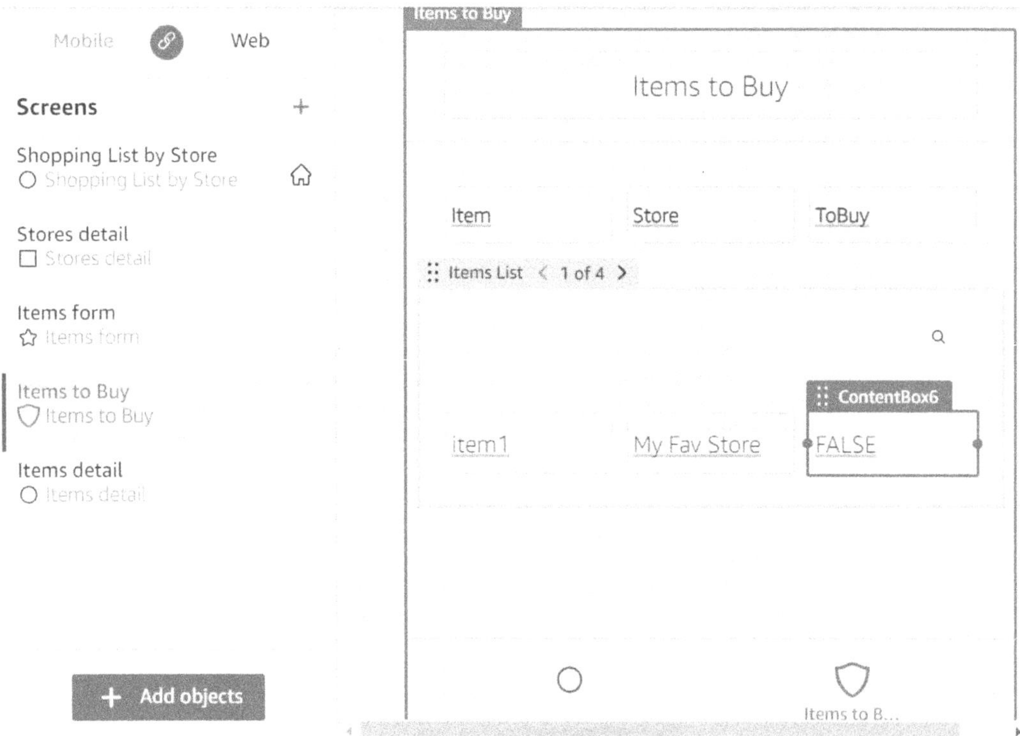

Figure 11.19 – Items to Buy screen

Now let's customize it in the following steps:

1. First, let's replace **ContentBox6** with a checkbox control similar to what we have in the items list we added to the **Store detail** screen.
2. Next, when shopping, we would like to see the **Items to Buy** on top and hide the items that have been bought or show them at the bottom of the list or as an entirely separate list. However, in the **Items List**, we have the entire **Items** table set as the source of this list, so let's fix that by setting the source as the following filter formula: `=FILTER(Items, "Items[Bought] = FALSE")`.
3. Lastly, we would also like to be able to add items to this list from this screen. So let's do that. Add a button in the screen's **Header** section and update the text to **+ Item**.
4. With the button selected, set the **On click:** property under **QUICK ACTIONS**, to navigate to the **Items form** screen.

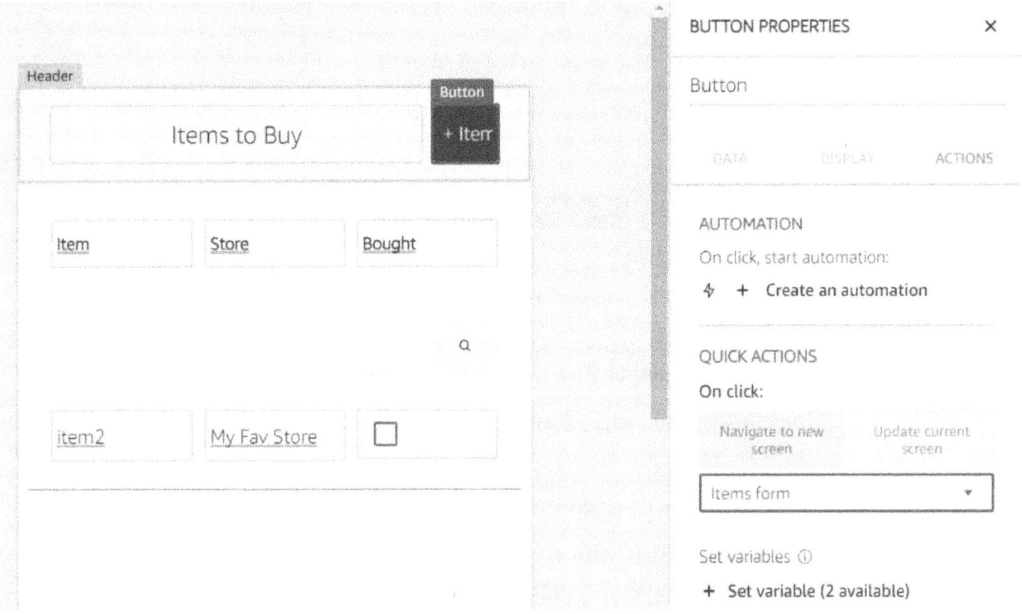

Figure 11.20 – Items to Buy screen after customization

> **Exercise 3**
> Show the list of bought items.

Items detail

As a result of completing the *To Do* from the previous section, you will also have the screen shown in *Figure 11.21* generated in your app:

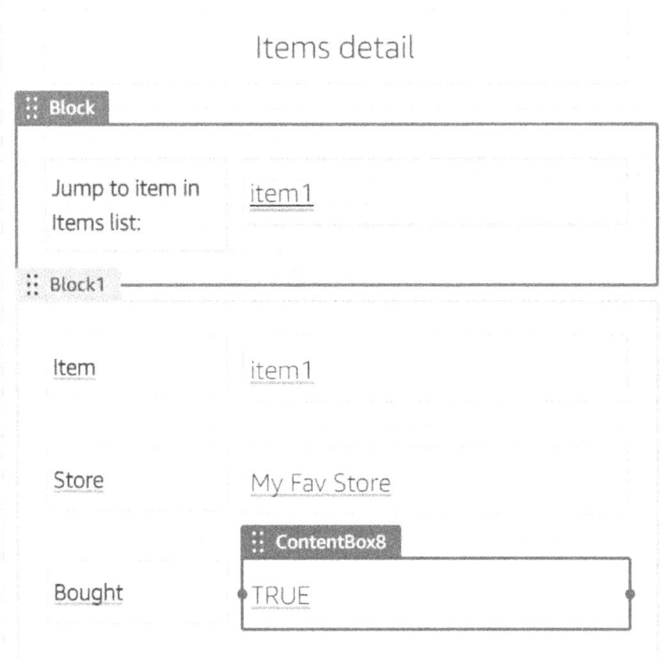

Figure 11.21 – Items detail screen

As such, the screen does not require any changes but here are some customizations I prefer to make:

1. Hide the **Block** by setting the visibility to `=FALSE`. I do this to remove the option to be able to select other items from the list but it is not a requirement.
2. Replace the **ContentBox8** with the **checkbox** control to make it consistent with our other screens.
3. Lastly, mark the data cells displaying **item1** and **My Fav Store** as editable to remove the need to create the edit screen or separate editable controls for the items.

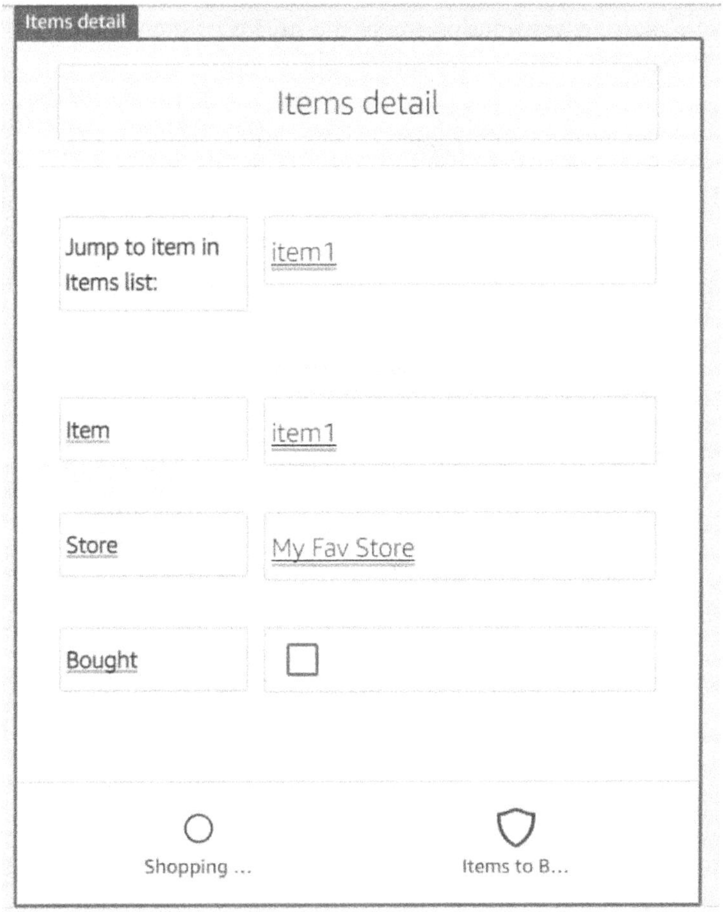

Figure 11.22 – Items detail screen after customization

And with that, our **Shopping List** app is ready, with one caveat – we do not have a way to delete an item or clear the list of bought items. Recall that we built the functionality to delete a task in *Chapter 5, Powering the Honeycode apps with Automations*, and therefore, I'm leaving the addition of a **Delete** button on the **Items detail** screen as a task for you to complete. Instead, let's see a different way to do this *clearing* in a bulk way in the next section.

Clearing the bought items list

To clear all the items in the bought list, we will add a button on the **Items to Buy** screen and configure its actions in the following steps:

1. Add a new block after the **Items List**, add a button inside this block, and update the text on the button to `Clear`.

2. Next, select **Create an automation** under the **ACTIONS** tab of **BUTTON PROPERTIES**.

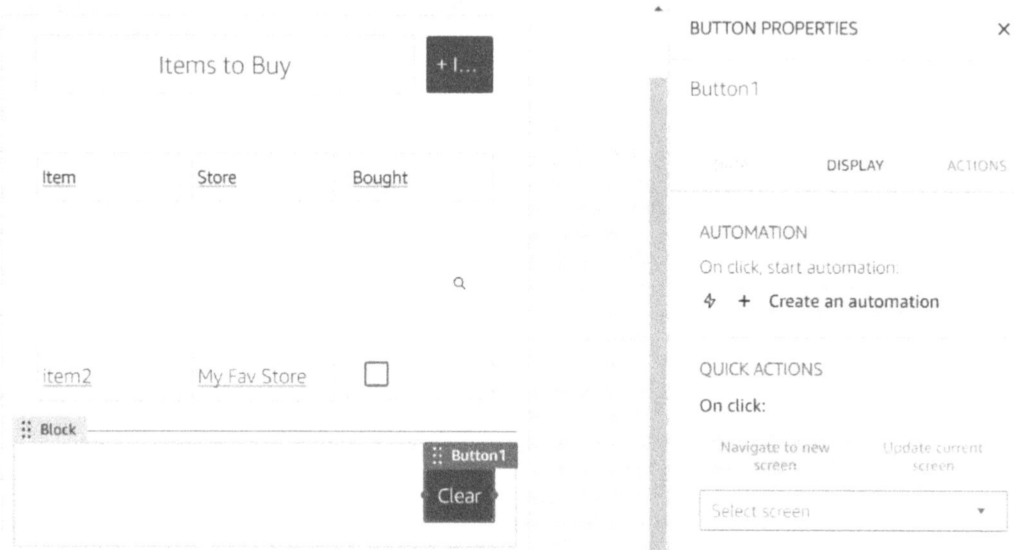

Figure 11.23 – Adding automation to the newly added Clear button

3. Click **Add actions** and select **Delete a row**.

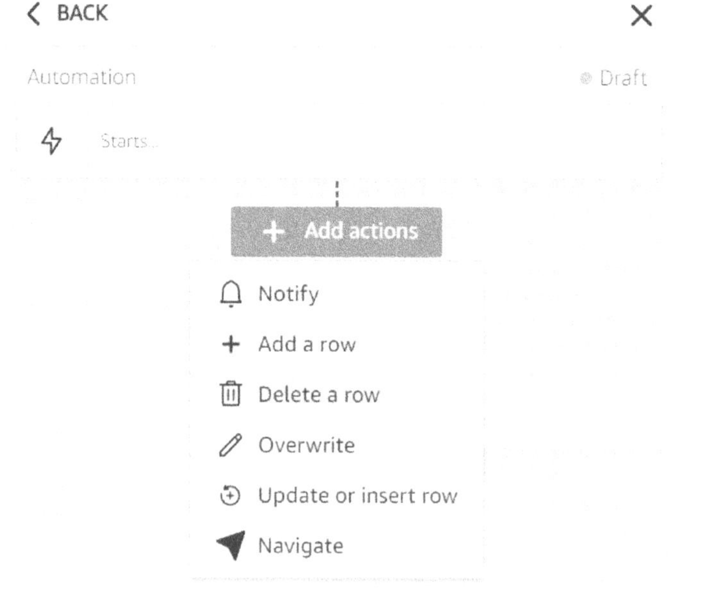

Figure 11.24 – Add actions | Delete a row

4. Within the **delete** block, select the option to delete **Specified row** and then set the following formula in the **Delete Row** field that shows up on the selection: `=Filter(Items, "Items[Bought]=TRUE")`.

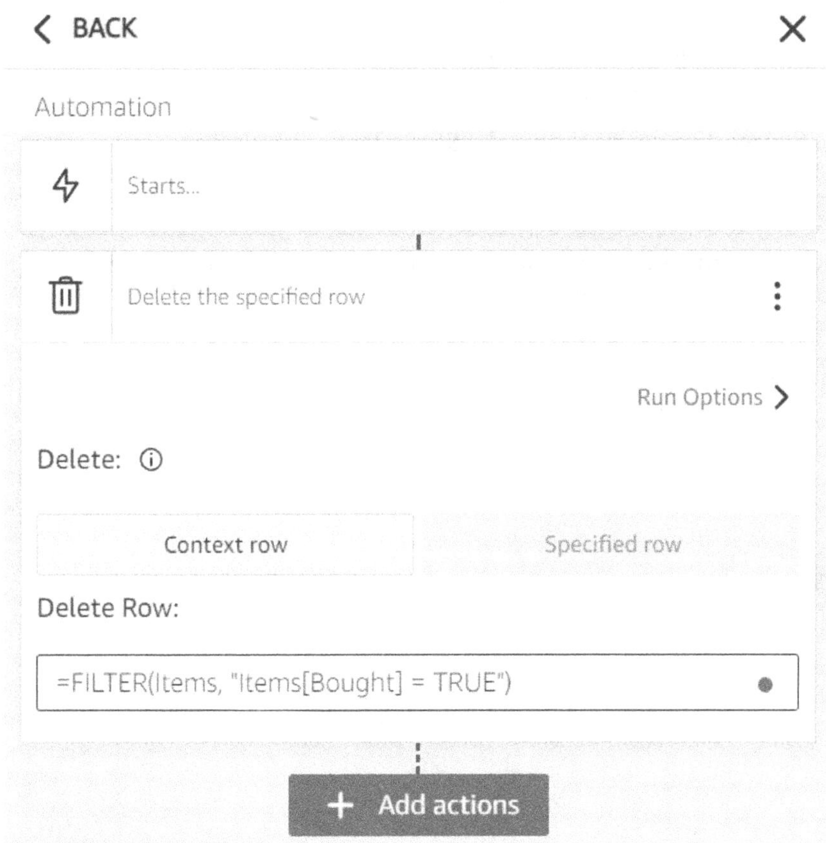

Figure 11.25 – Configuring the delete block of automation

5. Click **Back** and finish creating the automation.

In this automation, we essentially configured our **Clear** button to delete all the rows that have items marked as bought. We specified that using the `Filter` function we updated in *Step 3*. As you complete *Exercise 3*, you'll find that this is the same formula you will need to use as the source of the list to show all bought items.

Building an app using the Wizard

While creating our **To-Do** app, we saw how we can use **wizards** to create or bootstrap our app-building process. In this section, we will go through the key steps that could have helped set up our app even faster than what we covered in the last sections:

1. Click on the + icon to add a new app.
2. On the popup, select the tile **Use app Wizard**.
3. In the **Source** dropdown, choose the **Stores** table.

> **Note**
>
> Under the **Settings** tab, **Add a detail screen** and **Add a form screen** options will be selected as shown in *Figure 11.26*.

4. Click **Next**.

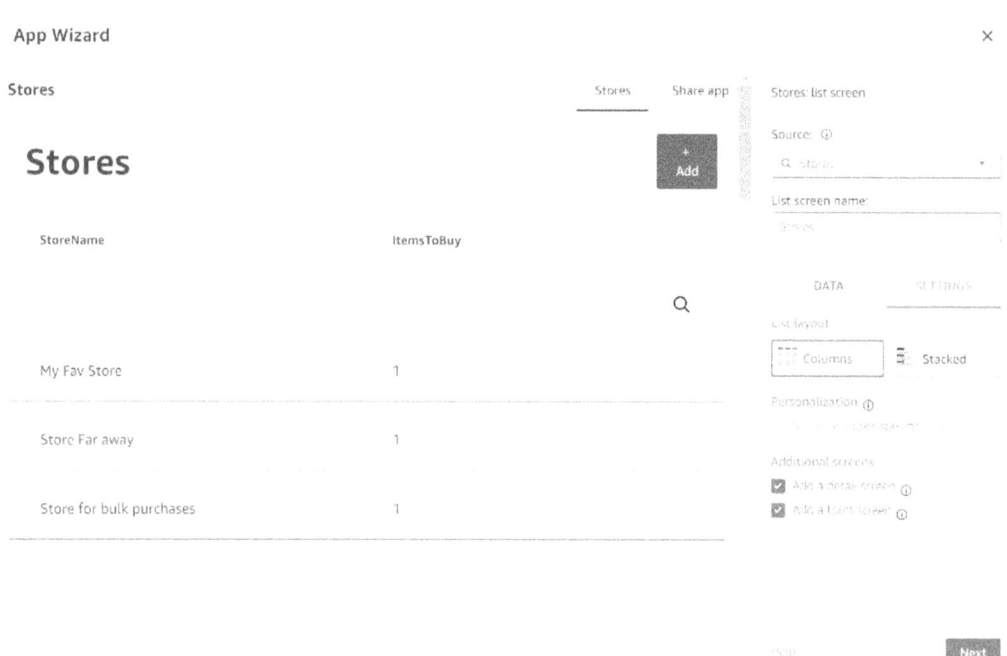

Figure 11.26 – Creating screens for the Stores table using the wizard

5. The wizard loads the **Stores detail** screen. Click **Next**.
6. The wizard then loads the **Stores** form. Click **Done**. The wizard completes the creation of three screens and shows the option to add more screens as shown in *Figure 11.27*:

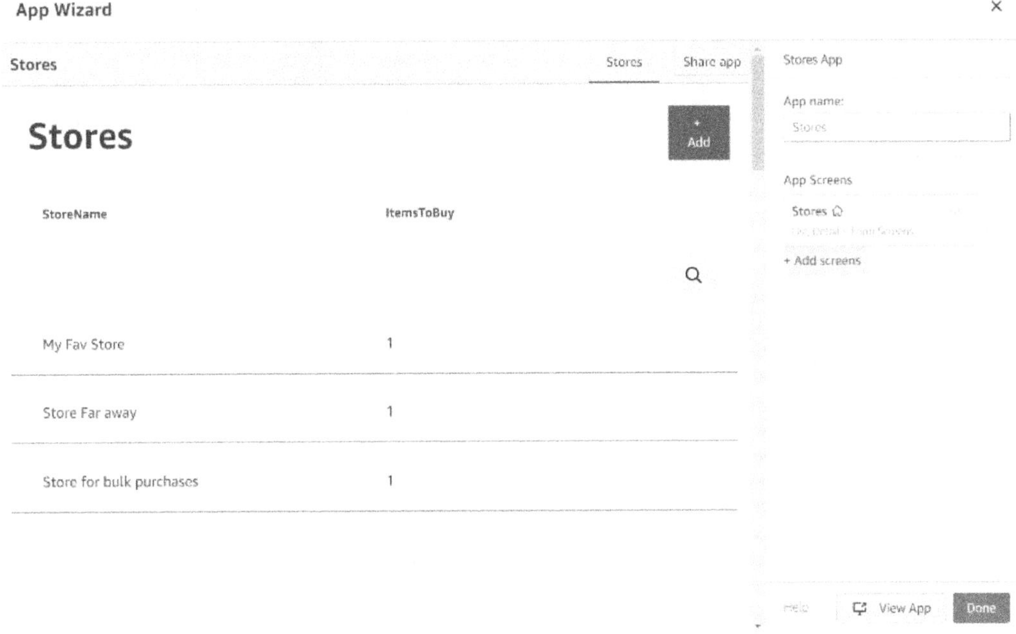

Figure 11.27 – Wizard after creating screens for the Stores table

7. Next, we click on **Add screens**, repeat *Steps 3 to 5* with the **Items** table as **Source** as shown in *Figure 11.28*. When on the **Items detail** screen, remember to mark the **Item** and **Store** fields on the screen as editable.

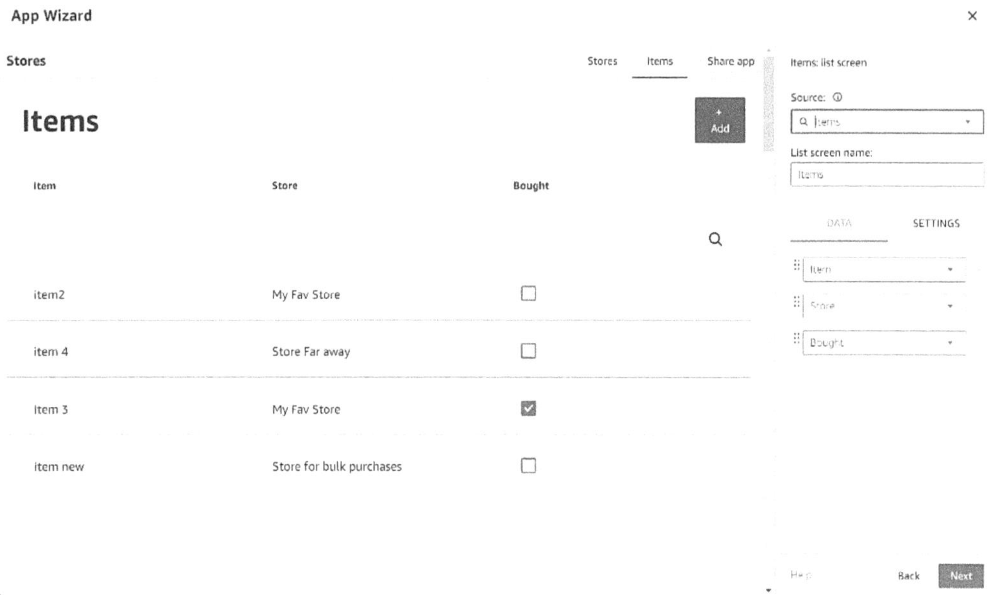

Figure 11.28 – Adding screens for the Items table using the wizard

8. Click **Done**.

Now wasn't that fast? Of course, the app isn't complete yet and there are a few edits needed to customize it and bring it to parity with what we built from scratch, but those are all marginal and covered in the previous sections. As a bonus, we even got a form for adding a new store from the app.

Summary

In this chapter, we built another app for an everyday use case of managing multiple shopping lists. We applied our learnings so far to define and set up our data model and then build the app from scratch. We learned how we can delete multiple rows in a single action as well as how we can use a column with filters to display the count later and use it to display the entire set of rows on the screen as per the need. Finally, we also used the wizard and learned how we can utilize it to expedite our building journey in the future.

However, wizards are not the only way to speed up the app building process in Honeycode. In the next chapter, we will make use of an existing template to build an app to enable submitting nominations and also vote for the nominees to choose a winner.

12
Building a Nominate and Vote App in Honeycode

In our day-to-day life, we all come across contests in various forms. In some, we are participating as a contestant, while in others, we are just viewers. In some, we may be judging, while in others, we are the organizer, the viewer, and might also be the judge. In an effort to engage viewers, a very common form of contest extensively used by media is to decide a winner by popular vote. A variant of this even allows for an audience to first nominate the candidates, which are then subject to voting for deciding a winner. We see similar contests also featuring in professional spaces, but a more common form of this problem is seen in the form of all-hands meetings that typically encourage employees to submit questions ahead of time.

Often, there are no specific tools available to quickly set up an easy tool or process for supporting this and you may find yourself using emails, Word documents, or spreadsheets to manage it. This typically results in common issues, such as how a question gets selected from the submissions, which is usually up to the organizer and may not be fully transparent to the team members, or as an organizer, there may be duplicate questions submitted that are not easy to spot and screen, and more.

In this chapter, we will see how Honeycode can help solve this everyday use case by building a nominate and vote app. More specifically, I will show you how to use the power of Honeycode templates by using an existing template and customizing it for our use case of running a contest where everyone can submit nominations and later vote for selecting a winner. We will also have some exercises to further your understanding as well as extend the functionality.

In this chapter, we're going to cover the following main topics:

- Defining the app requirements
- Translating requirements to app interactions
- Defining the data model
- Building the app

Technical requirements

To follow along with this chapter, you'll need to have access to Amazon Honeycode, which necessitates a laptop with a web browser, preferably Google Chrome, and optionally a mobile device running either a Honeycode supported version of Android (currently requires Android 8.0 and up) or iOS (currently requires iOS 11 or later).

Furthermore, we'll use the Honeycode terminology and refer to the components that we covered in *Chapter 2, Introduction to Honeycode*, *Chapter 3, Building Your First Honeycode Application*, *Chapter 4, Advanced Builder Tools in Honeycode*, and *Chapter 5, Powering the Honeycode apps with Automations*, and therefore recommend you complete those first.

Defining the app requirements

We first introduced this section in *Chapter 3, Building Your First Honeycode Application*, where, before building our app, we listed down the requirements or the use cases that we wanted the app to fulfill. And we made use of that list throughout the chapters as a guide for defining our data model, conceptualizing the application interface, and visualizing the interactions between various onscreen elements as well as the data displayed. Therefore, I'd encourage you to take 5 minutes to think about what your app should do and what it should look like, and then make a list.

> **To-Do**
>
> Take 5 minutes and list down how you would like your Nominate and Vote app to work.

Here is what my list looks like:

1. Contest organizers must be able to set one or more contests at the same time.
2. The app must mandatorily support an option for requesting details for each nomination.
3. App users must be able to submit nominations.
4. App users must be able to provide details regarding the nomination.
5. App users must be eligible to vote.
6. Optionally, voting must be restricted just to a panel of judges.
7. The app must make provision to restrict the number of votes a user can cast.
8. A user must only be able to see their votes and not the total number of votes against each nomination.
9. Apps must have a restricted organizer view to view total votes and ranks.

Your list may have some or all of these requirements and may have even more, and that is perfectly fine. You may find some of these requirements missing in the app we build in this chapter and may want to extend that for yourself after you finish the chapter.

Translating requirements to app interactions

Based on the requirements we listed in the previous section, our app will have four primary views:

1. View all ongoing contests.
2. View all nominations for a contest and vote (if allowed).
3. View the contest details and submit a nomination(s).
4. An organizer's view of the overall status of the contest with vote counts and details.

Users should be able to select the running contests, which should take them to a screen with all the submitted nominations. Clicking on the nominations should show the details of the entry, which is useful if, as part of the nomination, there is a requirement to provide additional context for the entry.

Against the nominations list, we would like to have a button to vote, and it should be visible or enabled only for people eligible to vote. The button should work as a toggle to vote and retract. If there is a limitation on the number of votes per person, buttons should be disabled when the limit is reached.

There will be another button, to submit a new nomination, which could also have a condition set so that it's either visible or enabled only for those who are allowed to submit nominations.

For the organizer, we could have a separate app, or the contest detail screen could have fields that are visible only to them.

Now that we have an understanding of the key interactions and screens of the app, let's define how our data model enables these screens and interactions.

Defining the data model

Our app has three distinct entities:

- Contests
- Nominations
- Votes

Therefore, we will need three separate tables for them. The nominations will need to have mapping to the contests they are part of, and the votes are mappings of voters against nominations.

If we decide to restrict the voting to a panel, we will also need a table to list down the members of that panel, and the same applies to permissions for organizing. If we are to limit that, we need a list of approved organizers also stored as a table.

Building the app

Alright, so now we know what our app should do, have a fair idea of what it should look like to enable the listed use cases, and know how we will structure and store the data to power the app.

If you have completed *Chapter 11, Building a Shopping List App in Honeycode*, you may notice that the first three views listed in the *Translating requirements to app interactions* section can be quickly generated through the **App Wizard** using a data source for listing contests as the main screen and with the **Add a detail screen** and **Add a form screen** options selected. Similarly, you can create three additional screens for nominations and then edit these generated screens to adapt some functionalities and delete the unwanted controls and screens. The only missing piece that you will not be able to generate with the wizard is the voting controls and the ability to separate the organizer's view of contests.

However, recall that we have another way of creating applications in Honeycode, which is through templates. In *Chapter 6, Introduction to Honeycode Templates*, we looked at all the templates that are present in Honeycode in 2021, and from that list, two templates offer voting capabilities:

- **Collaborative Brainstorming**
- **Instant Polls**

One major difference between the two is that **Instant Polls** has pre-defined options to vote for, while **Collaborative Brainstorming** allows for the addition of new ideas and is therefore much like the behavior we want to build for our app.

So, let's start building it up using the **Collaborative Brainstorming** template.

Creating the template app

We begin by creating the template app and associated workbook by following these steps. Go to the dashboard and create a workbook using the **Collaborative Brainstorming** template and name it `Honeycode Awards`.

> To-Do
>
> Before going to the next section, try out the template-created Brainstorming app if you haven't tried it out thus far.

And with that, we are ready to modify the template to build our Awards app.

Setting up the tables

The workbook is created by using the template loaded with five tables as shown in *Figure 12.1*:

- **Ideas**
- **Questions**
- **Status**
- **Upvote**
- **_README**:

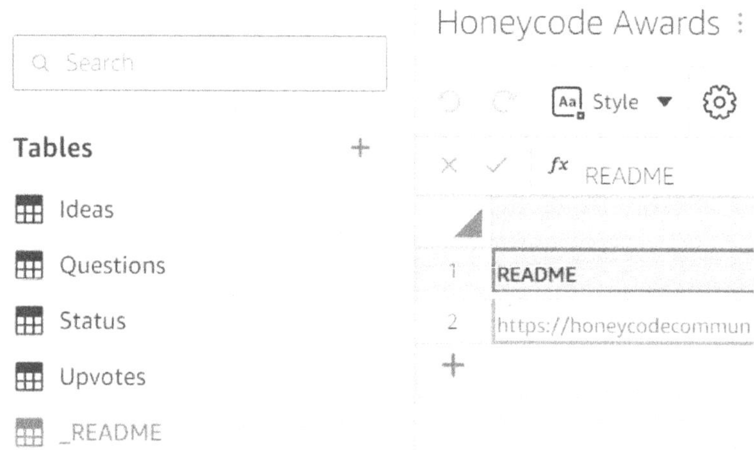

Figure 12.1 – Tables created by the template

In the previous section, *Defining the data model*, we made a note of three key entities, namely, Contests, Nominations, and Votes, that will require their own tables.

> **Did You Notice a One-to-One Relation between These Entities and the Tables Created by the Template?**
>
> Questions can be mapped to Contests, Ideas to Nominations, and Upvotes or Votes are the same.

Let's now review these tables and update them to meet our requirements.

Questions/Contests table

The **Questions** table, shown in *Figure 12.2*, consists of columns for the question, its context, and its metadata, such as the status, who created it, and how many ideas have been submitted:

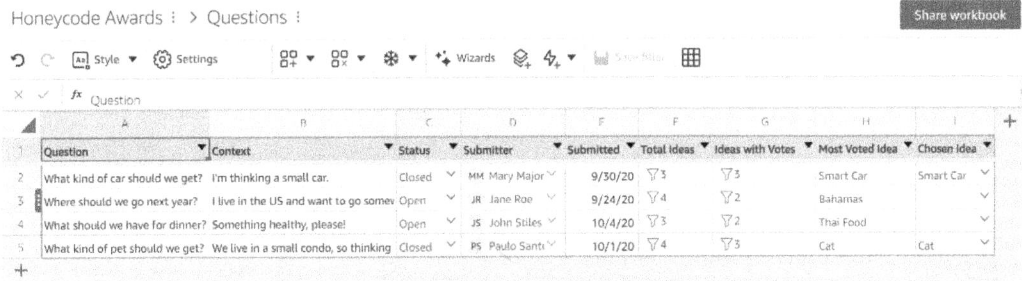

Figure 12.2 – Questions table

Similar to the mapping of tables, you may be able to visualize how these fields map to their counterparts when it comes to a contest. So, let's update this table to make it so:

> **Note**
>
> The renaming of tables and other entities that we will do in this and the other tables is intended to reduce the mental load of managing the mapping. This has no direct impact on the app's functionality.

1. The first step is to rename the **Questions** table **Contests**.
2. Next, we will rename the table columns as follows:

Current column name	Renamed column name
Question	Contest
Context	Contest Details
Total Ideas	Total Nominations
Ideas with Votes	Nominations with Votes
Most Voted Idea	Most Voted Nomination
Chosen Idea	Winner

Table 12.1 – Mapping for renaming the columns from the Questions table

3. Now insert a new column and name it `Max Votes Allowed`.
4. With **Max Votes Allowed** selected, click on **Settings** to open the right panel with column properties and set **Formula** as `=-1`. We will use **-1** as the default value of this column to represent no limit on the number of votes.

5. The updated **Contests** table is shown in *Figure 12.3*:

Honeycode Awards > Contests

Style Settings Wizards

fx Open

	Contest	Contest Details	Status	Submitter	Submitted	Total Nomination	Nominations with Votes	Most Voted Nomination	Winner	Max Votes Allowed
2	What kind of car should we get?	I'm thinking a small car.	Closed	MM Mary Major	9/30/20	3	3	Smart Car	Smart Car	-1
3	Where should we go next year?	I live in the US and want to go somewhere overseas.	Open	JR Jane Roe	9/24/20	4	2	Bahamas		-1
4	What should we have for dinner?	Something healthy, please!	Open	JS John Stiles	10/4/20	3	2	Thai Food		-1
5	What kind of pet should we get?	We live in a small condo, so thinking a small pet.	Closed	PS Paulo Santı	10/1/20	4	3	Cat	Cat	-1

Figure 12.3 – Contests table

Ideas/Nominations table

The **Ideas** table contains the submitted idea along with the details, including the mapping to the question it was submitted for, who submitted it and when, and how many votes it has. Similar to the previous section, let's update this table to make it our **Nominations** table:

1. Rename the table `Nominations`.
2. Next, we will rename the table columns as follows:

Current column name	Renamed column name
Idea	Nomination
Related Question	Submitted for Contest

Table 12.2 – Mapping for renaming the columns from the Ideas table

The updated **Nominations** table is shown in *Figure 12.4*:

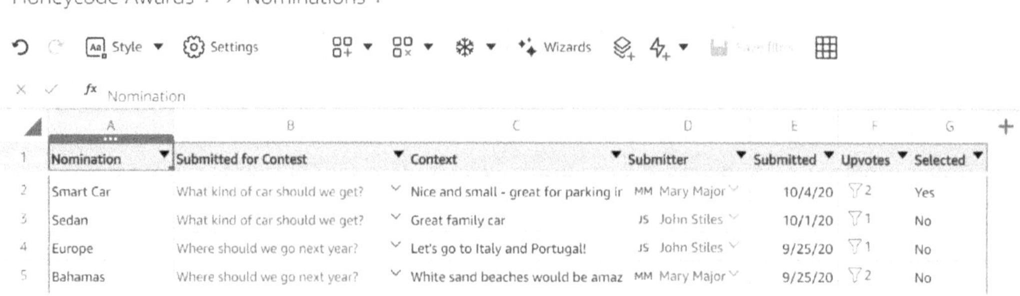

Honeycode Awards > Nominations

Style Settings Wizards

fx Nomination

	Nomination	Submitted for Contest	Context	Submitter	Submitted	Upvotes	Selected
2	Smart Car	What kind of car should we get?	Nice and small - great for parking ir	MM Mary Major	10/4/20	2	Yes
3	Sedan	What kind of car should we get?	Great family car	JS John Stiles	10/1/20	1	No
4	Europe	Where should we go next year?	Let's go to Italy and Portugal!	JS John Stiles	9/25/20	1	No
5	Bahamas	Where should we go next year?	White sand beaches would be amaz	MM Mary Major	9/25/20	2	No

Figure 12.4 – Nominations table

Upvotes table

The **Upvotes** table contains the voting records, with each row containing an upvoted idea, the question it is related to, and who upvoted it. To record the votes in a contest, we need the same information and just need to rename the columns to make them aligned with the rest of the tables.

Next, we will rename the table columns as follows:

Current column name	Renamed column name
Idea	Nomination
Question Title	Contest
User	Voter

Table 12.3 – Mapping for renaming the columns from the Upvotes table

The updated **Upvotes** table is shown in *Figure 12.5*:

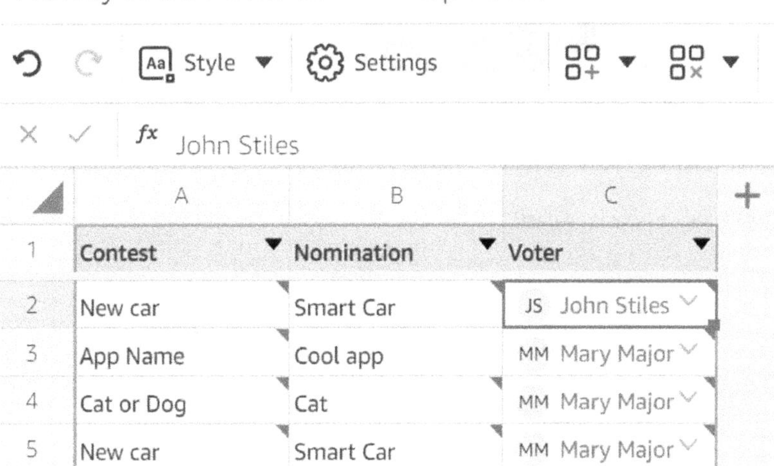

Figure 12.5 – Upvotes table with renamed columns

Now that we have set up the tables, let's take a look and update the app created by the template.

Editing the template app

The template generates the Brainstorming app, shown in *Figure 12.6*, to be used out of the box:

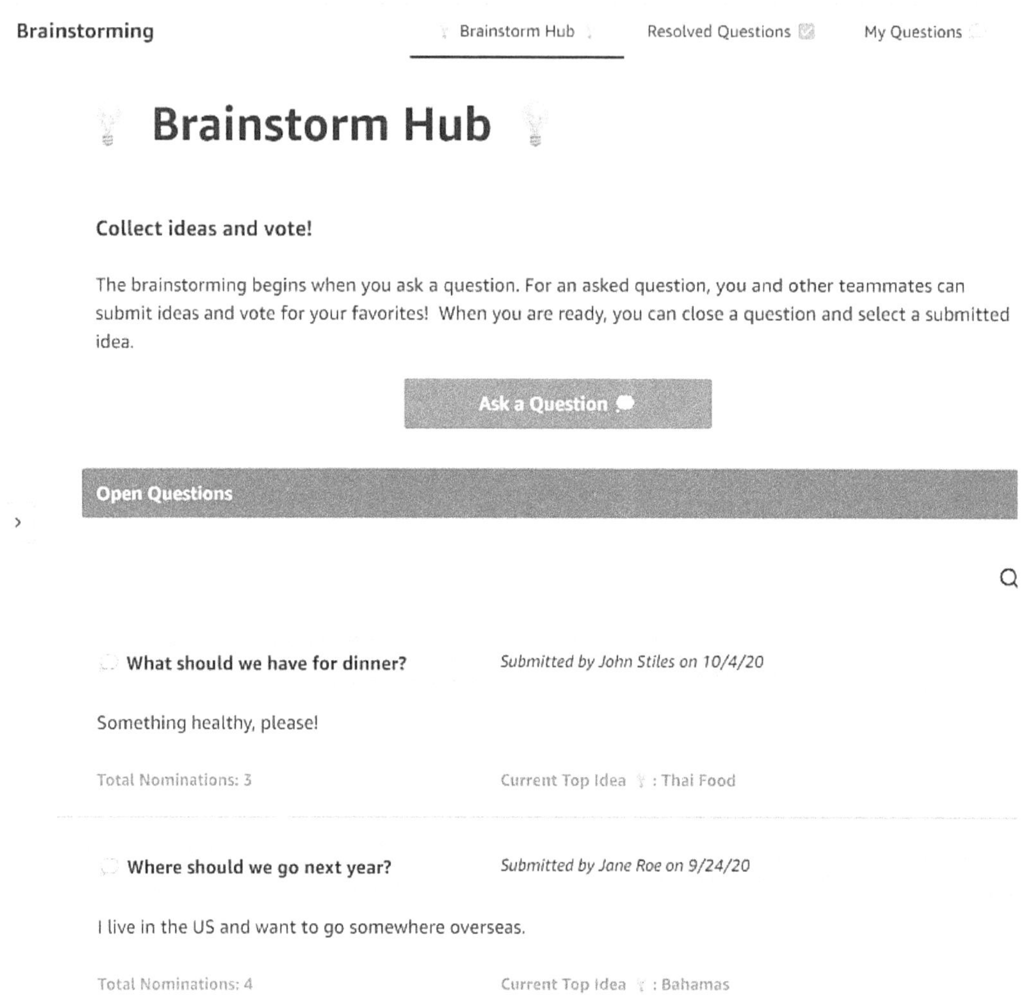

Brainstorming
Brainstorm Hub
Resolved Questions
My Questions

Brainstorm Hub

Collect ideas and vote!

The brainstorming begins when you ask a question. For an asked question, you and other teammates can submit ideas and vote for your favorites! When you are ready, you can close a question and select a submitted idea.

Ask a Question

Open Questions

What should we have for dinner?
Submitted by John Stiles on 10/4/20
Something healthy, please!
Total Nominations: 3
Current Top Idea : Thai Food

Where should we go next year?
Submitted by Jane Roe on 9/24/20
I live in the US and want to go somewhere overseas.
Total Nominations: 4
Current Top Idea : Bahamas

Figure 12.6 – Web view of the Brainstorming app generated by the template

The app comes with three visible screens and one hidden screen, shown in *Figure 12.7*, that we will review and update in the following subsections to adapt the app to our requirements:

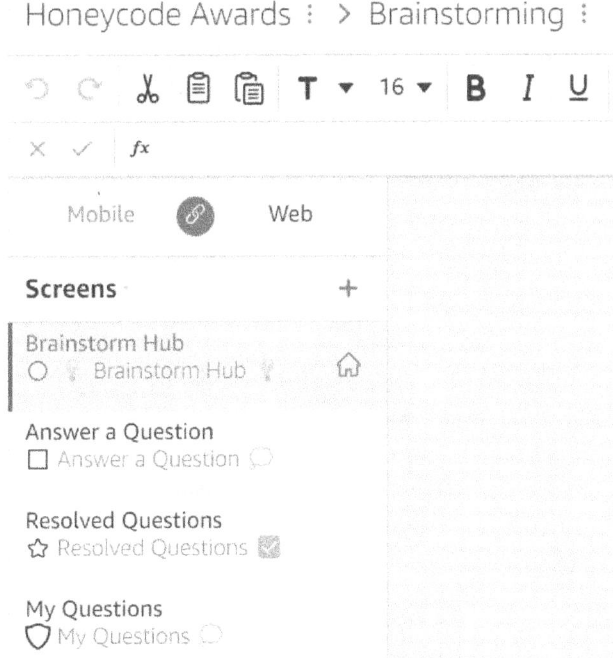

Figure 12.7 – Builder view of the Brainstorming app

Similar to the data model (tables and columns), we will do a fair amount of renaming, but we will also set visibility conditions or remove some controls as well as add additional controls for the missing functionalities. So, let's start updating our app by first renaming it `Honeycode Awards`.

Brainstorm Hub

This is the home screen of the app and lists down all the questions that are currently open for submitting ideas. It also allows the submission of new questions by using the **Ask a Question** button. The screen can be grouped into three parts:

- A top section that allows the addition of new questions
- A middle section, which is a form for collecting details for the new questions to be added
- A bottom section, which is the list of questions open for ideas

Now, recall that we mapped **Questions** to **Contests** when updating our data model and, by corollary, this will be our screen for listing all the ongoing contests. So, let's start the update process by renaming the screen **Contests**. Next, we will update each screen section.

Top section

This section consists of three objects: two textboxes with some text explaining the purpose of the app, and a button to allow a new question to be asked. See *Figure 12.8* for reference:

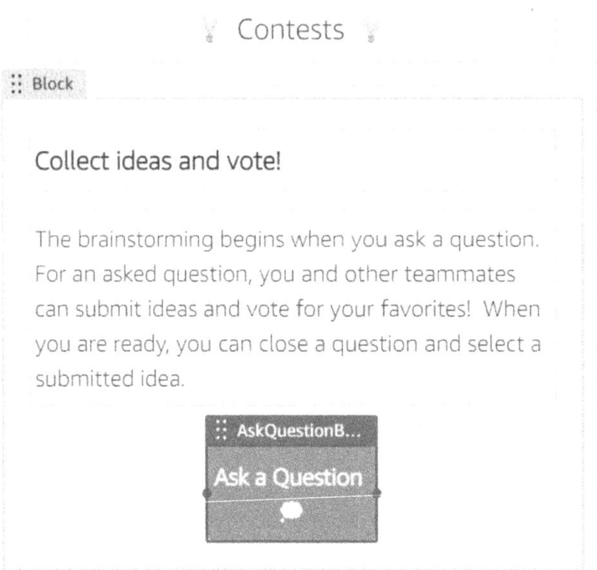

Figure 12.8 – Top section of the Contests screen

For the Awards app, we do not need any introduction text and therefore we will delete the two textboxes and update the text on the button to **Create Contest**. You may also want to rename the **AskQuestionButton** button to **CreateContestButton**.

Middle section

This section consists of the form for getting user input for adding a new question; see *Figure 12.9* (**a**). It also consists of debugging variables inside the boxes labeled **(b)**, as well as a hidden content box (labeled **(c)**) to display the confirmation message when a new question is successfully added.

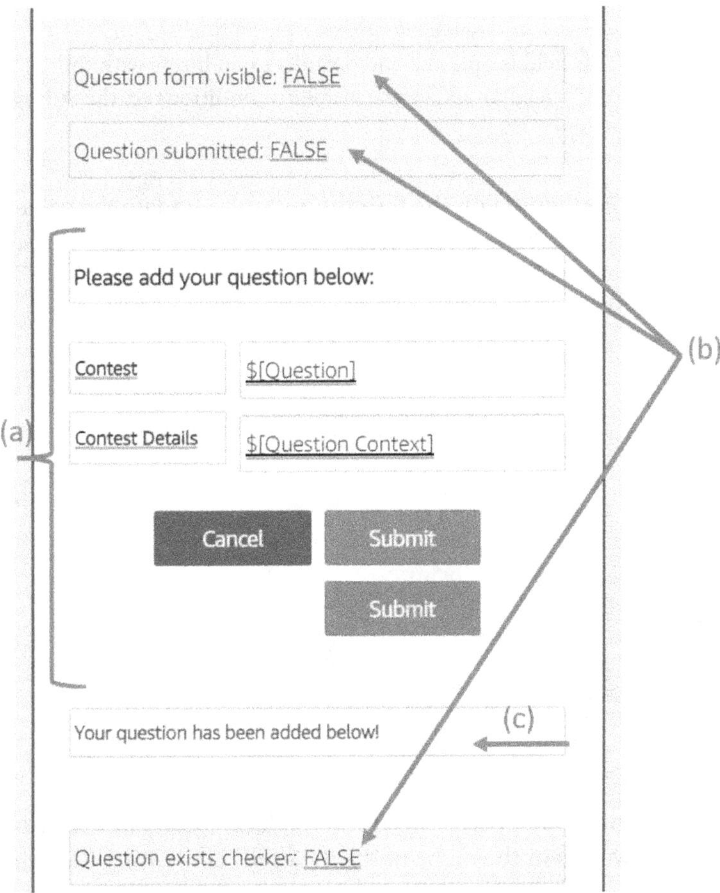

Figure 12.9 – Components of the middle section of the Contests screen

Let's update this section with the following steps:

1. Update the **Please add your question below** text to **Add contest details**.
2. You can update the **$[Question]** and **$[Question Context]** variable names by renaming their corresponding data cells **Contest** and **Contest details** respectively.
3. Since we added a column property to allow a configurable value for the number of votes permitted in the contest, let's add controls right above the form buttons and map them to the **Max Votes Allowed** column.

> **Note**
>
> Since the value of this field is optional and defaults to no limits with the value of `-1`, we do not need to update the visibility conditions on the **Submit** buttons.

4. Update the action configured on the **Submit** button to update the value of the column using this field, as shown in *Figure 12.10*:

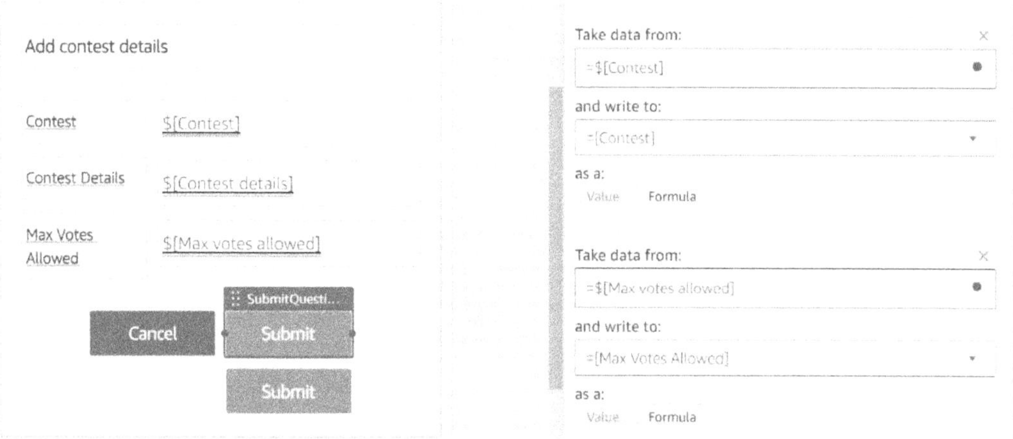

Figure 12.10 – Updating the action on the Submit button to save the Max Votes Allowed value

5. In the hidden yellow boxes, you may want to replace **Question** with **Contest**, both in the textbox as well as in the name of the data cells containing the debug variables.
6. Finally, update the hidden textbox message to **New contest created!**.

Bottom section

This section contains the list of open questions and their metadata. See *Figure 12.11*:

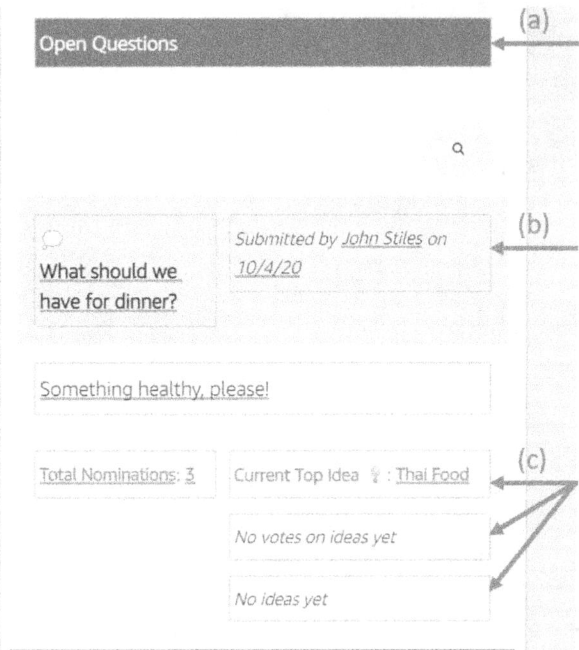

Figure 12.11 – Bottom section of the Contests screen

So, let's update them all to instead display contests and their data:

1. We have already named the screen **Contests**, so we can simply delete the textbox labeled **(a)** in *Figure 12.11*.
2. We do not need the information of who created the contest and so can delete the content box with information about who submitted the question, indicated by label **(b)** in *Figure 12.11*.
3. Lastly, we do not want to show the voting information to our users and will therefore delete the boxes labeled **(c)** in *Figure 12.11*.

The updated screen in the builder view is shown in *Figure 12.12*:

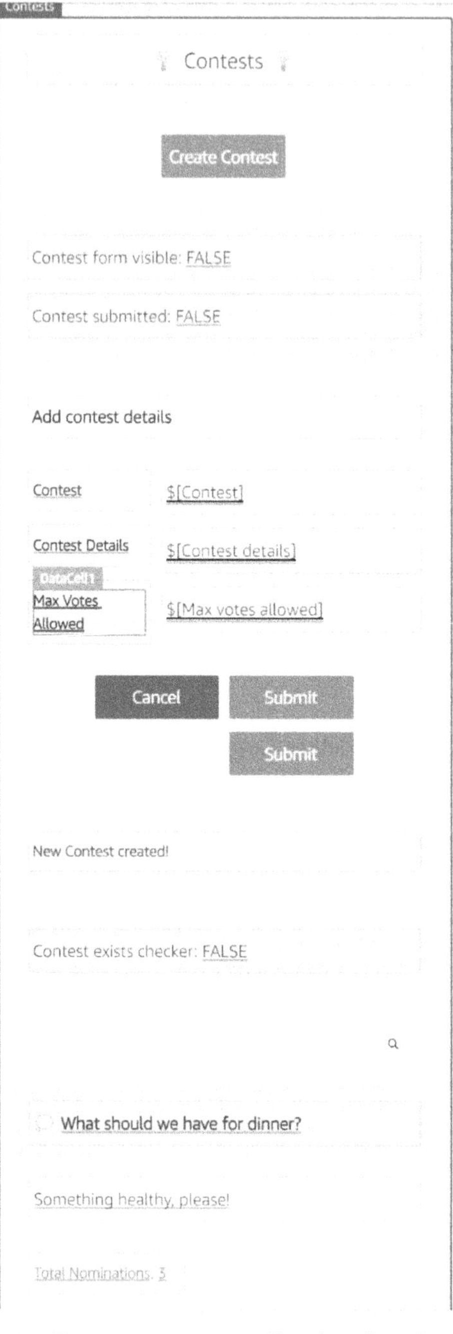

Figure 12.12 – Contests screen once all updates have been applied

> **Note**
>
> The template created the form and the list on the same screen. Given there were only a few fields, it may feel like creating a new screen was a waste. I will, however, encourage you to create a separate form screen for any input as it reduces the need to create hidden blocks on the screen and also separates concerns – read versus write – making the application more manageable.

Answer a Question

This is a hidden details screen that is accessed by navigating from a question, and is overloaded with additional functionalities. The screen provides details of the question asked, an option to submit an idea (an answer to the question), update the question's status, and list all the ideas submitted so far. Similar to the **Contests** screen, this screen also serves many purposes and can be grouped into four parts:

- A top section, which displays the details of the question
- A second section, which is a form for submitting an idea for the question
- A third section, which is a form for selecting the idea when closing the question
- A bottom section, which is a list of ideas submitted for the question

Now, recall that we mapped **Ideas** to **Nominations** when updating our data model and, by corollary, this will be our screen for listing all the nominations for the selected contest, adding a new nomination, and updating the contest status. So, let's start updating by renaming the screen **Contest details**. Next, we will update each screen section.

Top section

This section consists of a hidden block with a yellow background containing the variables for debugging and/or easy reference, along with a visible section displaying the details of the question. There are two sets of buttons with conditional visibility:

- One for adding an idea
- One for updating the status:

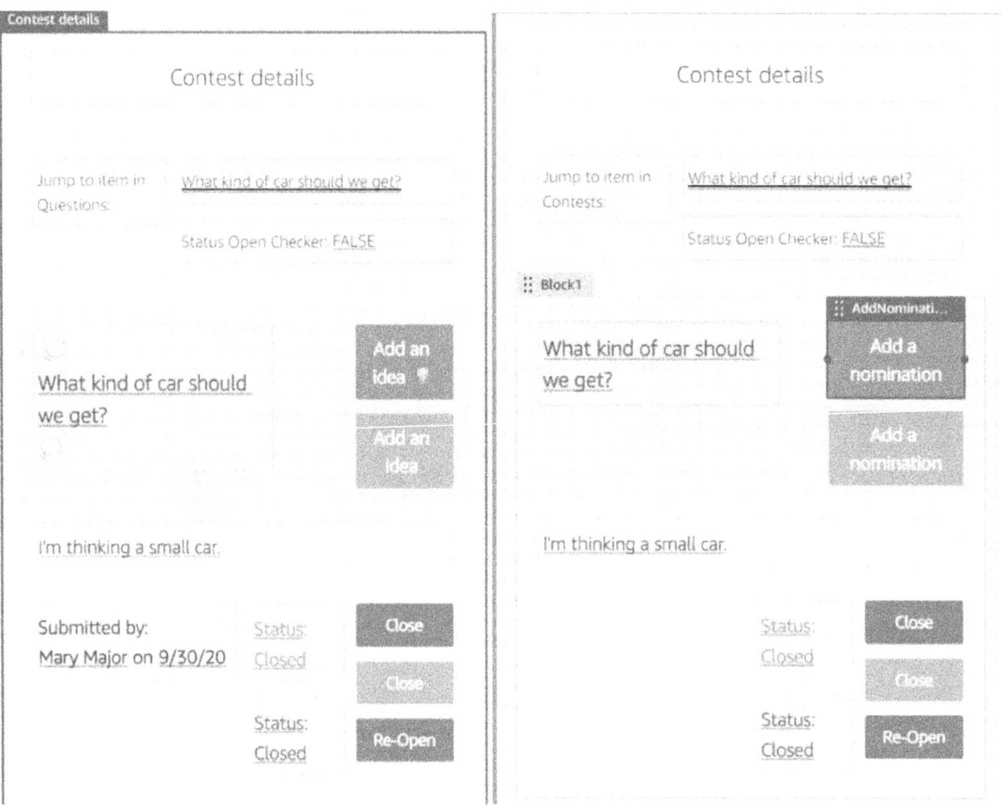

Figure 12.13 – Top section of the Contest details screen
(prior to editing on the left and after updating on the right)

Most of the information here is relevant for the contest's details, so we only need to make a couple of changes to arrive at the screen shown on the right of *Figure 12.13*:

1. Rename the button text from **Add an idea** to **Add a nomination**.
2. It is typically not important who created the contest, so we will delete the content box.
3. Lastly, update the text in the hidden block from **Jump to item in Questions** to **Jump to item in Contests.**

Second section

This section consists of the form for getting user input for adding a new idea and debugging variables contained inside the hidden block with a yellow background, as shown in *Figure 12.14*:

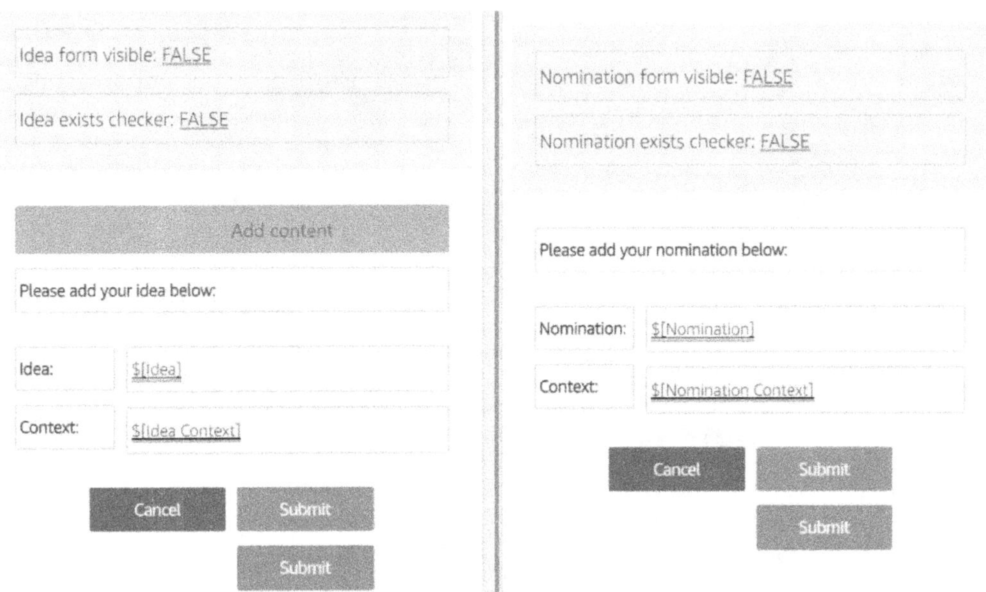

Figure 12.14 – Second section of the Contest details screen (prior to editing on the left and after updating on the right)

Let's update this section with the following steps to arrive at the screen shown on the right of *Figure 12.14*:

1. Let's start by replacing **Idea** with **Nomination** across this section, including the variables declared in the form.
2. The button with the gray background and text, **Add content**, is added at this point to create a visual separation when displaying the form in the app and can be removed.

Third section

This section consists of the form for closing the question by choosing an idea from the submissions and a hidden block with a yellow background containing the debugging variable, as shown in *Figure 12.15*:

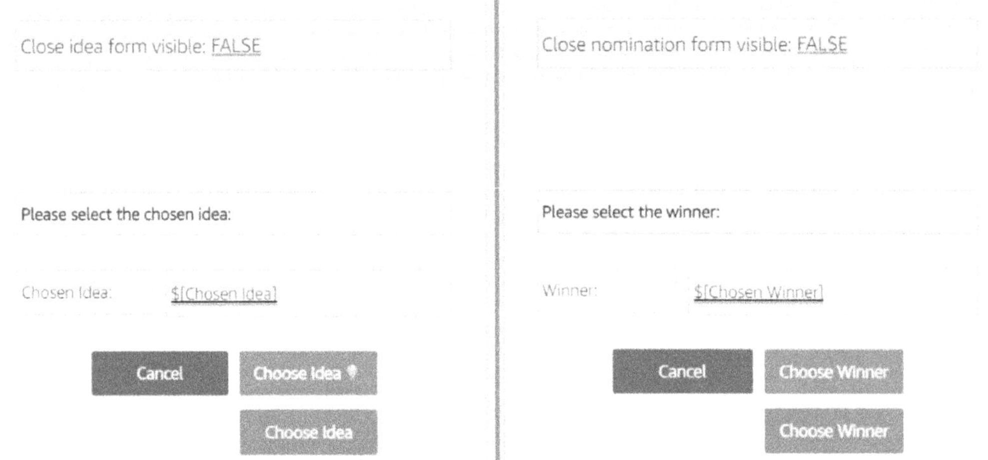

Figure 12.15 – Third section of the Contest details screen
(prior to editing on the left and after updating on the right)

Next, we will update this section with the following steps to arrive at the screen shown on the right of *Figure 12.15*:

1. Let's start by updating the hidden textbox and replacing **Idea** with **Nomination**.
2. Next, replace the **Chosen Idea** text with **Winner**. Also, update the name of the corresponding variable.
3. Finally, update the text on the button from **Choose Idea** to **Choose Winner**.

Bottom section

This section consists of two lists of submitted ideas with their metadata. The first list is visible for the open questions and includes the buttons to vote or unvote for the idea, while the second list is for closed questions and displays the selected idea. See *Figure 12.16*:

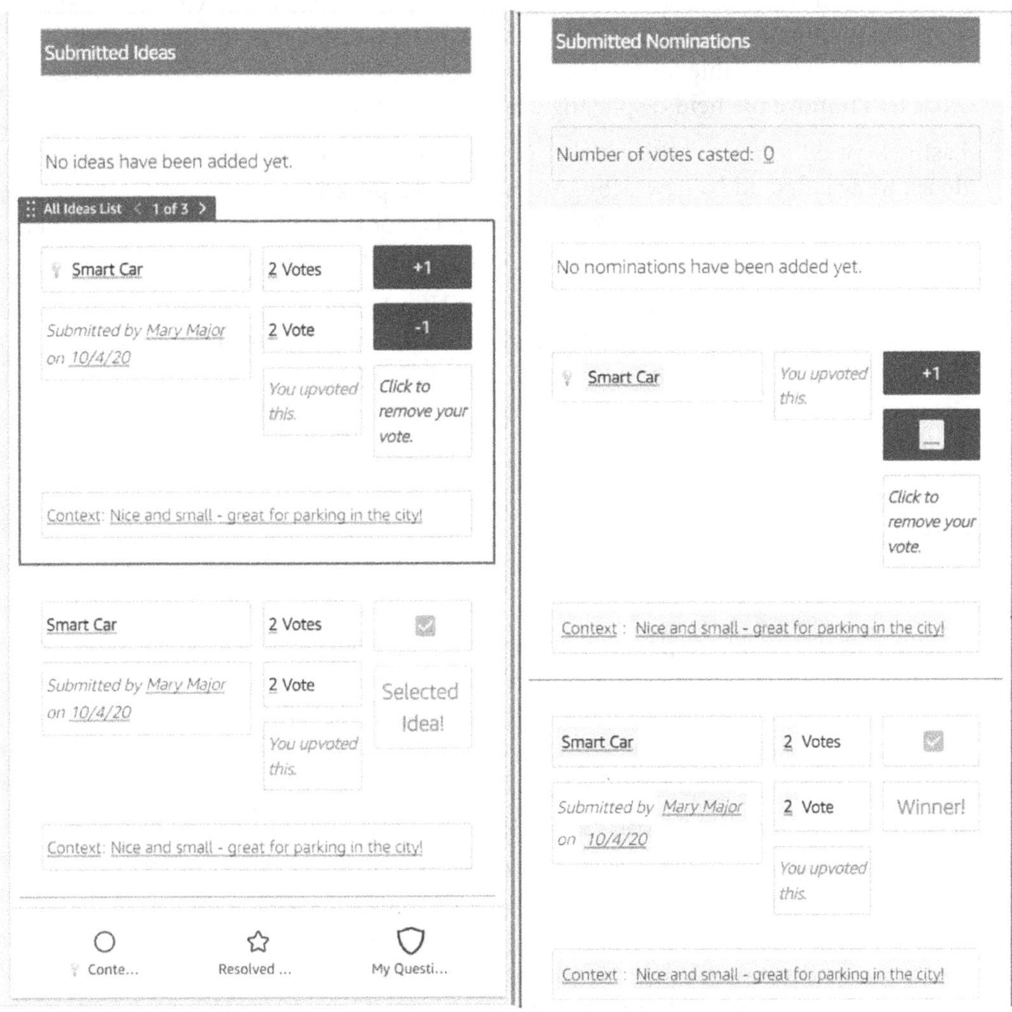

Figure 12.16 – Bottom section of the Contest details screen (prior to editing on the left and after updating on the right)

Let's update and customize them to display nominations and their data as per our requirements to arrive at the screen on the right in *Figure 12.16*:

1. Let's start by replacing **Ideas** with **Nominations** across this section.
2. Update the **Selected Idea** text to **Winner**.
3. As listed in the requirements, we do not want to display how many votes are cast against a nomination, so we will delete the two boxes displaying the number of votes from the list visible for open contests, but not from the list for closed contests. Also, let's remove the field displaying who submitted the nomination.
4. Lastly, we want to restrict the number of votes as per the contest configuration. To do so, we will first add a new debug variable for keeping a count of the number of votes cast so far. Add a new block and set its fill color to be the same as that of other hidden blocks on the screen, but do not set its visibility to `FALSE` yet.
5. Add a context box to the block, set the text as **Number of votes casted**, and also add a variable named **VotesCasted**. Set the initial value of this variable to `=Rows(FILTER(Upvotes,"Upvotes[Voter]=% AND Upvotes[Contest]=%",$[SYS_USER], $[InputRow]))`, as shown in *Figure 12.17*:

Figure 12.17 – Adding a debug variable for tracking the number of votes cast by the app user

6. Next, update the visibility condition on the **AddVoteButton** button by adding a condition to show it only when the number of votes cast is not equal to **Max Votes Allowed**. The updated visibility condition is `=AND(FILTER(Upvotes,"Upvotes[Voter]=% AND Upvotes[Nomination]=%",$[SYS_USER],[Nomination])=0, $[VotesCasted] <> $[InputRow][Max Votes Allowed])`.

7. Before we move forward, let's test our update.
8. Create a new contest using our app, and set **Max Votes Allowed** to `1`:

    I. Add a couple of nominations for the contest, as shown in *Figure 12.18*:

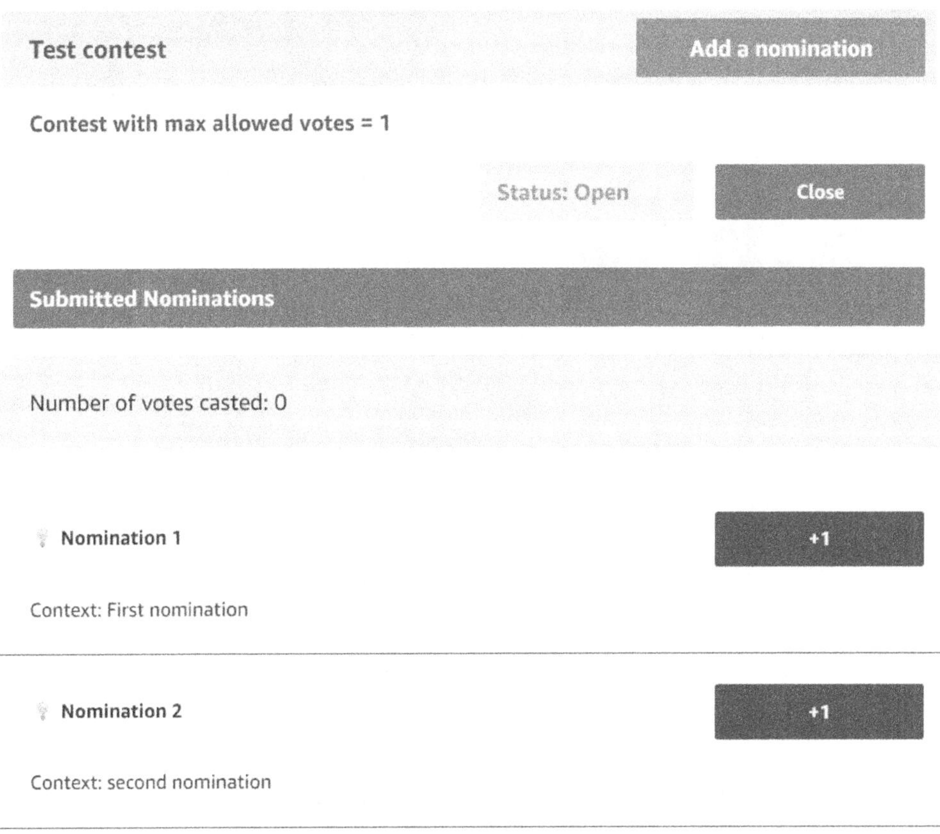

Figure 12.18 – Testing setup for the maximum votes allowed

II. Upvote one of the nominations. Verify that the upvote button against the other nomination is no longer visible and that **Number of votes casted** has changed to **1**. See *Figure 12.19* for reference:

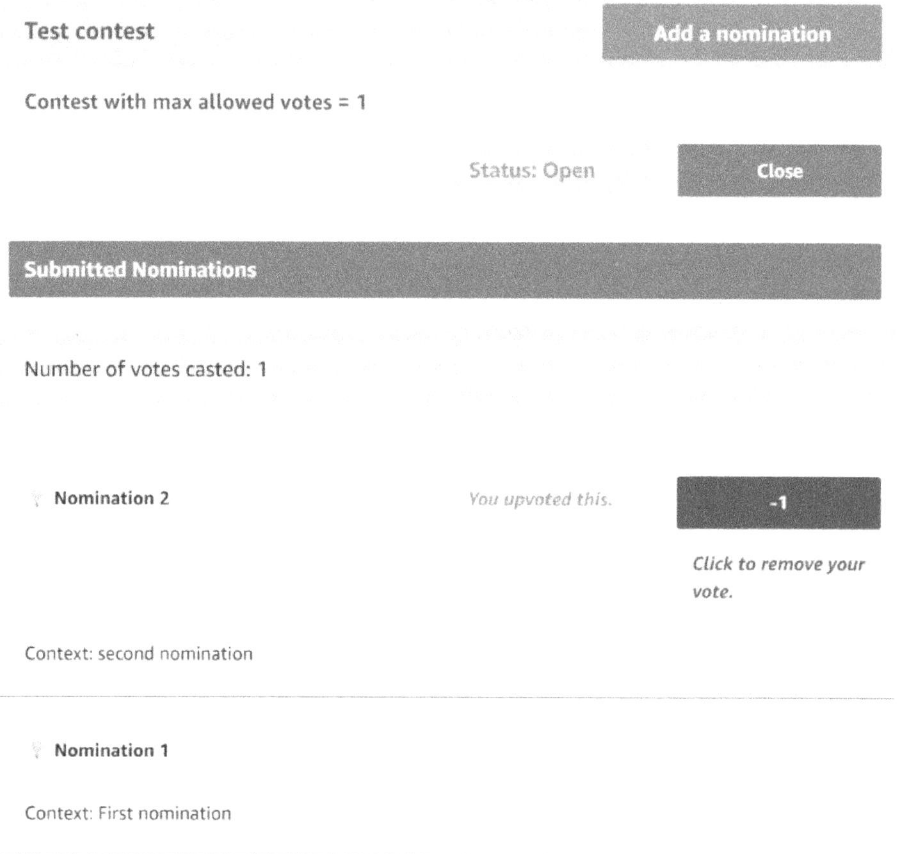

Figure 12.19 – Verifying upvote buttons are hidden after the maximum number of votes have been cast

9. Set the visibility condition of the block added in *step 4* previously as `=FALSE`.
10. Finally, we no longer need to sort the first nomination list according to the number of votes as the votes themselves are not shown. So, we update the source field of the list by removing the `Order by` clause to leave us with the following updated formula: `=FILTER(Nominations,"Nominations[Submitted for Contest]=%",$[InputRow][Contest])`.

> **Note**
>
> As previously noted in this section, the screen is overloaded with multiple functionalities that should ideally be spread across three separate screens:
>
> 1. Contest details: Displaying details of the contest and the nominations
> 2. Add a nomination: Containing the form for adding a nomination
> 3. Close the contest: Containing the form to choose the winner and close the contest.

Resolved questions

This is the screen for listing down the questions that are closed. Among other details, it also shows the selected idea for the question. Depending upon your requirements, you may not need this screen at all and can simply delete it or repurpose it for displaying past contests, as shown in *Figure 12.20*:

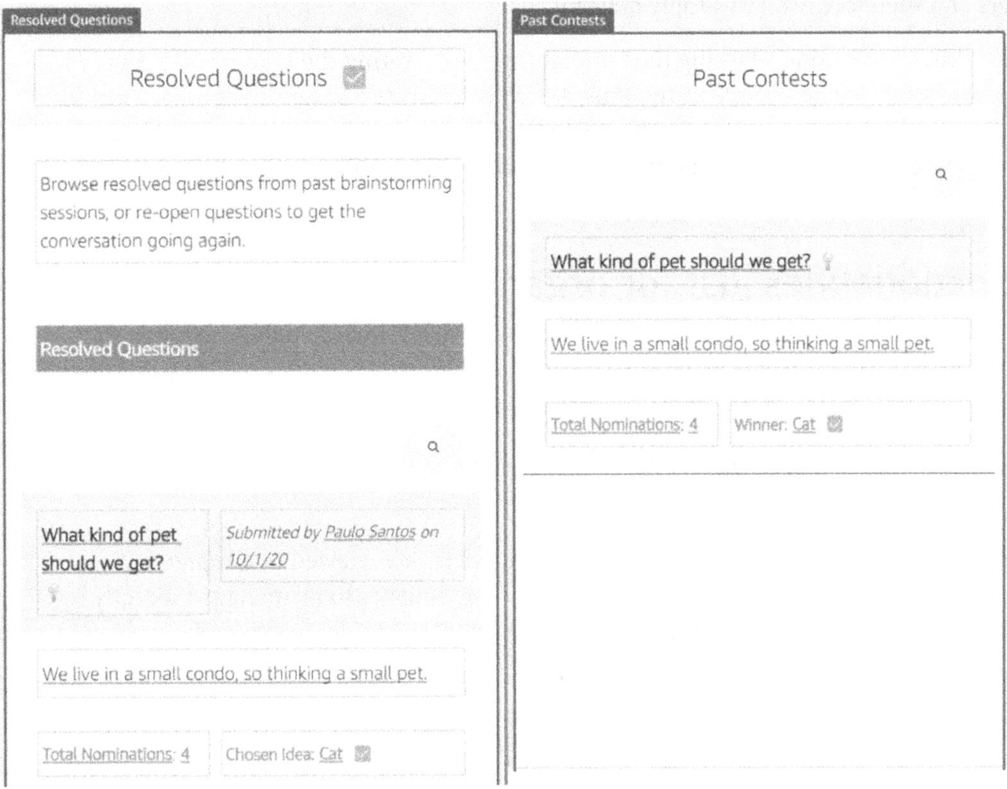

Figure 12.20 – The Resolved Questions screen on the left and Past Contests on the right

However, if you intend to keep it, let's make quick updates to convert it for our Awards app:

1. Let's start by renaming the screen **Past Contests**.
2. We do not need the content box at the top with general instructions and the green bar with the **Resolved Questions** text, so let's delete them.
3. The **Submitted by** information is not relevant to the contest, so we will delete that too.
4. Finally, update the **Chosen Idea** text with **Winner**.

My questions

This is the screen customized for listing down the questions that were created by each user. However, listing contests created by each user does not make much of a use case, especially since we want to restrict the rights to create contests to just a small subset of users and therefore, we can simply delete it.

With this, we are done adapting the template app for creating the **Honeycode Awards** app. However, one requirement remains, which is to create a separate set of features for organizers, including permissions to create the contest as well as view the number of votes against nominations for an ongoing contest and suchlike. This will be addressed in the next section.

Functionalities just for organizers

Typically, for any organization or group hosting contests, there is a fixed set of folks organizing the event and therefore, as also listed in the requirements, we do not want everyone to be able to create contests. We have three options here:

1. Build our app such that it allows new contests to be created only by the organizers.
2. Build a separate app for the organizers.
3. Similar to the event management template that we reviewed in *Chapter 9*, *Event Management Template*, we can leave this functionality to be managed directly by updating the relevant tables.

Now, you may have a team of organizers, but not everyone needs to have access to the data model or the app builder. However, if we go with the third option, we will need to provide them with access to the workbook, which is not desirable. Therefore, the recommended approach will be to create another app or build the app such that it allows additional controls for the organizers. In either case, the controls and objects we'll use in the app screen for creating the contest will largely be the same, with the major difference being the means of access control. When you have a separate organizer app, you can share the organizer app only with the organizing team. However, if we build the feature in a single app, we will need a way to identify the organizers.

Given that creating a separate app and sharing it with a smaller group is a simpler option, let's look into how we can separate the functionality in the same app.

Organizers table

In order to be able to customize the app to a set of users (organizers), we need to be able to identify them as such. To do so, we need to have a defined collection of them, and we can achieve that by having a separate table for this group and adding all the members to it:

1. Create a blank new table and name it **Organizers**.
2. Rename **Column1 Organizer**, and delete the other two columns.
3. Set the **Organizer** column's format as **Contacts**.
4. Add your organizers to this group, add yourself if the others are still not finalized so as to be able to test the functionality.

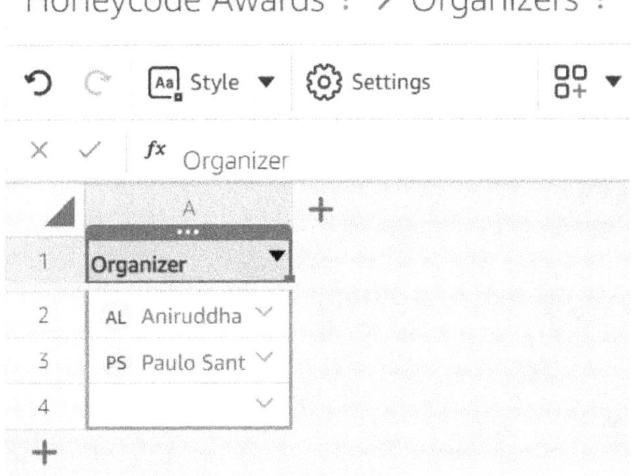

Figure 12.21 – Organizers table

Updating the app

Next, let's update the app to only allow organizers to create a new contest as well as be able to update the contest's status. We will do so by updating the visibility conditions of the controls for these functionalities:

1. Open the **Contests** screen of the app in the **Builder** view.
2. Update the visibility condition for **CreateContestButton** to check whether the app user is part of the organizer's group using the following formula: `=AND( NOT($[ContestFormVisible]), $[SYS_USER] IN Organizers[Organizer])`.
3. We will next add a similar check on the status change buttons in the **Contest details** screen. Select the **Close** button and update the visibility condition to `=AND($[IsOpen]=TRUE,$[CloseNominationVisible]=FALSE, $[NominationFormVisible]=FALSE, $[SYS_USER] IN Organizers[Organizer])`.
4. Similarly, select the **Re-Open** button and update the visibility condition to `=AND(NOT($[IsOpen]), $[SYS_USER] IN Organizers[Organizer])`.

And that's it. Remove yourself from the **Organizers** table and observe that you will no longer have the option to create a contest or update its status.

You can similarly update the visibility conditions on the number of votes for each nomination instead of deleting them while creating the **Contest details** screen in the earlier sections.

> **How about Creating a Separate View for the Organizers?**
>
> Creating a separate screen for the organizers can be a decent solution except that Honeycode does not support visibility conditions at the global level, which means we cannot hide the navigation menu based on the app user. You may choose to create such a view, which will be an empty screen for anyone other than the organizers.

Restricting voting to a panel of judges

We had another optional requirement to enable judging by only a selected panel, and that will have a similar choice and implementation as we had for the organizers. One thing we already have for enabling such a setup is that organizers can simply set **Max Allowed Votes** to **0**, and that will disable the button to upvote for all users. Then, you can simply enable it for a specific set of users (judges) within the same app or, if you have a separate app, then no update is required.

> **Exercise 1**
>
> Add the feature to allow votes only by a panel of judges.

Summary

In this chapter, we built another app for a relatively common use case in organizations. We achieved our goal not by building from scratch, but by making use of an existing template to speed up our development time.

While doing so, we learned how we can adapt an existing template to fit our requirements and then add more features on top of it to further enhance the experience and functionality. Finally, we also reviewed the trade-offs on how a functionality can be implemented in different ways, including the creation of a separate app, and learned how it can be built into a single app.

In the current and the previous chapter, our examples were more focused on app creation and the app builder, with limited use of automation. In the next chapter, we will build an app to conduct periodic business reviews and see how automation can help power our use cases.

13
Conducting Periodic Business Reviews Using Honeycode

In most businesses, there are several periodic review meetings at different levels of the organization, and they all share a common structure in terms of the data being reported and reviewed. Typically, this data is collaboratively generated and often shared and updated over emails, requiring someone to collect all threads and compile them for the final report.

In this chapter, we'll build an app that lets teams collaboratively work on a single source of truth and, therefore, always be able to see the latest information.

We'll also automate the process of sending reminders for updates as well as take a look at generating a sample report from all the updates.

In this chapter, we're going to cover the following main topics:

- Defining the app requirements
- Translating requirements to app interactions
- Defining the data model
- Building the app

Technical requirements

To follow this chapter, you'll need to have access to Amazon Honeycode, which requires a laptop with a web browser, preferably Google Chrome, and optionally, a mobile device running either Honeycode's supported version of Android (currently Android 8.0 and up) or iOS (currently iOS 11 or later).

Furthermore, we'll use the Honeycode terminology and refer to the components that we covered in *Chapter 2, Introduction to Honeycode*; *Chapter 3, Building Your First Honeycode Application*; *Chapter 4, Advanced Builder Tools in Honeycode*; and *Chapter 5, Powering the Honeycode apps with Automations.* The instructions in this chapter for various tasks are provided with the assumption that you are familiar with those operations and, therefore, I recommend you to complete the referenced chapters before we start with this chapter.

Defining the app requirements

We first introduced this section in *Chapter 3, Building Your First Honeycode Application*, where before building our app, we listed down the requirements or the use cases that we wanted the app to fulfill. We made use of that list throughout the chapter as a guide to define our data model and conceptualize the application interface and visualize the interactions between various on-screen elements as well as the data displayed. Therefore, I'd encourage you to take 5 minutes to think about what your app should do and what it should look like, and then make a list of it.

> To Do
>
> Take 5 minutes and list how you would like your app to work.

Here is what my list looks like:

1. I must be able to view all the recent updates.
2. I must be able to easily identify the recent updates and know which sections have not been updated.

3. I must be able to submit my updates under relevant sections.
4. I must be able to edit and update the submitted notes.
5. The application must be able to assign the contacts for different sections and subsections of the review.
6. The owners must receive a reminder to update their sections.
7. All the stakeholders must receive the review update once completed.

Your list might have some or all of these requirements or might have even more, and that is perfectly fine. You might find some of these missing in the app we will build and might want to extend that for yourself after you finish the chapter.

Translating requirements to app interactions

Based on the requirements we listed in the previous section, our app will have two primary views:

- A summary view for reviewing the updates in the meeting
- A personalized view for the owners to facilitate easy updates

The summary view should allow for navigation to drill into the details and, depending on the setup, this navigation can have multiple levels. In some cases, you may even want to build an alternate view with different navigation for the same set of data.

For example, while the summary view can be created to track and review updates on organization goals or OKRs, the other view could be created for reviewing the same information in the context of different teams due to a many-to-many relation between the organizational goals and team deliverables. It is possible that a single team can deliver many goals, and many teams can deliver a single goal.

In some organizations, viewing a summary at the OKR level could become too abstract and not provide enough depth in the update or visibility. In such cases, we can consider building another view to list the OKRs and allow navigation to drill down, but for weekly reviews, we use the next level of abstraction or grouping under the OKRs.

Defining the data model

For this chapter, we will consider a tech company to be our organization with organizational goals as the top-level north star that is served via multiple epics or themes. Each of the epics will have one or more features that need to be built for delivering the parent epic towards achieving the mapped organizational goals.

Our app, therefore, has four distinct entities requiring tables for each of the following:

1. Organization goals or OKRs
2. Epics
3. Features
4. Teams

Then, we need to be able to show the state (such as not started, in progress, and done), and the status (such as on schedule and behind schedule) for each of the deliverables as well as those on the aggregate level and, thereby, require a table for listing the state and status values.

Often, we may come across the scenario where a single epic will deliver against more than one OKR and we would like to capture that in our app. The most scalable and automated way of achieving that is to maintain the OKR-epic mapping in a separate table.

Finally, we need to be able to capture the updates against each deliverable. Here, if we want to preserve the historical updates, we'll need an archive table, otherwise, we can simply have the update field against each row of the entity that will be overwritten at each review.

Building the app

Alright, now we know what our app should do, have a fair idea of what it should look like to enable the listed use cases, and how we will structure and store the data to power the app.

So, let's start building the app.

Creating a new workbook

By now you must be familiar with the process of creating a workbook. So, go ahead and create an empty workbook from **Dashboard** and name it `Weekly Business Reviews`.

Creating tables

In the previous section, we noted the eight tables that we will need. So, let's set them up.

OKRs table

An organizational **OKR** typically consists of an objective or a goal and the metric used for measuring it. Typically, the start and the target value of the metric are also provided. And last, it has a mapping to the list of items that, when delivered, will help in achieving the goal as measured by the stated metric. For the ease of presentation in the app, we could also have a short code for each of the OKRs. So, with this data in mind, let's create the table by following these steps:

1. We have the default table created with the workbook, so let's rename that `OKRs`.
2. Create the following columns in the table by renaming the existing columns and then by adding new columns:

    A. `Short code`

    B. `Objective`

    C. `Metric`

    D. `Start value`

    E. `Target value`

    F. `Epics`

3. Set the following formula on the `Epics` column: `=Filter(OKRs_Epics,"OKRs_Epics[OKR]=%",THISROW())`.
4. Fill in some of your OKRs or dummy data in the table.

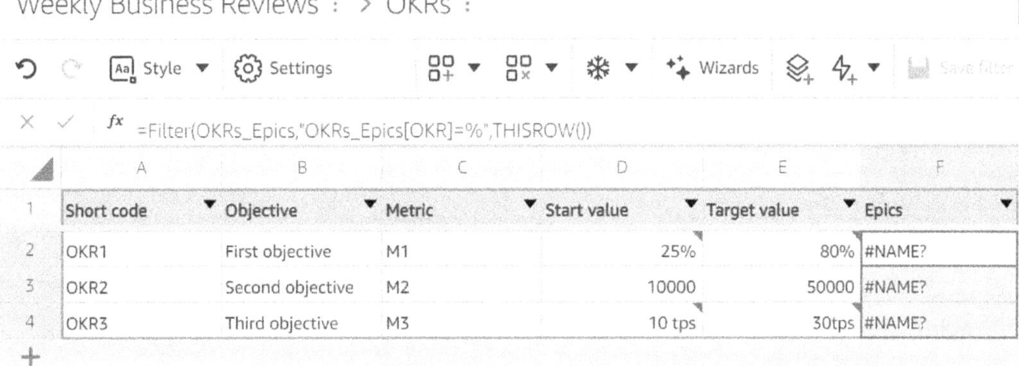

Figure 13.1 – OKRs table

Your `OKRs` table will look similar to *Figure 13.1*.

> **Do You Recall Why the Cells in the `Epics` Column Show the #NAME? Error Code and How It Gets Fixed?**
>
> Not sure? Then review the setup of the **Stores** table in *Chapter 11, Building a Shopping List App in Honeycode.*

Epics table

An epic will have a title, a short description, an expected or actual start date, and an expected or actual end date. Other information associated includes the team or the person primarily responsible for the epic's deliverability and the progress status. So, let's create this table:

1. Create a new table and rename it `Epics`.
2. Create the following columns in the table by renaming the existing default columns and then by adding new columns:

    A. `Title`

    B. `Description`

    C. `Start date`

    D. `Target completion date`

    E. `Owner`

    F. `Current state`

    G. `Status`

    H. `Features`

    I. `OKRs`

    J. `Updates`

    K. `Last updated`

3. Set the following formula on the `Features` column: `=Filter(Features,"Features[Parent Epic]=% ORDER BY Features[Status] ASC",THISROW())`.
4. Set the following formula on the `OKRs` column: `=Filter(OKRs_Epics,"OKRs_Epics[Epic]=% ",THISROW())`.
5. Set the following formula on the `Status` column: `=Findrow([Features])[Status]`.

> **Define Status for an Epic**
>
> We are using the definition that the status of the epic is defined by the worst of the statuses of the features needed to deliver the epic. That is, if any of the features needed to deliver the epic is **Red**, the entire Epic is **Red**. If no **Red** is present, then the Epic is **Yellow** if any feature is **Yellow**, otherwise it will be **Green**.

> **Can You Explain the Formulas in the Features and Status Columns?**
>
> In order to understand these formulas, first, complete the next section and set up the `Status` table. Once done, you'll realize that sorting the features in ascending order of their status will bring the feature with the worst status to the top of the filtered list because of the data set up in the `Status` table. And then, the `Findrow` function will pick the topmost row from this filtered list and, therefore, derive the epic's status from the feature with the worst status.

6. Format the `Start date`, `Target completion date`, and `Last updated` columns as **Date**.
7. Format the `Owner` column as **Rowlink & Picklist** with the source set as the **Teams** table.
8. Format the `Current state` column as **Rowlink & Picklist** with the source set as the **States** table.

> **Note**
>
> While you are able to set formulas with non-existing table names in *steps 3* to *5*, you cannot set non-existing tables as the source for the picklist. Therefore, you need to have the required tables created before completing *steps 7* and *8*.

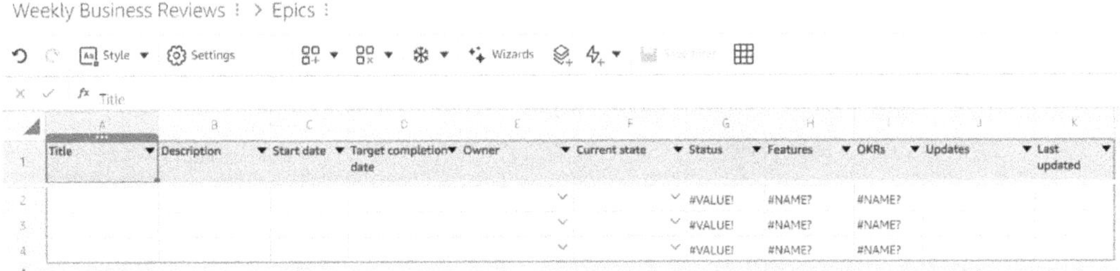

Figure 13.2 – Epics table

Since we will need the **Status**, **States**, and **Teams** tables for the **Features** table too, let's create them first to not have to park the table setup as we had to do for the `Epics` table. Your `Epics` table, upon the complete setup of the three tables and after finishing *steps 7* and *8*, will look similar to *Figure 13.2*.

Status table

The Status table, as discussed in the earlier section, is needed for maintaining a fixed list of values that can be assigned to an epic or a feature, and has a shared understanding across the organization. These values are typically custom to every organization and for this chapter, we'll use **Red**, **Yellow**, and **Green** as the status values. So, let's set up this table:

1. Create a new table and rename it `Status`.
2. Create the following columns in the table by renaming the existing default columns:

   A. `Id`

   B. `Status`

   C. `Description`

3. Fill the table with the values as shown in *Figure 13.3*:

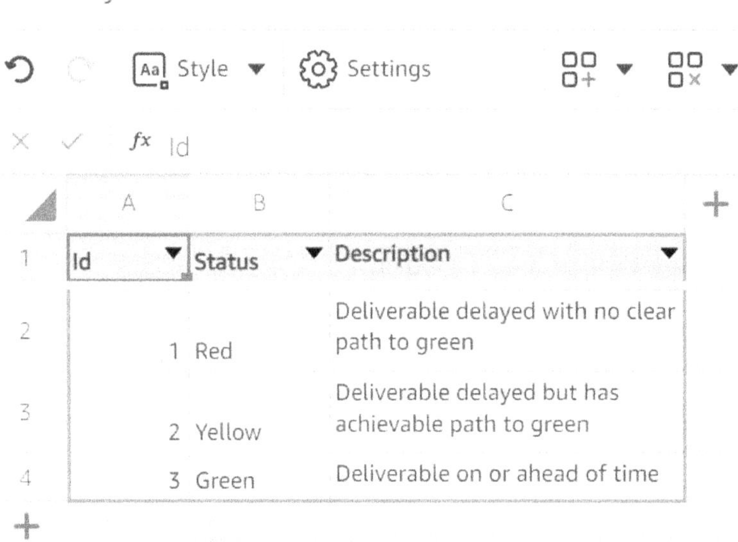

Figure 13.3 – Status table

> **Why Do We Need the Id Column?**
>
> We added the `Id` column because, for most common use cases, we would like to have the projects ordered by their status and, typically, they are either listed from `Red` to `Green` or vice versa. The `Id` field, as assigned in *Figure 13.3*, will enable such a sorting.

States table

Similar to the `Status` table, a **States** table is also needed for maintaining a fixed list of values that can be assigned to an epic or a feature, and has a shared understanding across the organization. These values are typically specific to every organization and for this chapter, we'll use `Not started`, `In progress`, `Ready for release`, and `Done` as the state values. So, let's set up this table:

1. Create a new table and rename it `States`.
2. Create the following columns in the table by renaming the two existing default columns and deleting the third column:

   A. `State`

   B. `Id`

3. Fill the table with the values, as shown in *Figure 13.4*:

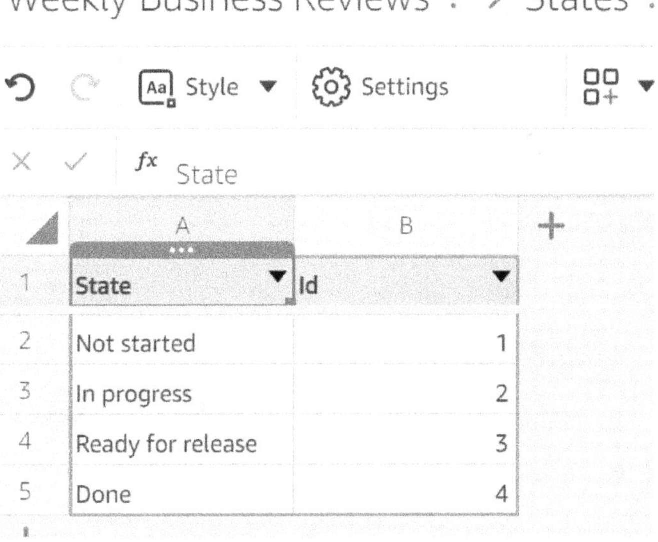

Figure 13.4 – States table

> **Note**
>
> Similar to the `Status` table, we have an `Id` column here to make the sorting experience on the UI better, where the items can be sorted in the logical order of the state they are in. But, this is less important in comparison to that of the `Status` table and is, therefore, not the first column, allowing for a simpler setup for representation on the app at a later stage.

Teams table

A **team** is typically responsible for delivering an epic or a feature, and we need to be able to show that information in our app as well. Moreover, we need to be able to provide a single point of contact for this team for any external party and, therefore, need to be able to quickly access that information. A team manager or a lead is a default choice for most cases, but in some product-driven organizations, epics can also be the responsibility of a product manager. Now, let's set up this table allowing for keeping the information of both the manager and product manager:

1. Create a new table and rename it `Teams`.
2. Create the following columns in the table by renaming the existing default columns:
   A. `Team name`
   B. `Manager`
   C. `Product Manager`
3. Format the `Manager` and `Product Manager` columns as **Contacts**.
4. Add your teams, managers, and product managers to the table.

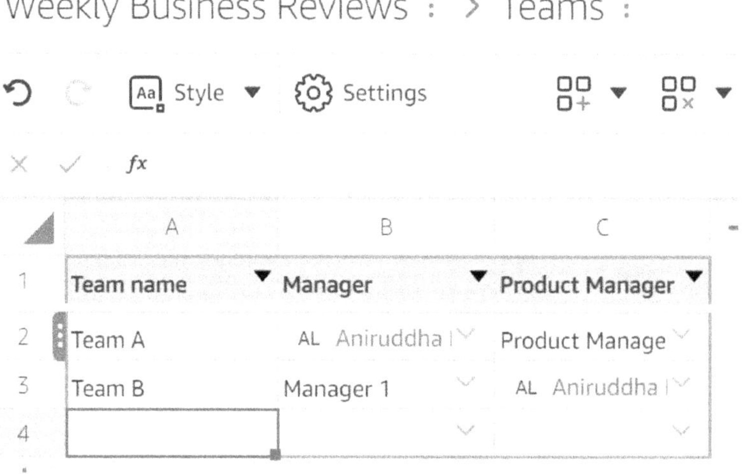

Figure 13.5 – Teams table

For the purpose of this chapter, I'm going to fill this table with some dummy data.

Features table

A **feature** is a smaller unit of work than an epic but will have nearly the same fields, such as a title, a short description, start and end date, owning team, and status. We could have, therefore, used the same table for listing epics and features by adding another column to indicate whether the row item is an epic or a feature, but it is a neater and cleaner representation to separate them since we do need to have a backlink from features to its parent epic. So, let's create this table:

1. Create a new table and name it `Features`.
2. Create the following columns in the table by renaming the existing default columns and adding new columns:

    A. `Title`

    B. `Description`

    C. `Start date`

    D. `Target completion date`

    E. `Owner`

    F. `Current state`

    G. `Status`

    H. `Parent Epic`

    I. `Updates`

    J. `Last updated`

3. Format the `Start date`, `Target completion date`, and `Last updated` columns as **Date**.
4. Format the `Parent Epic` column as **Rowlink & Picklist** with the source set as the `Epics` table.
5. Format the `Current state` column as **Rowlink & Picklist** with the source set as the `States` table.
6. Format the `Status` column as **Rowlink & Picklist** with the source set as the `Status` table.

7. Format the `Owner` column as **Rowlink & Picklist** with the source set as the `Teams` table.

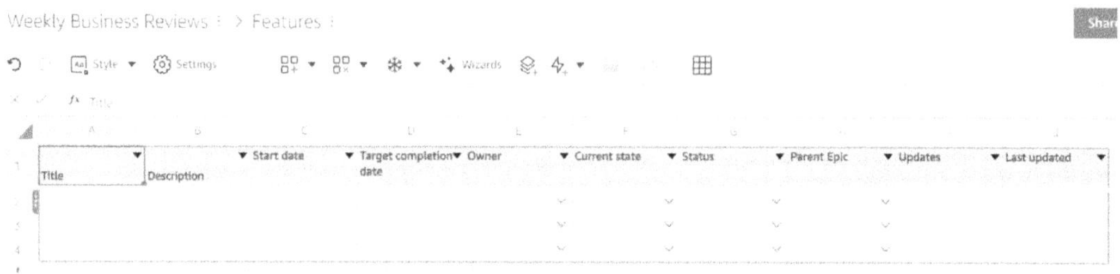

Figure 13.6 – Features table

Your `Features` table will look similar to *Figure 13.6*.

OKRs_Epics table

As discussed in the previous section, a single epic can contribute to more than one organizational goal and we do not know in advance how many goals it may contribute to. So, unlike how a feature is always linked to a single epic and, therefore, it suffices to have a single column for the backlink, we cannot follow the same approach for epics to OKR mappings. You might say that, theoretically, an epic is unlikely to contribute to more than three OKRs, and therefore, we can create three columns in the `Epics` table for linking the OKRs. A more scalable solution is to create a mapping table by following these steps:

1. Create a new table and rename it `OKRs_Epics`.
2. Create the following columns in the table by renaming the existing default columns and then deleting the last column:

   A. `OKR`

   B. `Epic`

3. Format the `Epic` column as **Rowlink & Picklist** with the source set as the `Epics` table.
4. Format the `OKR` column as **Rowlink & Picklist** with the source set as the `OKRs` table.

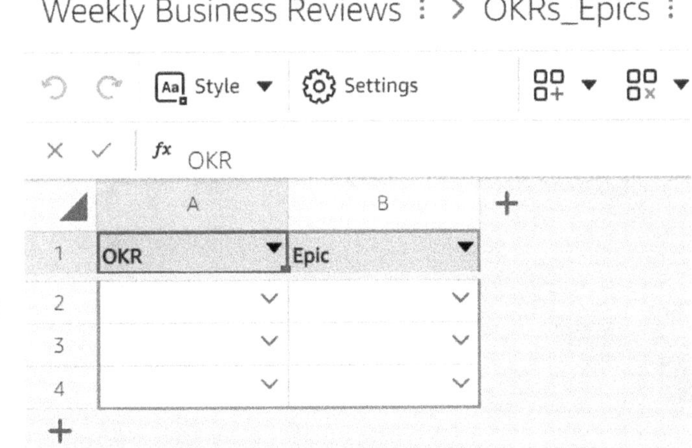

Figure 13.7 – OKRs_Epics table

Your `OKRs_Epics` table will look similar to *Figure 13.7*.

Past_Updates table

Earlier, we mentioned that some organizations might want to maintain the historical updates and statuses of various projects for later analyses of trends or to derive learnings for future planning. To achieve this, they may occasionally need to reference historical updates for a feature or an epic, requiring us to preserve a copy of the status and updates after each review. So, let's also set up a table for this:

1. Create a new table and rename it `Past_Updates`.
2. Create the following columns in the table by renaming the existing default columns and adding new columns:

   A. `Review date`

   B. `Epic`

   C. `Feature`

   D. `Update`

   E. `Status`

3. Format the `Review date` column as **Date**.
4. Format the `Epic` column as **Rowlink & Picklist** with the source set as the `Epics` table.
5. Format the `Feature` column as **Rowlink & Picklist** with the source set as the `Features` table.

6. Format the `Status` column as **Rowlink & Picklist** with the source set as the `Status` table.

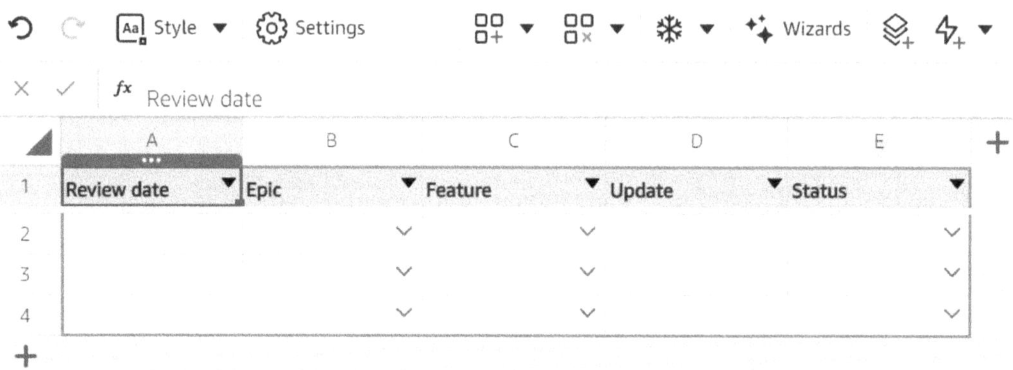

Figure 13.8 – Past_Updates table

If you want to capture any other information that can change between reviews, such as `State` and `Target completion date`, you can also add columns for those values.

With that, we now have our seven tables set along with an additional one for building the feature to save the update history. We are now ready to build our app.

Creating the app

In the previous sections of this chapter, we have covered what use cases our app should support and then discussed what views, screens, and interactions we will need in our app to enable them. We then set up our data model to support that. Now, let's follow the steps to build our app. We will be jump-starting the process using the app wizard:

1. Click on the + icon to add a new app, and on the popup, select the **Use App Wizard** tile.
2. Choose the `Epics` table as the source and change the layout to **Stacked** under the **SETTINGS** tab. We'll update the default stacked layout representation later for a better visual experience.
3. Remove fields such as `Description`, `Owner`, and `Start date` that do not need to be present in the summary view on the home screen and add fields for `Updates` and `Status` using the **+ Add column** button.
4. Change the list screen name from **Epics** (default value) to `Review`.

5. Click **Next** to create the detail screen. Make sure to mark the fields such as `Current State`, `Updates`, and `Target completion date` that will be updated for reviews as editable. You can always change editability later on in the builder after exiting the wizard but it is simpler to do it here.
6. Click **Next** to create the epic form screen, and remove the fields such as `Updates` and `Last updated` that are not needed to be filled in when creating a new epic. Also, rename the screen as `Add Epic`.
7. Click **Done** to load the next screen, as shown in *Figure 13.9*. Rename the app from **Epics** (default name) to `WBR`.

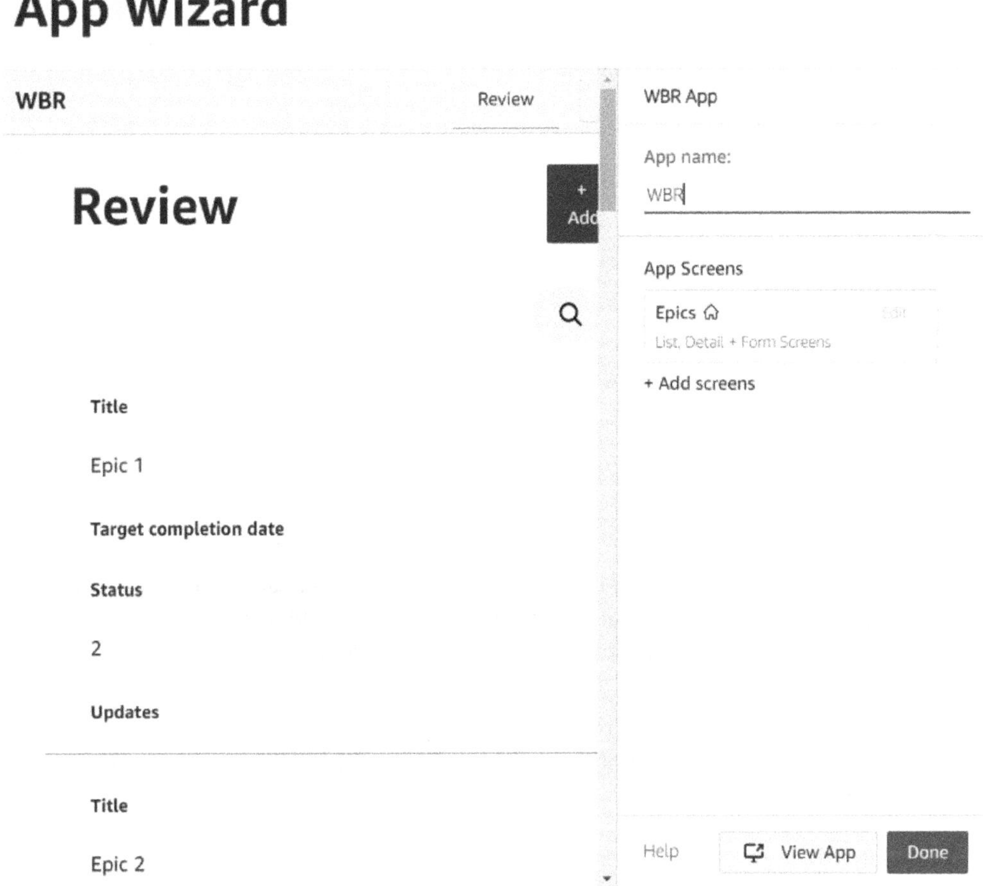

Figure 13.9 – Wizard after creating screens for the Epics table

8. Next, we click on **+Add screens** and repeat the preceding steps to create list, details, and form screens for the `Features` and `OKRs` tables.

9. Click **Done** to exit the wizard to load the newly created app in the builder view, as shown in *Figure 13.10*:

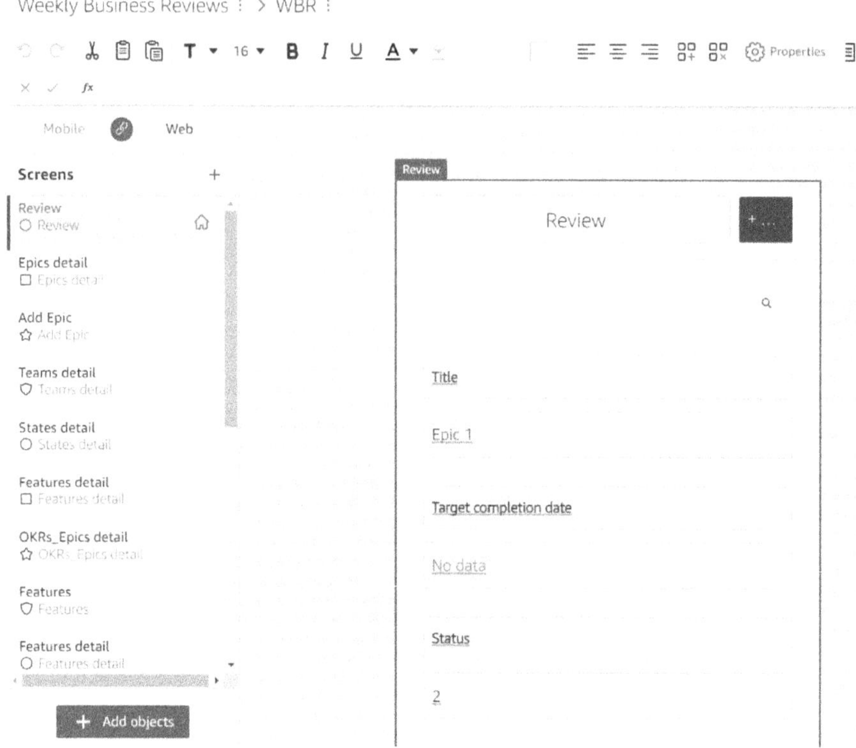

Figure 13.10 – WBR app in builder view after exiting the app wizard

> **Note**
>
> Along with the nine screens we created with the wizard (three each for the three tables – `Epics`, `Features`, and `OKRs`), there are additional screens created by the wizard such as **States details** and **Status details**.

Now that we have the basic screens created, let's customize them to fit our needs. But, before that, let's reduce our list of screens by deleting the following screens that are not required:

- **Teams detail**
- **States detail**
- **Status detail**
- **OKRs_Epics detail**

> **Note**
> One of the side effects of using a wizard is that there may be more than one screen with the previous names; delete them all.

Now that we have trimmed down our screens to the main screens, let's start customizing them.

Review screen

For a review, the most important pieces of information are the status of the projects (epics), their due date (or target completion date) for determining the criticalness of the status, and the updates since the last review. Furthermore, an organization may be more inclined to review the updates for deliverables that are with the **Red** status, followed by **Yellow**, and then **Green**, so a sorted list in that order of statuses will be preferred. Now, let's update the screen and its layout to achieve these requirements by following these steps:

1. Select the list control on the screen and update its source from `=Epics` to `=Filter(Epics, "ORDER BY Epics[Status] ASC")`.
2. Select the segment displaying the title of the epic. Style the text to bold and set the fill color to green. Under **SEGMENT PROPERTIES**, use **CONDITIONAL STYLING** to set the fill color corresponding to the status, as shown in *Figure 13.11*:

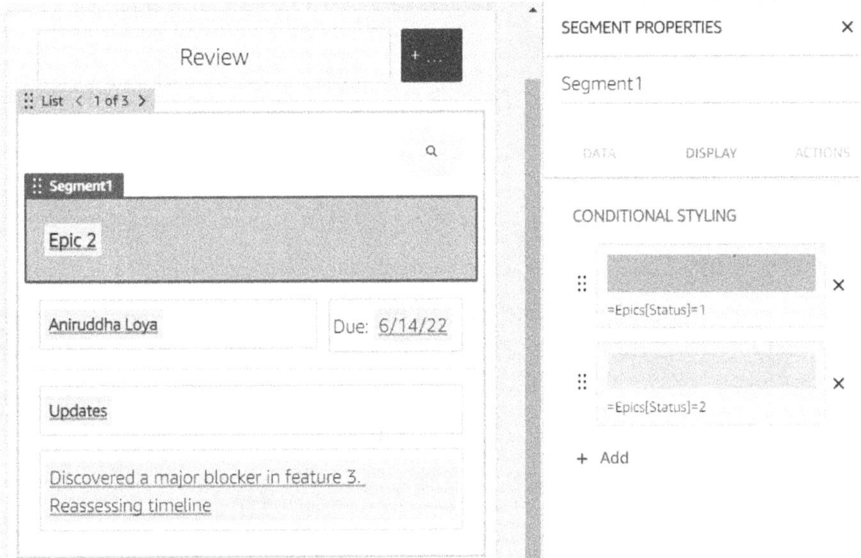

Figure 13.11 – Styled segment displaying the epic title and updated layout

I also rearranged the different content boxes as per my display preference and updated the button text from + **Add** to + **Add Epic**.

3. Next, we'll add a **data cell** outside of the list. Rename it `ShowUpdates`, set its type as **Variable** with the initial value as `=False`, and also set its visibility to `=False`.
4. Now, add a button and set its text as `Show Updates`. Set its visibility condition as `=NOT($[ShowUpdates])` and under **Quick Actions**, set the `ShowUpdates` variable's value to `=TRUE`, as shown in *Figure 13.12*:

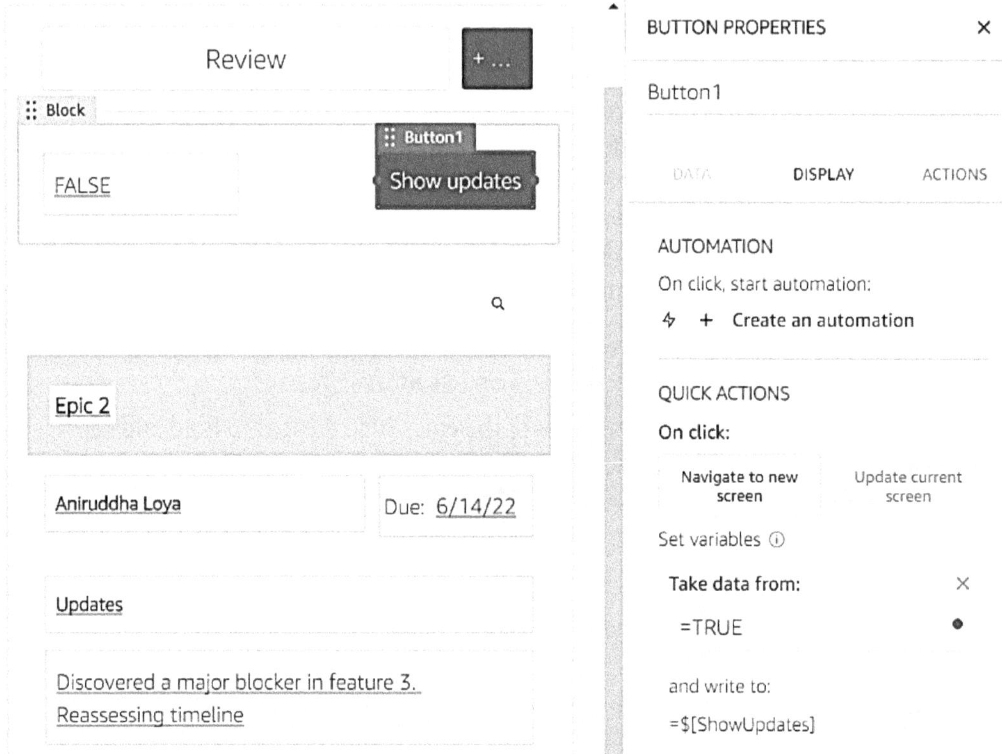

Figure 13.12 – Updating the value of the ShowUpdates variable

5. Add another button under the one created in *step 4*. Set its text as `Hide Updates`, set the visibility condition as `=$[ShowUpdates]`, and under **Quick Actions**, set the `ShowUpdates` variable's value to `FALSE`.

6. Finally, set the visibility condition on the three content boxes displaying the updates and due date as `=$[ShowUpdates]`.

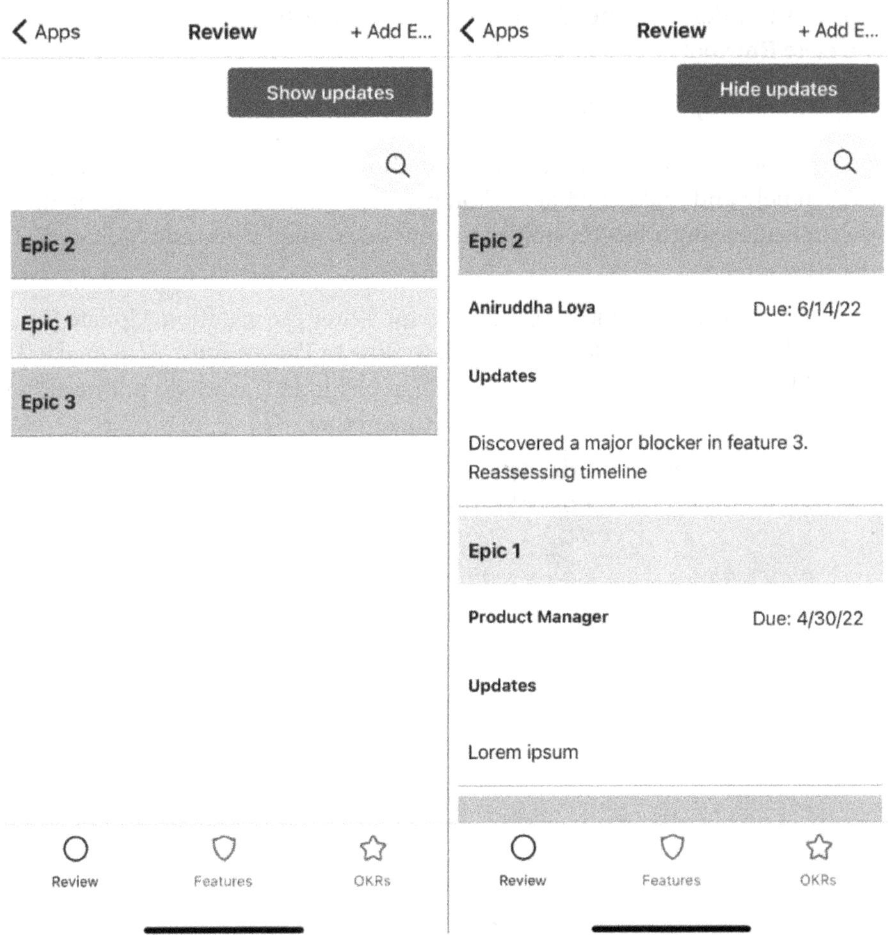

Figure 13.13 – Screenshots of the completed review screen

Figure 13.13 shows the screenshots of the completed review screen.

Epics detail screen

On this screen, we would like to have three main functions enabled:

1. The ability to update details of an epic
2. The ability to review all the features under this epic
3. The ability to add new features to the epic

However, similar to duplicates of deleted screens, we have a duplicate of the **Epics detail** screen too. Both display the same data, except one has a button to delete the epic and the other does not. You may decide to keep them both or reduce the duplication and maintenance effort by deleting one. Our recommendation is to delete the one that does not have a **Delete Row** button.

Now, let's update the only remaining **Epics detail** screen and enable these functions:

1. If you had not selected the editable fields on the screen in the wizard, you can now go to each field and make it editable. You may also choose to enable hiding the epic metadata fields using a button similar to how we enabled show and hide updates in the **Review** screen.
2. Next, we rearrange and remove some fields for better presentation. Update the text of the button at the bottom from **Delete Row** to **Delete Epic**. Also, make the topmost block with the `InputRow` variable to be hidden, as we do not need the functionality to jump from one detail page to another.
3. Update the shared source for the data cell displaying **Status** to `= [Status] [Status]` to display the status color instead of the ID.

Figure 13.14 – Epic detail screen's top half after completing steps 1 to 3

4. For the features listed under the epic, we'll update the list to mimic the behavior of the **Review** screen to show and hide the details of the feature as well as show the status by appropriately coloring the title bar.

> Tip
>
> You can copy and paste the entire block we created in the **Review** screen for showing and hiding updates to this screen. You can even copy the content boxes and update their data sources appropriately.

5. Lastly, we can copy the + **Add** button from the **Features** screen to here to allow creating a new feature directly from this screen. After copying, update the text to **+Feature,** and under the **Actions** tab, set the value of the **Parent Epic** variable to `=THISROW()`, as shown in *Figure 13.15*:

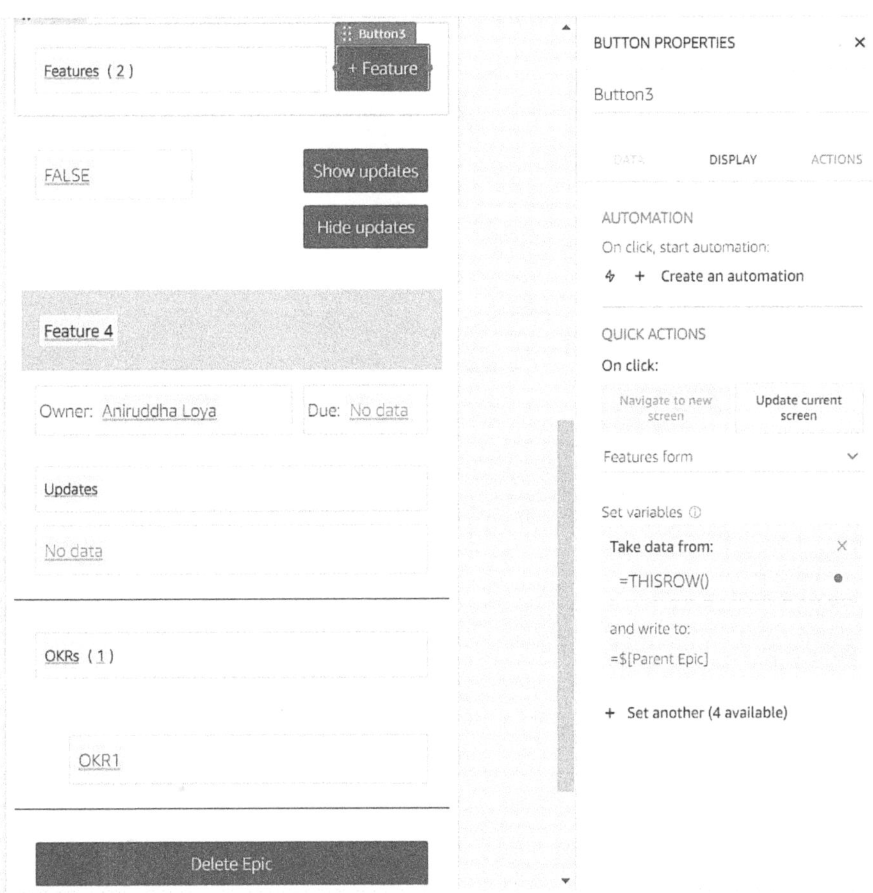

Figure 13.15 – Epic detail screen's bottom half with updates from steps 4 and 5

Our screen is now ready, however, we are missing one functionality – updating the **Last updated** date whenever there is a new update added to the epic. We achieve that by creating a workbook automation with these steps:

1. Click on the + icon to add a new automation and name it `Epic updates`.
2. Select **Column Changes** under **Automation trigger**, select the `Epics` table, and the `Updates` column.
3. Set `=NOT($[PREVIOUS] = [Updates])` as the condition to run the automation, as shown in *Figure 13.16*:

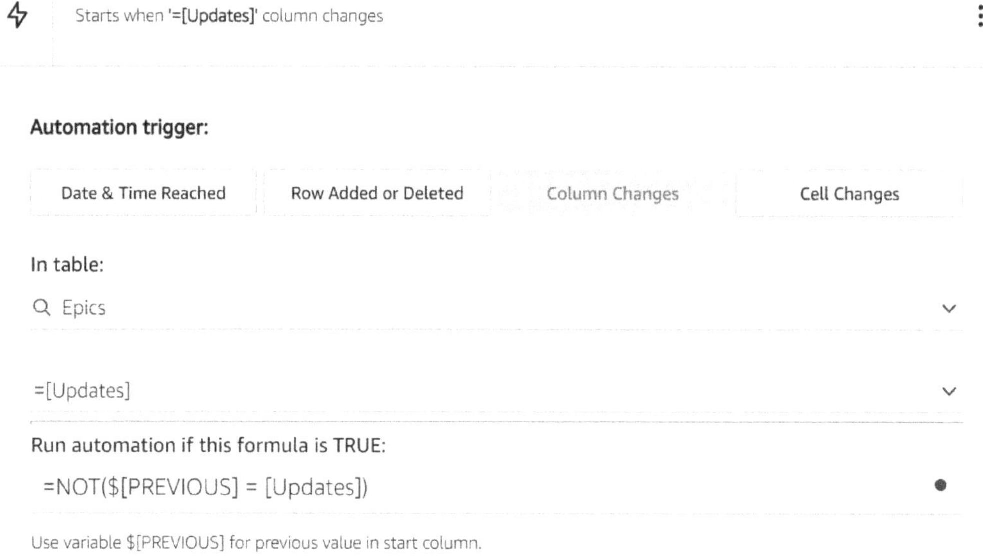

Figure 13.16 – Automation trigger block of the automation

4. Next, add the overwrite action and set the value of the **Last Updated** column to `=Today()`, as shown in *Figure 13.17*, and then publish the automation.

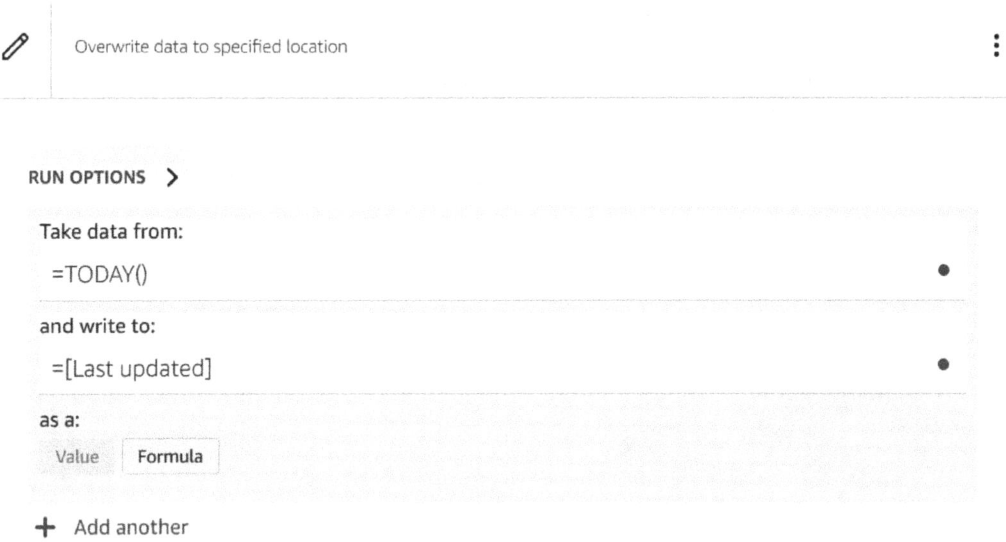

Figure 13.17 – Overwrite block of the automation

With this, our **Epics detail** screen is now complete.

Features detail screen

Similar to the **Epics detail** screen, we have a duplicate **Features detail** screen too. The recommendation will be to delete the one that does not have a **Delete Row** button but in this case, before deleting the duplicate screen, you will need to update the navigation on the features list in the **Epics detail** screen to navigate to the **Features detail** screen that we'll keep.

In terms of layouts and functions, there is a similarity to details displayed for epics and, therefore, we can keep the screens consistent in layouts. Next, you can rearrange the controls on the screen for better presentation and also update the text of the button from **Delete Row** to **Delete Feature**.

Lastly, to update the **Last updated** value for feature updates, we need automation similar to **Epic updates**, which I will leave as an exercise for you to complete.

> **Exercise 1**
> Create the automation to update the **Last updated** value for feature updates.

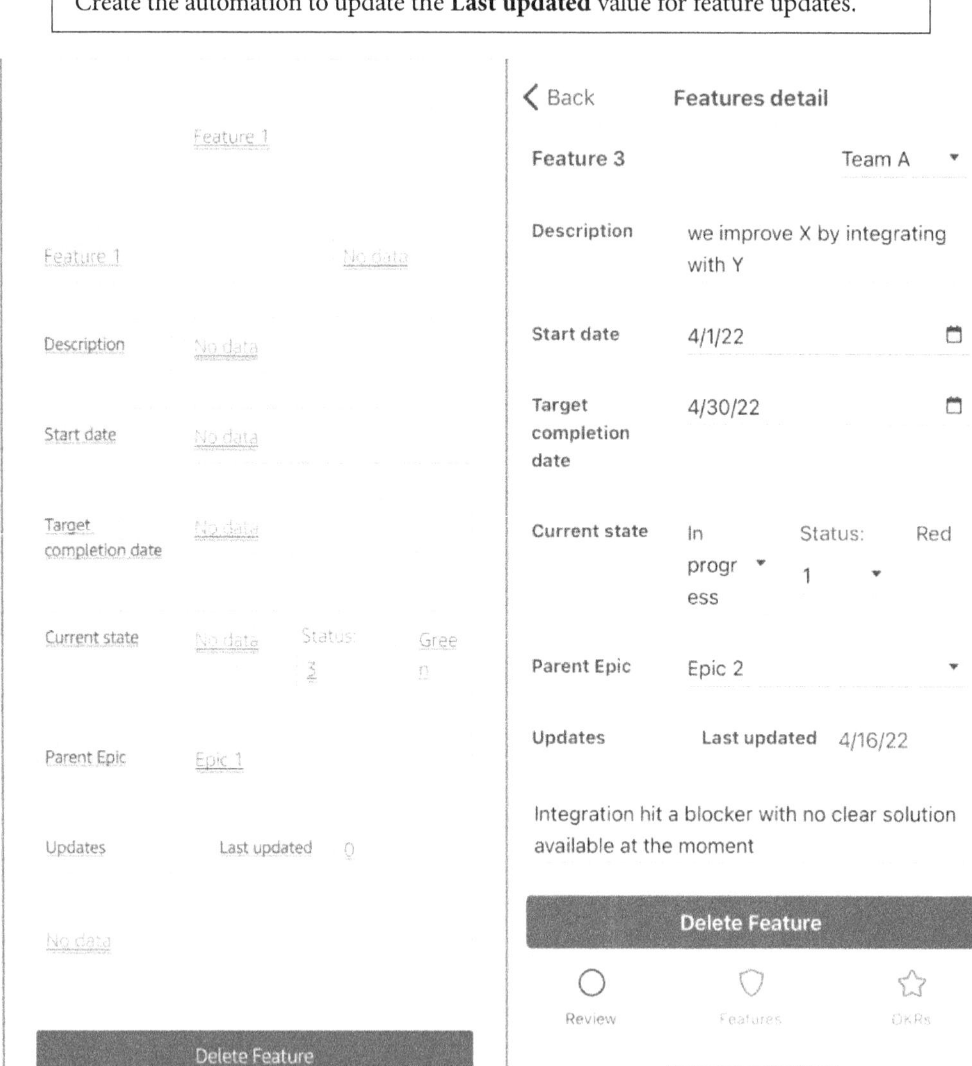

Figure 13.18 – Features detail screen in builder view on the left and a screenshot of the app on the right

Figure 13.18 shows the completed screen in builder view as well as the screenshot of the mobile app.

OKRs screen

Next, we take a look at the **OKRs** screen, which is for simply listing the organization's objectives and providing navigation to view the details about it. With some simple visual and layout changes, I have my screen looking like *Figure 13.19*:

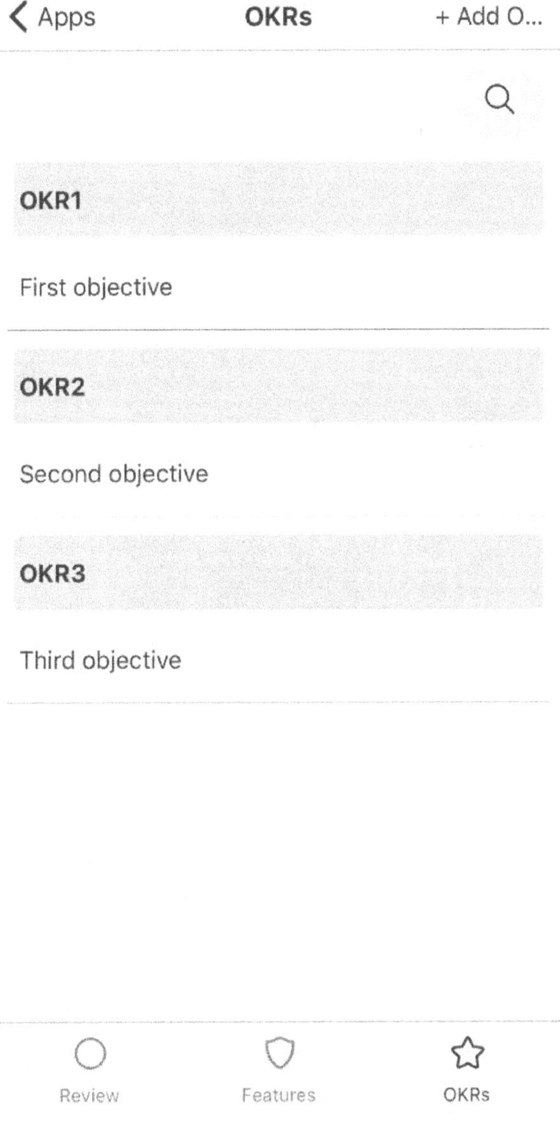

Figure 13.19 – Screenshot of the OKRs screen

OKRs detail screen

Similar to the **Epics detail** screen, this screen provides the details about the OKR and lists all the epics that are mapped to it. We will update the list of epics to display the epic title and owner and conditionally style it to display the status, as shown in *Figure 13.20*:

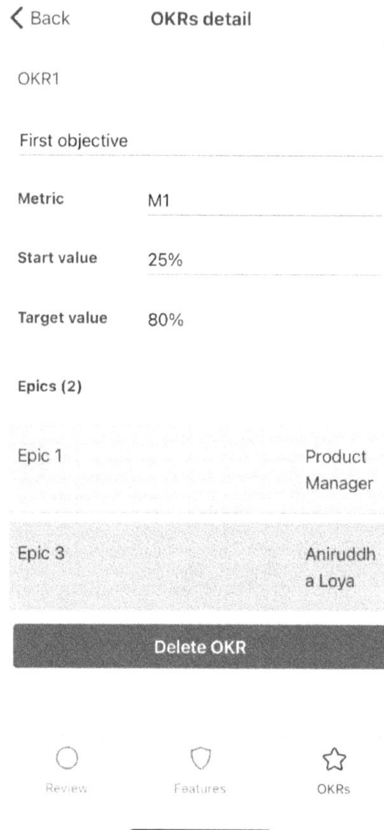

Figure 13.20 – Screenshot of the OKRs detail screen

Features screen

The **Features** screen created with the wizard lists all the features but it is of no great use in its current form simply listing all the features. We will repurpose this screen to allow the feature owners to quickly see which of their features they have updated and which they have not. And, from there, they can quickly navigate and add the updates. So, let's start with renaming the screen as `My Features`.

We will use the columnar layout for this screen, so if you had selected **Stacked layout** in the wizard, you can replace the existing list control with a new **Column list** control.

Now, let's add the functionality to see all features or see features filtered to only list the ones that the app user owns. We would also like to make that filtered view to be the default:

1. Duplicate the setup we did for showing and hiding updates in the **Review** screen.
2. Rename the data cell as `ShowAll` and update the button texts to `Show All` and `My Updates` in accordance with the action they perform.
3. Add two more data cells, name them `AllFeatures` and `Filtered`, and make them hidden.
4. Mark them as variables and set their initial value as `=FILTER(Features, " Order By Features[Last updated] ASC")` and `=FILTER(Features, "Features[Owner][Manager]=% Order By Features[Last updated] ASC",$[SYS_USER])`, respectively.
5. Update the data source on the feature list to `=IF($[ShowAll], $[AllFeatures], $[Filtered])`.

The completed `My Features` screen is shown in *Figure 13.21*:

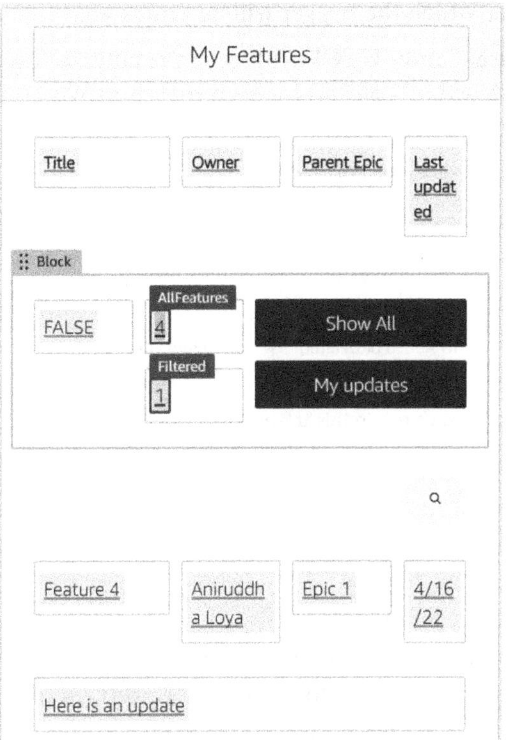

Figure 13.21 – My Features screen

With this, all our app screens are complete. However, we do have a few requirements that are incomplete, namely sending reminders for the update, sending review updates to all, and archiving the updates from the last review. Let's build those in the next sections using automation.

Sending review updates to all and archiving the updates

To be able to automatically send the updates, we need to have a trigger based on the review schedule. Since we named the app `WBR` (short for `Weekly Business Review`), we will build this automation to trigger every week by following these steps:

1. Create a table and name it `WBR_Schedule`.
2. Rename the first column as `Review date`, the second column as `Updates`, and delete the last column.
3. Create an automation and name it `WBR updates`.
4. Select **Date & Time Reached** under **Automation trigger** and choose the `WBR_Schedule` table for which the trigger will be evaluated.
5. Configure the **Date & Time (UTC)** field to set the automation to execute 1 day after the date in the **Review date** column for each row, as shown in *Figure 13.22*:

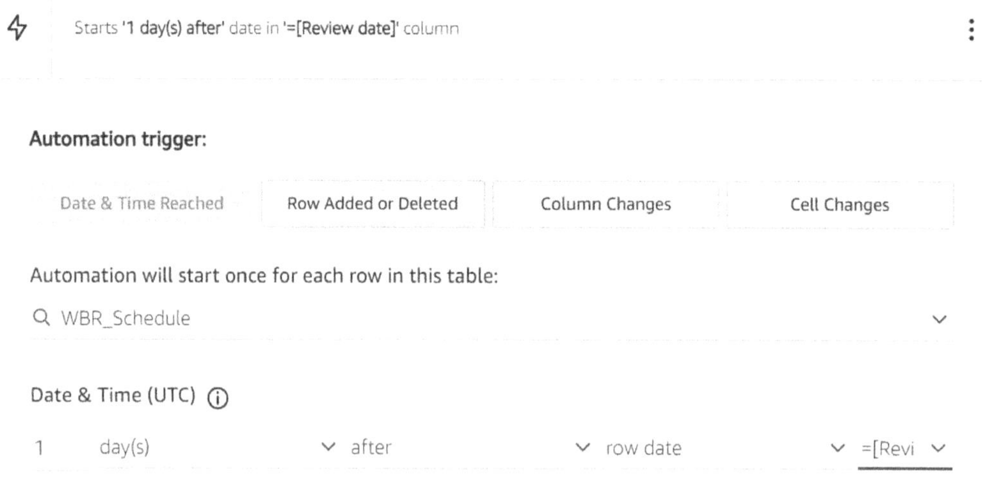

Figure 13.22 – Automation trigger configuration block

Next, we will copy all the updates to the `Past_Updates` table:

6. Add action to **Update or insert row** and select the `Past_Updates` table as the table to update or add rows to.

7. Under **RUN OPTIONS**, set `=FILTER(Epics,"7>%",DATEDIF(Epics[Last updated],TODAY(),"D"))` to select the epics that were updated in the last 7 days and, therefore, need to have their updates be copied to.
8. Since we always want to add new rows, select the matching condition to **Custom** and set it to `=FALSE`, as shown in *Figure 13.23*.
9. Finally, configure the fields to be copied from and to as follows:

Take data from:	and write to:
`=Today() - 1`	`=[Review date]`
`=Thisrow()`	`=[Epic]`
`=Epics[Updates]`	`=[Update]`
`=Epics[Status]`	`=[Status]`

Table 13.1 – Mapping for renaming column from the Questions table

The completed block for copying the updates to the `Past_Updates` table is shown in *Figure 13.23*:

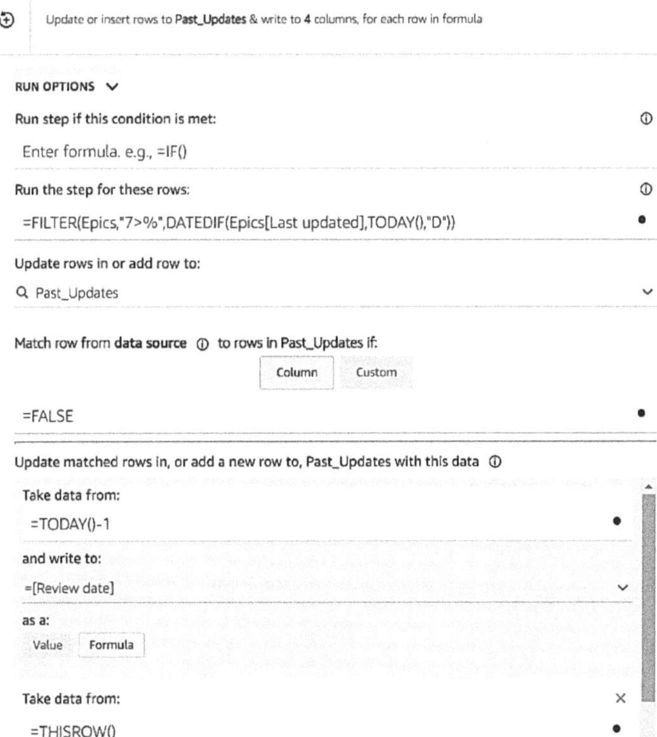

Figure 13.23 – Automation block for copying updates to the archive

Next, we need to concatenate all the updates into a single cell so that we can use it to send the notification in the email.

10. We start with adding a column to the `Past_Updates` table to concatenate the values for each row. Add a column to the table, name it `Concatenated update`, and set the following formula: `=CONCATENATE([Epic],"-",[Status] [Status],"-",[Update])`.
11. Now, add the **Overwrite** action block and set **RUN OPTIONS** to select the rows on which this step should execute using `=FILTER(Past_Updates,"Past_Updates[Review date]=%",TODAY()-1)`.
12. Set the source and destination fields as follows:

    Take data from: `=CONCATENATE(Findrow(WBR_Schedule,"WBR_Schedule[Review date]=%",TODAY()-1)[Updates]," ",Past_Updates[Concatenated update])`

    and write to: `=Findrow(WBR_Schedule,"WBR_Schedule[Review date]=%",TODAY()-1)[Updates]`

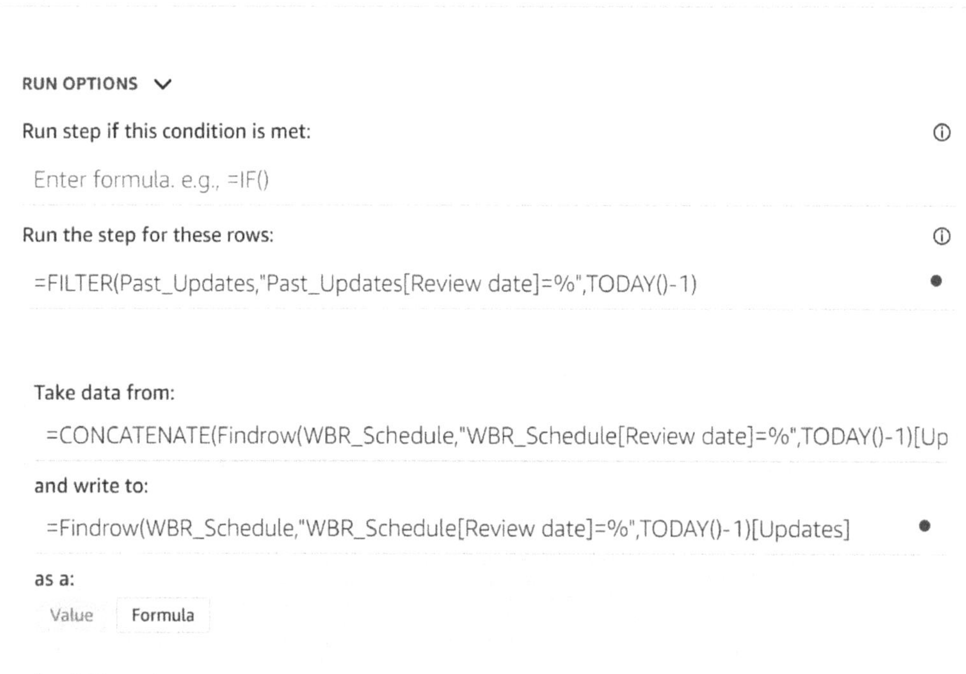

Figure 13.24 – Automation block for collecting all the updates and storing them in a single cell

Next, we add a new row to the `WBR_Schedule` table to set the next review date, creating the next trigger for the automation:

13. Add an action to **Add a row** and select `WBR_Schedule` as the target table.
14. Set the source and destination fields as follows:

    Take data from: `=Today()+6`

    and write to: `= [Review date]`

Figure 13.25 – Automation block to add a new row in the schedule with the date of the next review set

And finally, we will send the notification to the relevant audience:

15. Add the **Notify** action and add the email addresses in the **To** field.
16. Set **Subject** as `=[Review date] WBR updates`.
17. Set **Message** as: `This week's updates: =[Updates]`.

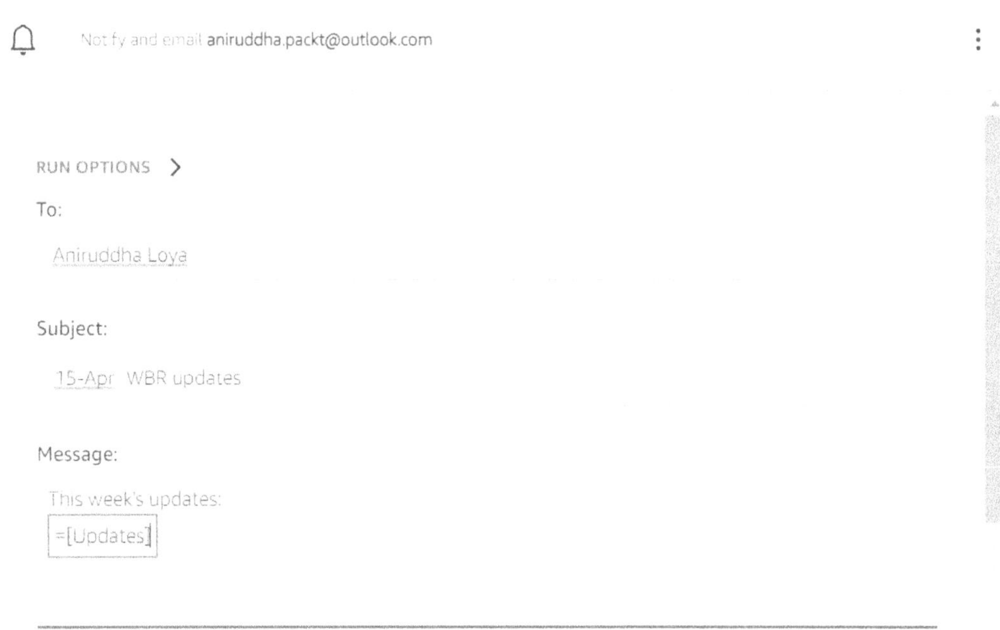

Figure 13.26 – Notify block of the automation configured for sending all the updates from the week

18. Attach the link to the app and publish the automation.

This completes our automation to archive the updates and send the notification.

> **Tip**
>
> To debug the automation, instead of *step 5*, set the trigger to be 1 minute after the current UTC and publish the automation. Once you are satisfied with the execution, you can revert it to *step 5*.

> **Note**
>
> The automation does not copy updates from the `Features` table. If you need those to be part of the summary notification or the archives, you can add another block to copy them over, similar to what we did for updates on the `Epics` table.

> **Note**
>
> Once archived, optionally, we would like to clear the updates from the `Epics` table and that block can also be added to this automation.

Sending reminders for providing updates

In the previous section, we created the `WBR_Schedule` table and set the date for the next review. We can use that as the trigger to send reminders 2 days before the review with the following steps:

1. Add a new automation and name it `Update reminder`.
2. Configure the **Date & Time reached** trigger to execute 2 days before the review date.
3. Add the **Notify** block to send a reminder to all the owners. You can simplify identifying the owners by sending the notification to all managers and product managers of the teams by setting the **To** field as `=Teams[Manager]` and `=Teams[Product Manager]`.
4. Attach the link to the app and publish the automation.

And, with that, the automation to send update reminders is complete, and so is our application.

Summary

In this chapter, we built an app for another common use case in organizations, conducting periodic business reviews. We made use of the App wizard to jumpstart our development and then customized the screens to meet our visual and functional requirements. In the process, we furthered our knowledge of wizard use as well as customization.

The most important aspect of the chapter, however, was the use of automation to power up our use cases. We learned how we can make a self-triggering automation loop using Date and Time triggers. And, in the other automation we built, we made use of the trigger for changing the value of a column to make updates to another field.

So far, we have covered use cases served through a single app, albeit at varying levels of complexity. But, not every use case can be served by a single app, and we have seen a couple of examples when reviewing the templates.

In the next chapter, we will build a use case requiring multiple apps and personalization to finish up this book.

14
Solving Problems through Multiple Apps within a Workbook

Real-world problems are often complex, require different access levels for different users, or may involve providing different functionality or views of the same underlying data. Sometimes, these requirements can be fulfilled in a single app, but at times, it is more prudent and simpler to split the responsibilities and functionalities in different apps and share them with relevant users.

In this chapter, we'll build apps that enable a realtor to interact with their clients. A realtor, at any given time, can be working with multiple clients, which can be buyers or sellers. Each buyer comes with varying requirements and preferences, and therefore, the realtor shares a curated list of properties personalized for each buyer to review and respond with their decision to request a viewing, put in an offer, or pass on the offer. With sellers, the realtor can book viewings based on the available time slots provided by the seller, as well as collect and share the offers on a property for the consideration of the seller.

This problem has two well-defined user personas and two very different sets of requirements and actions, and therefore, we will build separate apps to enable this use case. The first app will be for the realtor to manage the client profiles and preferences, share a list of properties or offers, and so on. Another app will be used by the clients to review the properties or offers and take action.

In this chapter, we're going to cover the following main topics:

- Defining the app requirements
- Translating requirements to app interactions
- Defining the data model
- Building the Realtor app
- Building the Client app

Technical requirements

To follow this chapter, you'll need to have access to Amazon Honeycode, which requires a laptop with a web browser, preferably Google Chrome, and optionally a mobile device, running either Honeycode's supported version of Android (it currently requires Android 8.0 and up) or iOS (it currently requires iOS 11 or later).

Furthermore, we'll use Honeycode terminology and refer to the components that we covered in *Chapter 2, An Introduction to Honeycode*, *Chapter 3, Building Your First Honeycode Application*, *Chapter 4, Advanced Builder Tools in Honeycode*, and *Chapter 5, Powering the Honeycode apps with Automations*. The instructions provided in this chapter for various tasks assume that you are familiar with those operations, and therefore, you should consider reading the referenced chapters first.

Defining the app requirements

We first introduced this topic in *Chapter 3, Building Your First Honeycode Application*, where, before building our app, we listed down the requirements or the use cases that we wanted the app to fulfill. We made use of that list throughout the chapter as a guide to define our data model, conceptualize the application interface, and visualize the interactions between various onscreen elements as well as the data displayed. Therefore, I'd encourage you to take 5 minutes to think about what your app should do and what it should look like, and then make a list.

> **To-Do**
>
> Take 5 minutes and list how you would like your app to work.

Here is what my list looks like:

1. The realtor must be able to add new buyers and capture their details, requirements, and preferences for the property.
2. The realtor must be able to add new sellers and capture their details and the property being sold.
3. The realtor must be able to add a new property to the app that is being sold by other realtors for their buyer clients.
4. The realtor must be able to quickly identify the preference or profile changes made by a buyer or seller. This can be either through a visual indicator on the app, a notification, or a combination of both.
5. The realtor must be able to assign properties to specific buyers based on preferences.
6. The realtor must be able to share the offers placed with the sellers.
7. The realtor must be able to book times with sellers to allow interested buyers to view the property.
8. The sellers must receive a notification on any new offers as well as any viewings booked for the property.
9. A property can be assigned to multiple buyers for consideration.
10. Whenever a new property is mapped to a buyer, they must receive a notification for it.
11. Whenever a buyer takes an action on a property, the realtor must receive a notification for it.
12. The client app must ensure that the buyers are only able to review properties assigned to them and not those assigned to other buyers. Similarly, sellers should only see their properties and offers, not those of other sellers.

Your list may have some or all of these requirements and may have even more, and that is perfectly fine. You may find some of these missing in the app we will build and may want to extend the list for yourself after you finish the chapter.

Translating requirements to app interactions

Based on the requirements we listed in the previous section, let's discuss how and what our apps will display and what the interactions will look like on different screens.

The Realtor app

For a realtor, the primary actions are as follows:

- Review the buyer client's profile and assign or remove properties based on their current or updated preferences and actions.
- Book viewings for properties being sold by seller clients and also share the details of offers being placed on them.
- Manage the list of properties for sale in the market – add new ones and remove those that are no longer on the market.
- Manage client profiles – add new ones and remove or deactivate those whose requirements are met or are no longer a client.

Reviewing these interactions, we can infer two primary views for a realtor:

- A list of clients that navigates to the details view when clicked and the ability to either filter by client type or have separate views for each.
- A list of properties that navigates to the details about the property. Again, this view should provide the toggle between the properties for sale on the market and those that are being sold by their clients.

The details page of buyer clients and properties for sale on the market should allow for a mapping functionality (the realtor can assign clients to a property from the property detail page and vice versa).

Also, the details page of seller clients and the properties being sold by seller clients should allow viewings to be booked as well as updates to be added about the offers.

Finally, there can be an optional third view of a feed of actions that were taken by the clients so that the realtor is quickly able to review the updates and act wherever needed.

The Client app

While listing the requirements, we noted two client personas, *buyers and sellers*, with different interactions and requirements.

For a seller, the primary actions are as follows:

- Review the offers on the properties and accept or reject them.
- Provide time slots for a realtor to book viewings.

To enable these actions, the primary set of views can be the following:

- A list of offers grouped by the property being sold and quick buttons to accept or reject them. Optionally, there can be a field to add a comment for the realtor.
- A list of time slots is provided for viewing with an indicator of whether it has been booked or not. There is also support for adding new time slots.
- Lastly, a view of the list of properties being sold that has navigation to a detailed view, with functionality to edit the property details, review and act on the offers, and also review and add viewing time slots.

For a buyer, the primary actions are as follows:

- Review the properties suggested by the realtor and take action on them.
- Review their buying preferences and update them as needed.

To enable these actions, the primary set of views can be the following:

- A list of properties with a visual indicator or ordering to indicate new additions or updates. The list navigates to a detailed page, allowing for actions to request a viewing, put in an offer, or pass on the offer.
- A screen to review and update the profile and buying preferences.

Reviewing the actions as well as the views for the two client types, we do not see any overlap, except that both will have a view with a list of properties navigating to a detailed page. However, the list itself will have a different visual representation and so will the detailed page. Therefore, the optimal solution is to build two separate apps.

> **Note**
> If you were to build an app using some other platform or by writing code, you can easily get by with a single app that can serve both client types. However, Honeycode does not support conditional visibility for the primary screens, and therefore, your combined app will have empty screens based on the client type as well as a cluttered navigation panel.

Defining the data model

Based on the requirements and app interactions discussed in the previous sections, you should be able to identify these four primary entities for your data model:

- **Clients**
- **Properties**
- **Viewing slots**
- **Offers**

> Note
>
> The preceding list is with an implied assumption that there is only one realtor. If we were to build this app for a realtor firm, we would also have a separate entity for the realtor.

Apart from these primary entities, we also need to be able to capture the different actions, types of clients and properties, and so on. In the previous chapters, we learned the benefit of keeping them as defined options in a table, and therefore, we will have to create tables for each of those.

Lastly, a realtor can map a single property on sale to multiple buyers. Similar to how we maintained OKR-Epic mapping in *Chapter 13, Conducting Periodic Business Reviews Using Honeycode*, in a separate table for scalability, we will apply the same rationale and store the property-buyer mapping in a separate table.

Building the app

Alright, so now we know what our app should do, have a fair idea of what it should look like to enable the listed use cases, and also know how we should structure and store the data to power the app.

Let's start building it up.

Creating a new workbook

By now, you must be familiar with the process of creating a workbook. So, go ahead and create an empty workbook from the Dashboard and name it `Honeycoded Realtors`.

Creating tables

In the previous section, we discussed the entities and tables that we will need. So, let's set them up.

> **Note**
>
> While creating tables in *Chapter 13, Conducting Periodic Business Reviews Using Honeycode*, we learned that in order to completely set the tables for primary entities with **picklists**, we would need to create the tables for secondary entities first. So, here we will start with creating tables for secondary entities.

The Client_Types table

Create a table, named `Client_Types`, with a single column, named `Type`, containing the following values for client types:

- `Buyer`
- `Seller`

The Property_Types table

Create a table, named `Property_Types`, with a single column, named `Type`, containing the following values for property types:

- `Condo/Apartment`
- `Townhouse`
- `Independent`
- `Commercial`

The Buyer_Actions table

Create a table, named `Buyer_Actions`, with a single column, named `Action`, containing the following values for the actions that a buyer can take:

- `Request viewing`
- `Make an offer`
- `Not interested`

The Seller_Actions table

Create a table, named `Seller_Actions`, with a single column, named `Action`, containing the following values for the actions that a seller can take:

- `Accept`
- `Reject`

The Clients table

Create a table, named `Clients`, with the following columns:

- `Client` – Format the column as **Contacts**.
- `Type` – Format the column as a **Rowlink and picklist** with the source as the `Client_Types` table.
- `Present address`
- `Phone number` – Format the column as **Number** with `0` decimal places
- `Area of search`
- `Maximum property price` – Format the column as **Currency**.
- `Minimum number of rooms` – Format the column as **Number** with `0` decimal places.
- `Minimum property area`

> **Note**
>
> Setting the column format as **Number** with 0 decimal places affects how the number is displayed but not how it is stored. To ensure that the value provided in the column is in the correct format, the validation has to happen in the app.

The Properties table

Create a table, named `Properties`, with the following columns:

- `Id` – Set the column formula as `=GETROW(Properties,-1,THISROW())[Id]+1` and then set the value for the cell in the second row as `1`, as shown in *Figure 14.1*.
- `Type` – Format the column as **Rowlink and picklist** with the source as the `Property_Types` table.
- `Address`

- `Asking price` – Format the column as **Currency**.
- `Number of rooms` – Format the column as **Number** with `0` decimal places.
- `Property area`
- `Year of construction` – Format the column as **Number** with `0` decimal places.
- `Other details`
- `Seller client` – Format the column as **Rowlink and picklist** with the source as `=Filter(Clients,"Clients[Type][Type]=%","Seller")`.

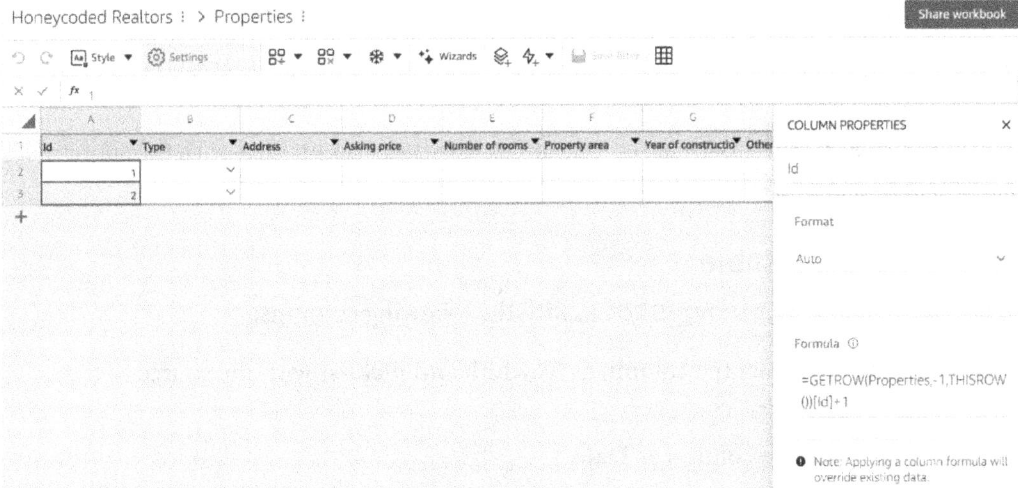

Figure 14.1 – The Properties table displaying the setup of the Id column

> **The GETROW Function**
>
> The `GetRow` function takes an ordered set of table rows as a source, an integer as an offset, and a context row to return a row relative to the provided row from that collection.

> **Exercise 1 – Explain the Formula Set on the Id Column?**
>
> Now that you understand the `GetRow` function, can you explain the purpose and use of the formula set in the `Id` column? Is there an alternative way to achieve the same outcome?

The Offers table

Create a table, named `Offers`, with the following columns:

- `Property` – Format the column as **Rowlink and picklist** with the source as the `Properties` table.
- `Buyer client` – Format the column as **Rowlink and picklist** with the source as `=Filter(Clients,"Clients[Type][Type]=%","Buyer")`.
- `Offered amount` – Format the column as **Currency**.
- `Offer valid until` – Format the column as **Date and time**.
- `Additional comments`
- `Seller client` – Format the column as **Rowlink and picklist** with the source as `=Filter(Clients,"Clients[Type][Type]=%","Seller")`.
- `Seller action` – Format the column as **Rowlink and picklist** with the source as the `Seller_Actions` table.

The Viewing_Slots table

Create a table, named `Viewing_Slots`, with the following columns:

- `Property` – Format the column as **Rowlink and picklist** with the source as the `Properties` table.
- `Date` – Format the column as **Date**.
- `Start time` – Format the column as **Time**.
- `End time` – Format the column as **Time**.
- `Is booked` – Format the column as **Checkbox**.

The Property_Recommendation table

Create a table, named `Property_Recommendation`, with the following columns:

- `Property` – Format the column as **Rowlink and picklist** with the source as the `Properties` table.
- `Buyer client` – Format the column as **Rowlink and picklist** with the source as `=Filter(Clients,"Clients[Type][Type]=%","Buyer")`.
- `Buyer action` – Format the column as **Rowlink and picklist** with the source as the `Buyer_Actions` table.

With this, we now have our tables created and set up with appropriate formats to enable our next step of creating the apps.

Creating the Realtor app

In the previous sections of this chapter, we have covered what use cases we would like to support and identified the need for separating apps for a realtor and their clients. We then discussed what views, screens, and interactions we will need in our app for a realtor. Now, let's follow the steps to build our app:

1. Launch **app wizard** to create a new app.
2. Choose the `Properties` table as the source and select the columns you would like to display in this list.
3. Click **Next** to create the detail screen and update the screen name to `Property detail`. Mark the fields for **Asking price**, **Other details**, and **Seller client** as editable, leaving the others as-is because we do not expect those to change once set. **Seller client** is left editable in case the client is acquired after the property was already added to the app.
4. Click **Next** to create the form screen and update the screen name to `Add property`.
5. Click **Done** to load the next screen, as shown in *Figure 14.2*. Rename the app from **Properties** (the default name) to `Realtor`.

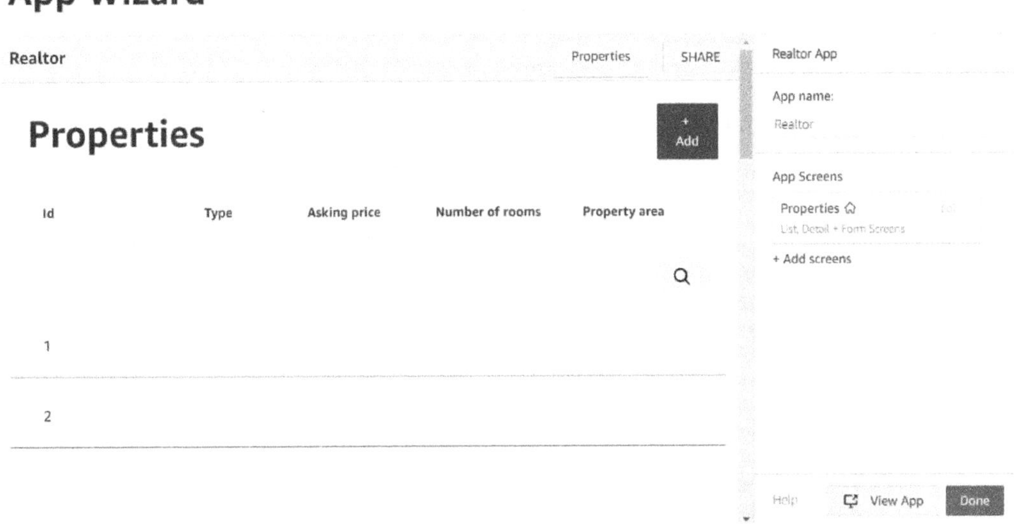

Figure 14.2 – App wizard after creating screens for the Properties table

6. Next, we click on **+Add screens** and create a list, details, and form screens for the `Clients` table.
7. Click **Done** to exit the wizard to load the newly created app in the builder view, as shown in *Figure 14.3*:

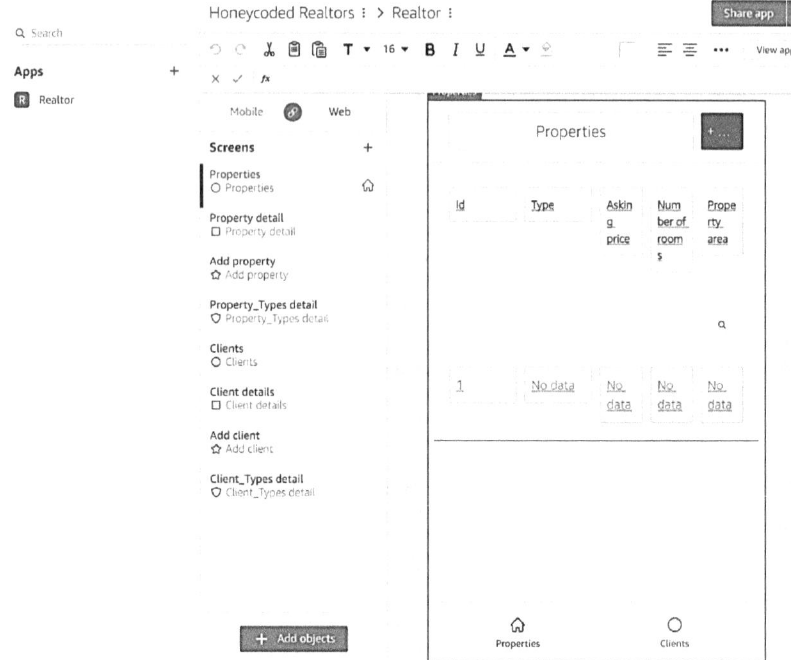

Figure 14.3 – The Realtor app in the builder view after exiting the wizard

8. Delete the following screens, as they are of no use in this app:
    - **Property_Types detail**
    - **Client_Types detail**

Now that we have our main screens created, let's start customizing them and adding missing interactions to complete the app.

The Properties screen

The screen currently lists all the properties; however, as discussed in the previous section, a realtor would like to be able to quickly access the properties their clients are selling and also filter the properties to be able to quickly identify the ones based on client criteria and assign to them. So, let's update the screen to enable that:

1. Add a **data cell** outside of the list. Rename it `ShowAll`, set its type as `Variable`, with the initial value set to `=False`, and visibility set to `=False`.
2. Now, add a **button** and set its text as **All Properties**. Set its visibility condition as `=NOT($[ShowAll])`, and under **Quick Actions**, select **Update current screen** and set the **ShowAll** variable's value to `=TRUE`.
3. Add another button below the one created previously. Set its text as **Client Properties**, set the visibility condition as `=$[ShowAll]`, and under **Quick Actions**, select **Update current screen** and set the **ShowAll** variable's value to `FALSE`.
4. Update the source of **List** to `=IF($[ShowAll], Filter(Properties), FILTER(Properties, "NOT(ISBLANK(Properties[Seller client]))"))`.
5. Finally, under the **DISPLAY** tab for **LIST PROPERTIES**, check the **Filter** option and select **Type**, **Asking price**, and other fields that the realtor would like to filter on.

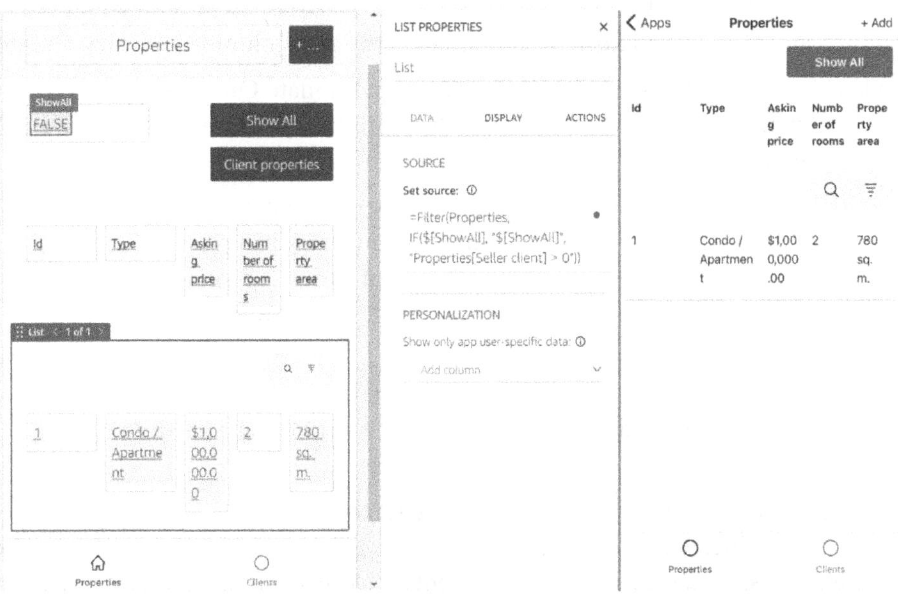

Figure 14.4 – The Properties screen in the builder view on the left and an app screenshot on the right

The updated **Properties** screen is shown in *Figure 14.4*.

The Property detail screen

In our section on defining the app interactions, we noted that we should be able to map properties to buyers, book a viewing, or add or update offers. So, let's add that to the screen:

1. Add a **form** object. Select the **Button + form** screen option and set the `Property_recommendation` table in the **Add form data to** field.
2. Under **display field**, remove the `Buyer action` column and click **Create**.

   This will add a new button and a new **Property_Recommendation form** screen.
3. Update the text on the button and the new screen's name to `Recommend to client`.
4. The button's **Quick Actions** already has a navigation set to the newly created form screen. Let's also pass the property row to pre-fill the value by setting the property variable of the form screen to take data from `=$[InputRow]`.
5. Add a **list** control to show all the buyers that the property was recommended to and their actions. Unselect the **search** option under the **Display** tab and update its data source to `=Filter(Property_Recommendation, "Property_Recommendation[Property] = %", $[InputRow])`.
6. Add another **Button + form** screen, with data to be added to the `Offers` table. Remove the `Seller client` and `Seller action` columns and click **Create**.
7. Update the new button's text to `Add offer`, and update **Quick Actions** to pass the current property as context to be pre-filled in the form screen. Also, update the form screen name to `Add offer`.
8. The realtor will only need to add offers for properties that belong to their seller clients and therefore needs this function only for such properties. So, let's hide the button when there is no seller client by setting the visibility condition of the box containing the button to `=NOT(ISBLANK($[InputRow][Seller client]))`.
9. Add another list control to show all the offers on this property. Remove the search option and update the source of the list to `=Filter(Offers, "Offers[Property] = %", $[InputRow])`.
10. Add one more **Button + form** screen, with data to be added to the `Viewing_Slots` table. Update the button's text to `Add viewing slot` and update **Quick Actions** to pass the current property as context to be pre-filled in the form screen. Also, update the form screen name to `Add viewing slot`.

11. Add another list control to display all the viewing slots for the property. Remove the search option, replace the data cell for the `Is_Booked` column with a checkbox, and update the source to `=Filter(Viewing_Slots, "Viewing_Slots[Property] = % AND Viewing_Slots[Date] >= %", $[InputRow], TODAY())`.
12. Finally, update the text on the **Delete Row** button to `Delete Property` and move it to the bottom of the screen. Also, set the visibility to `False` for the block at the top of the screen that contains the `InputRow` variable.

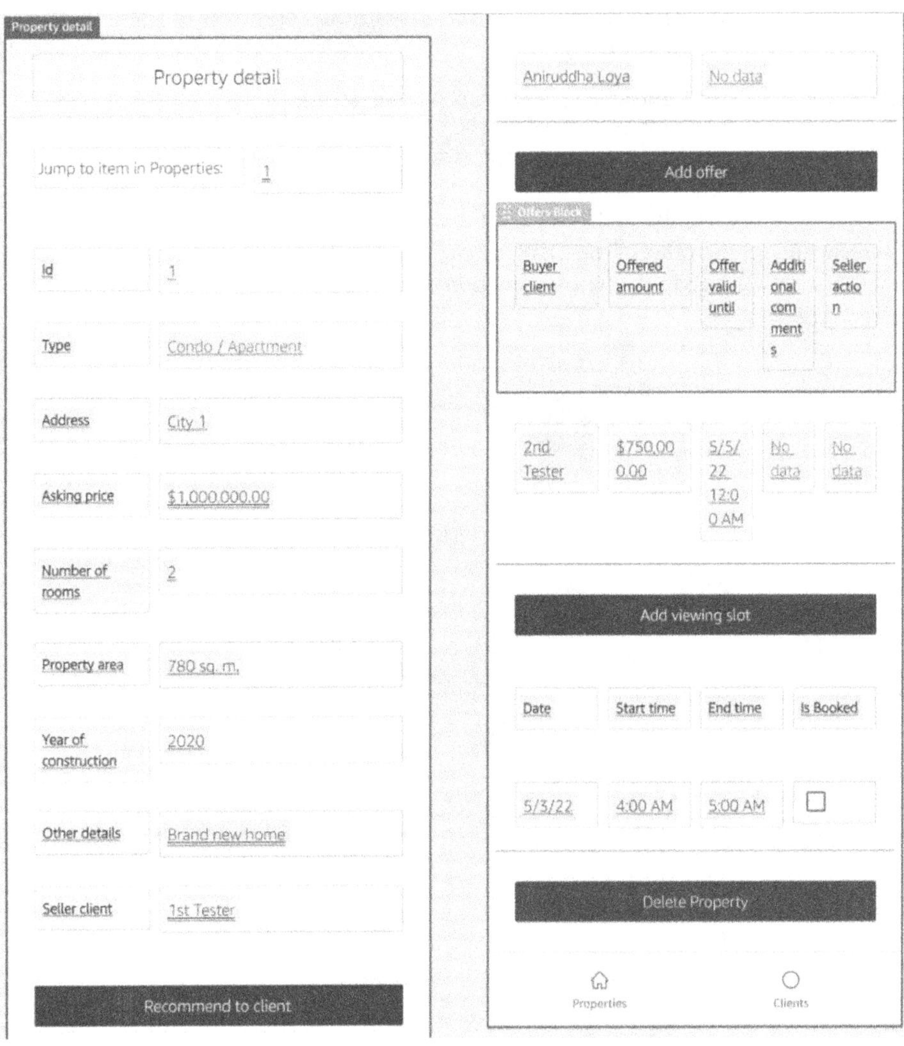

Figure 14.5 – The builder view of the completed Property detail screen, split into two halves

The completed **Property detail** screen in the builder view is shown in *Figure 14.5*.

The Clients screen

The screen currently lists all the clients; however, as discussed in the previous section, a realtor may want to be able to review their clients by type and also like to know which clients have recently updated their profile. So, let's update the screen to enable that:

1. Enable the filter on the list control and select the `Type` column to allow filtering.
2. To identify which clients have been updated, we need the information on when the last update was made on the profile and when was the realtor last reviewed it. So, let's first add `Last updated` and `Last reviewed` columns to the `Clients` table and format them as **Date**.
3. Select the list control and click **Create an Automation**.

> **Note**
> There should be a **Quick Actions** configured to navigate to the **Client details** screen, and set the value of the `InputRow` variable. That action will be automatically added to the new automation that we are creating, as shown in *Figure 14.6*.

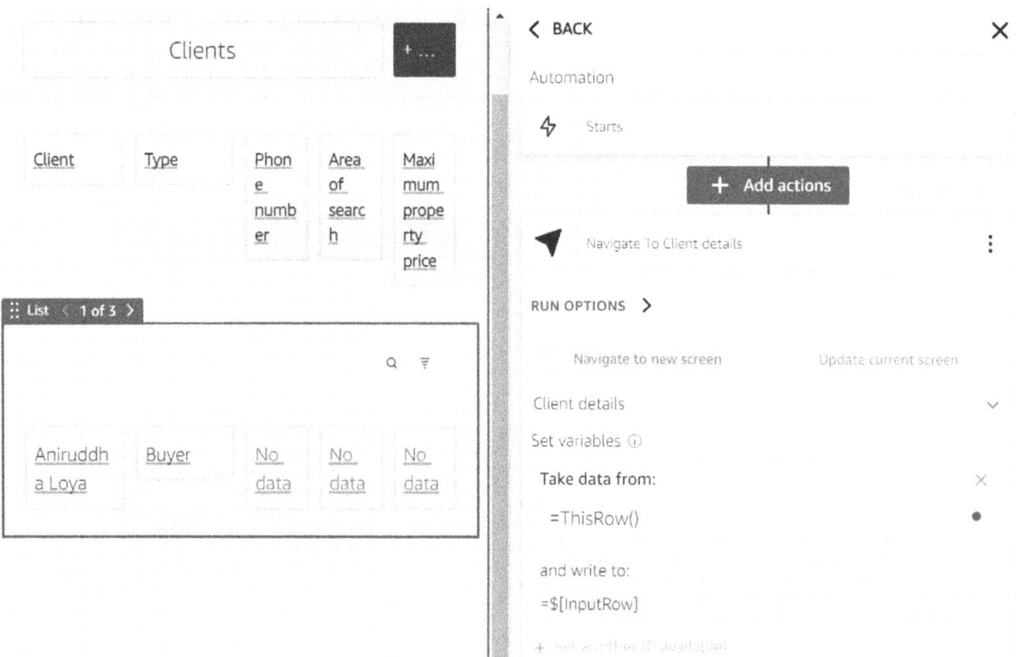

Figure 14.6 – The Quick Actions configuration, auto-configured in the automation

4. Add an **Overwrite** action. Set the source and destination fields as follows:

   - **Take data from**: `=TODAY()`
   - **and write to**: `=ThisRow() [Last reviewed]`

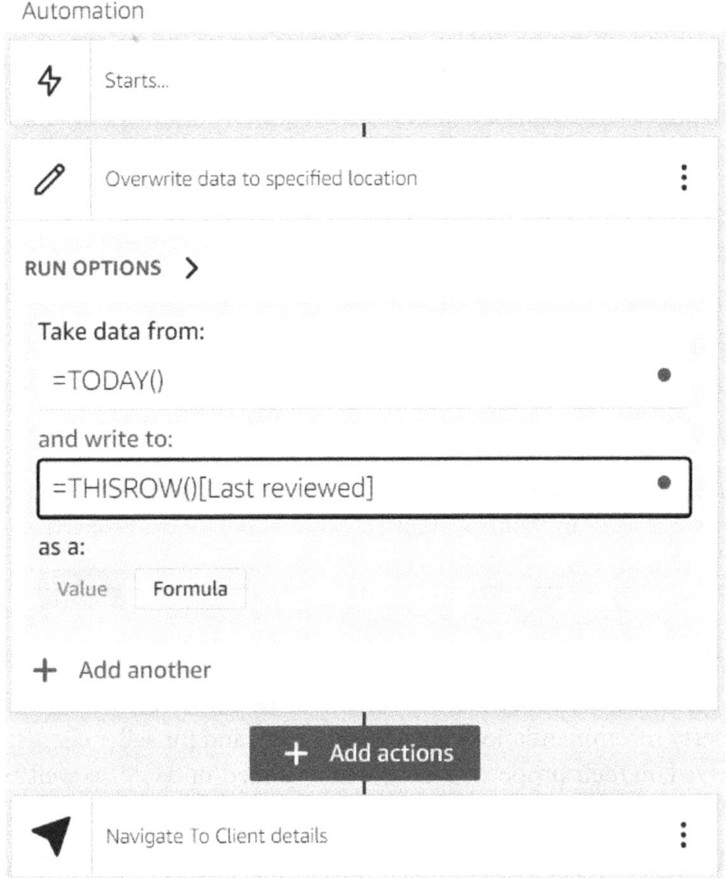

Figure 14.7 – Overwriting the last reviewed value for the row

5. Finally, we add conditional styling to the **segment** inside the list. Set the condition as `=Clients[Last updated] > Clients[Last reviewed]` and use **Style** to fill it with a green color or any other visual clue you may want to use.

The completed **Clients** screen will look similar to *Figure 14.8* on a web app:

Realtor

Properties Clients SHARE

Clients

+ Add

Client	Type	Phone number	Area of search	Maximum property price
Aniruddha Loya	Buyer	1231231230		$1,000,000.00
1st Tester	Seller	2323454566		
2nd Tester	Buyer	9900998877		$800,000.00

Figure 14.8 – A screenshot of the Clients screen on a web app

> **Note**
>
> We only enabled the **Add On** filter for the list, but you can implement the list toggle behavior using the buttons, similar to what we did for the **Properties** screen.

The Client details screen

As discussed in the interactions of a realtor, the **Client details** screen should allow the addition of property recommendations for a buyer client, and for seller clients, it should allow offers received on their properties for sale to be added/updated, as well as bookings for viewings.

As you may realize, all these additions are the same as what we did for the **Property details** screen, and the update to this screen is therefore left as an exercise.

> **Exercise 2**
>
> Complete the Client details screen.

The completed **Client details** screen should look similar to the screenshots shown in *Figure 14.9*:

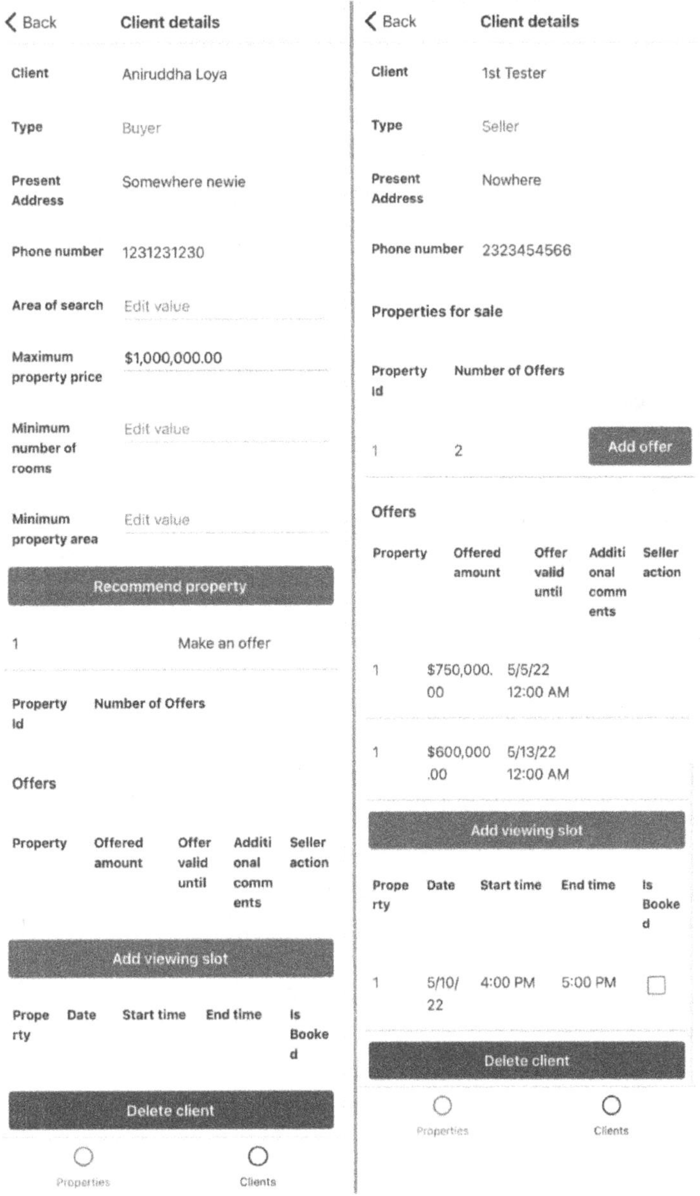

Figure 14.9 – Screenshots of the Client details screen for the buyer (left) and seller (right) clients

No changes are required on the other app screens, and with this, our Realtor app is complete. Next, we will build the client apps, starting with the one for buyer clients.

Creating the Buyer app

Similar to our Realtor app, let's make use of app wizard and build the view we need.

1. Launch the app wizard to create a new app.
2. First, use the `Properties` table as the source, and create the list and details view only. Remove the `Seller client` column from the detail view and ensure that it is not added to your list view.
3. Rename the app from **Properties** (the default name) to `Buyer`.
4. Next, click on **+Add screens** and create the list and details screens for the `Clients` table.
5. Remove the `Type`, `Last Updated`, and `Last Reviewed` columns from the list and details screen. Make all the fields, except **Client**, editable in the details view.
6. Click **Done** to exit the wizard to load the newly created app in the builder view, as shown in *Figure 14.10*:

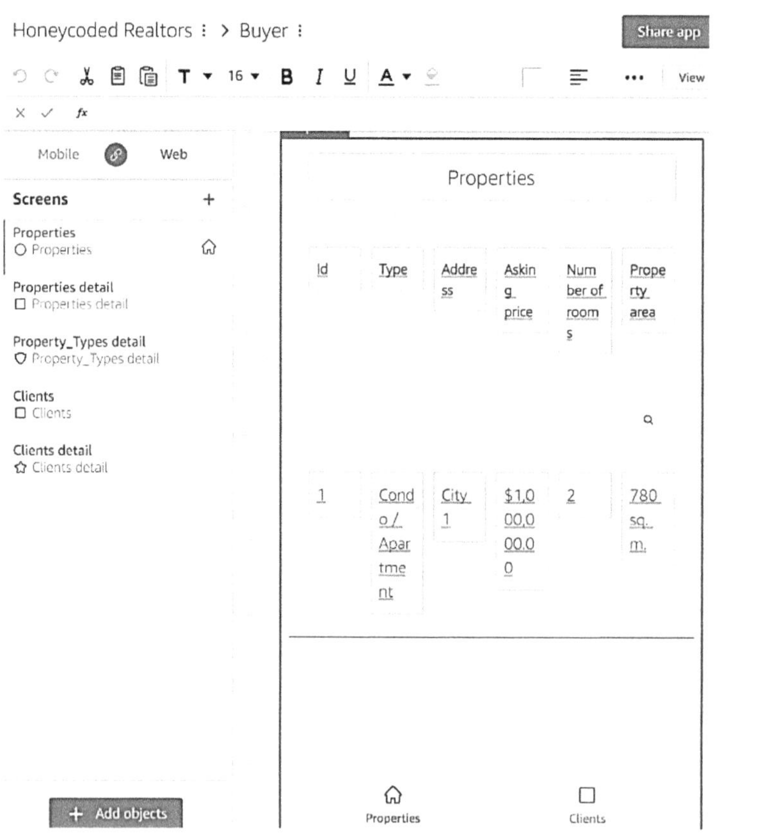

Figure 14.10 – The Buyer app in the builder view after exiting the wizard

7. Delete the following screens, as they are of no use in this app:

   - **Property_Types detail**
   - **Clients**

> **Why Are We Deleting the Clients Screen?**
> Because in the **Buyer** app, there is no need to show are other clients of the realtor, and it also maintains privacy.

8. Deleting the **Client** screen makes **Properties** the only navigable screen, but we need the buyer clients to be able to access their profiles. So, using **APP NAVIGATION PROPERTIES**, we move the **Client details** screen from the **HIDE FROM NAVIGATION** section to the **SHOW IN NAVIGATION** section to arrive at the screenshot shown in *Figure 14.11*:

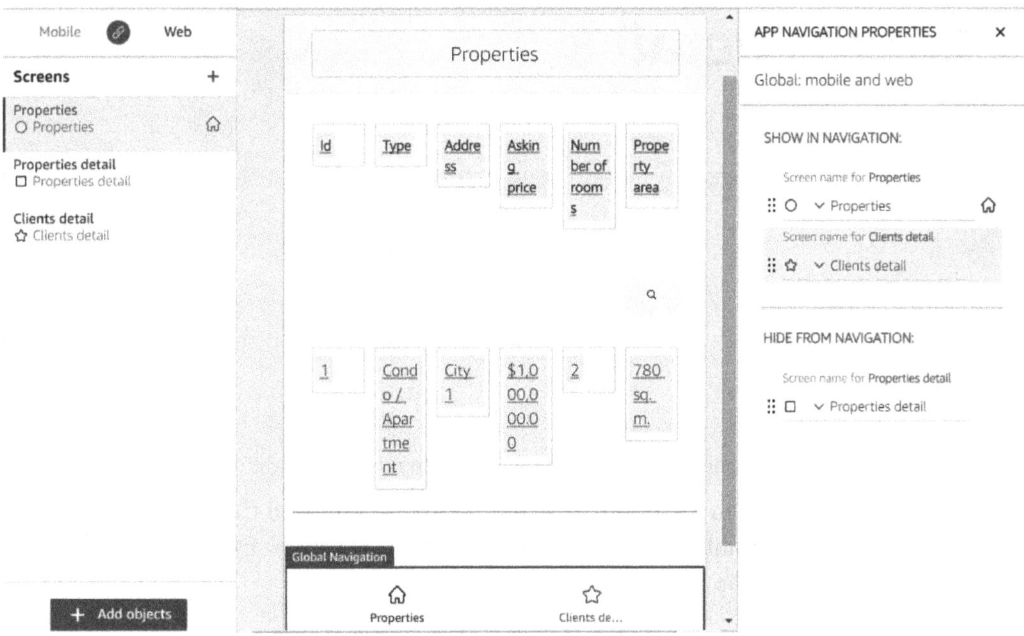

Figure 14.11 – The Buyer app after deleting the screens and updating navigation

Now that we have our main screens created, let's start customizing them and add missing interactions to complete the app, starting with the **Client details** screen to continue with the changes we made in this section.

The Clients detail screen

A detail screen is typically meant for navigation from a list view to show more information. A list screen provides the context of the row that was clicked to set up the variable, named `InputRow` by default, which is then used as a source of the detailed block and dereferenced to show various values.

However, in the previous section, we deleted the list view of the `Clients` table, given we only have to display details of a single client, the app user. We also made the screen directly navigable by making it appear in the navigation, but there is no source list to set the `InputRow` variable. So, if you open your app at this stage, the client details screen will appear empty, with a dropdown to select a client, as shown in *Figure 14.12*:

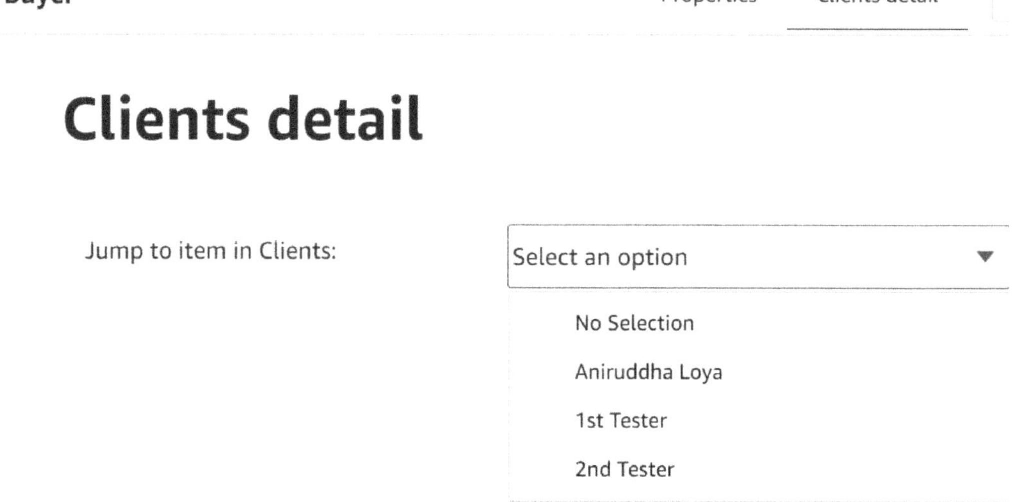

Figure 14.12 – The empty Client details screen due to missing navigation context

As shown in the screenshot in *Figure 14.12*, even though we removed the list screen for clients, there is still a possibility to navigate to other clients. Let's now fix the issue as well as update this screen with the following steps:

1. We start by changing the visibility of the **block** containing the visible dropdown and setting it to `FALSE`.
2. Next, we set the `InputRow` variable to always show the details of the app user by setting the initial value of the variable to `=FINDROW(Clients, "Clients[Client] = %", $[SYS_USER])`.

3. Finally, we rename the screen `Profile` to arrive at the stage shown in *Figure 14.13*:

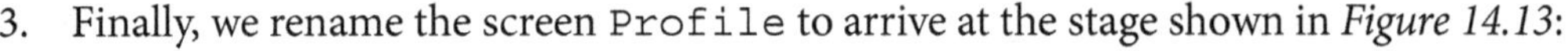

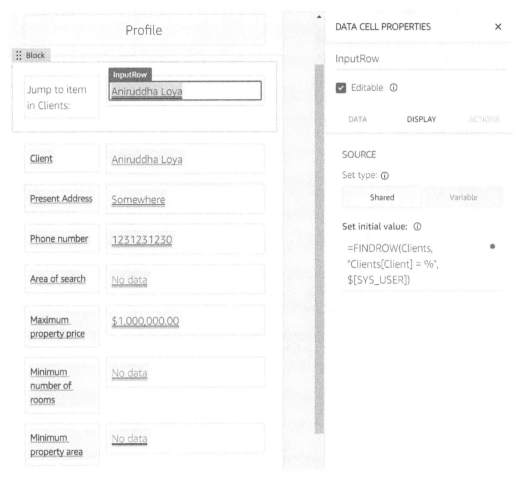

Figure 14.13 – The Client details screen, updated to the Profile screen

But we are not finished with this screen yet. Recall that we customized the **Clients** screen of the Realtor app to show a visual clue whenever there is a new update to the client profile since the last review. We set the `Last reviewed` column, but so far, we haven't set the value for the `Last updated` column.

To do so, we need to update the `Last updated` column value if any change is made to the profile by the client. This can be done in two ways, and we'll apply one of those here and leave the other as an exercise:

1. Add a data cell to the screen. Rename it `IsEditable` and set its type as variable, with **initial value** set to `=False` and visibility set to `=False`.
2. Now, add a button and set its text as `Edit`. Set its visibility condition as `=NOT($[IsEditable])`, and under **Quick Actions**, select **Update current screen** and set the `IsEditable` variable's value to `=TRUE`.
3. Add another button below the one created previously. Set its text as `Done`, set the visibility condition as `=$[IsEditable]`, and under **Quick Actions**, set the `IsEditable` variable's value to `FALSE`.
4. Create a copy of all the **content boxes** that are displaying values from the client's profile, except the one displaying the name, and uncheck the **editable** checkbox for each of them.
5. For all the newly added content boxes in the previous step, set **visibility** to `=NOT($[IsEditable])` while setting the **visibility** condition on the original content boxes, except the one displaying the name, as `=$[IsEditable]`.
6. Select the **Done** button added in *step 3* and create an automation for it. Verify that the previously configured **Quick Action** is now transferred automatically to the automation.
7. Add an **Overwrite** action. Set the source and destination fields as follows:
   - **Take data from**: `=TODAY()`
   - **and write to**: `=$[InputRow][Last updated]`

> **Note**
>
> There is still a small detail here to keep in mind. In the preceding setup, we have the editable data cells configured as shared, which means that even if we do not press **Done**, the data is already updated.
>
> This can be fixed by changing them to **Variable** and then adding another section in the **Done** button automation to update the values in the table.

With this, our **Profile** screen is now complete; the builder view of the screen is shown in *Figure 14.14*:

Figure 14.14 – The Builder view of the completed Profile screen

> **Exercise 3**
>
> Provide the alternative solution for updating the `Last updated` value in the event of any change to the profile.

The Properties screen

This screen currently lists all the properties; however, we need to display only the properties recommended to the client. This can be done in two ways; the one using **personalization** we'll cover in this section while leaving the other as an exercise. So, let's update it to meet our requirements:

1. Add a column to the `Property_Recommendation` table, name it `Client`, and set **column formula** to `= [Buyer client][Client]`.

> **Why Do We Need the Additional Column?**
>
> In Honeycode, the *personalization* feature works on columns that contain **Contacts** as the data. However, the `Buyer client` column appears to show as a Contact, which is just because the first column of the `Clients` table has **Contacts** as a format. The underlying value of the cells in that `Buyer client` column is a **Rowlink** to the `Clients` table. Dereferencing the value to the new column sets the value of that column to be of the `Contact` type and, therefore, can be used for personalization.

2. Select the list on the **Properties** screen and change its source from the `Properties` table to `Property_Recommendation`.
3. Under the **PERSONALIZATION** property configuration, select the **Client** column we created in *step 1*, as shown in *Figure 14.15*:

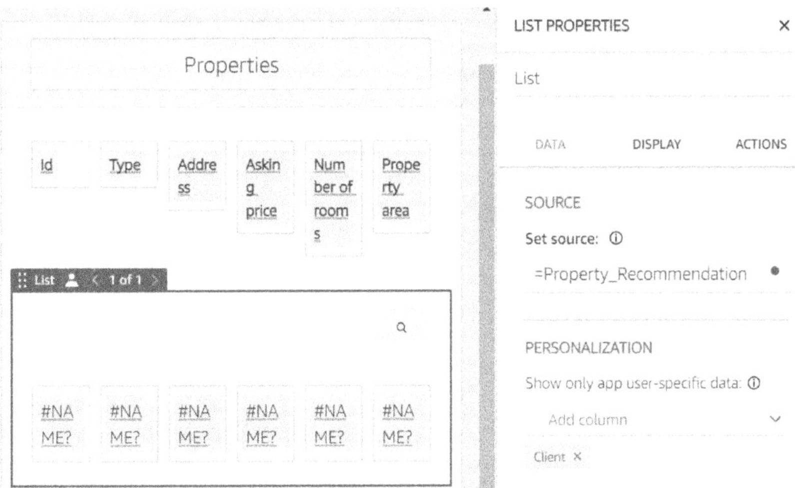

Figure 14.15 – Setting the column to use for personalization on the screen

4. Changing the source will break the display on the list, as shown in *Figure 14.15*, so let's fix it with a **double dereference** using the `Property` column. What that means is that for all the displayed fields, select their data cells and update the shared source by injecting `[Property]` between = and the existing dereference value. For example, update the source of the data cell for the `Id` column from `= [Id]` to `= [Property] [Id]`, as shown in *Figure 14.16*:

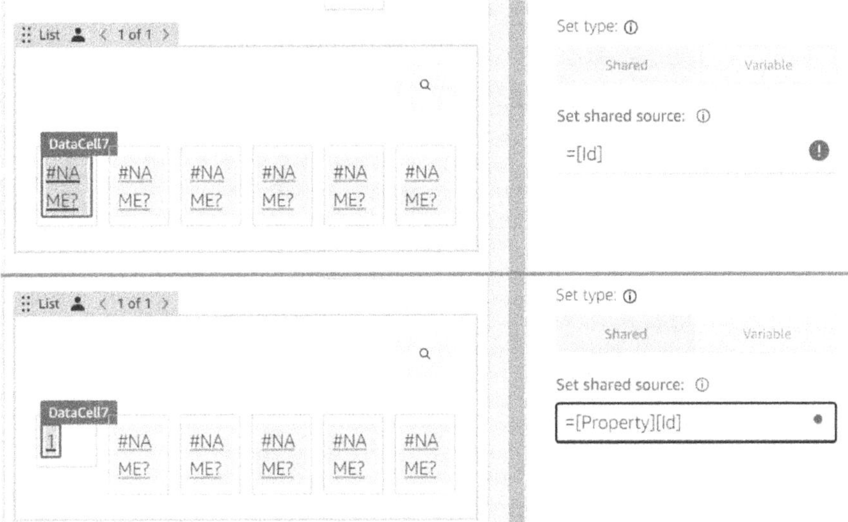

Figure 14.16 – The Id column's data cell before (above) and after (below) double dereferencing

> **Exercise 4**
>
> Can you explain why this simple update to double dereference works?

5. Optionally, you can add another column to show the action (if any) taken by the client.

And that is all we need to do on this screen.

> **Exercise 5**
>
> Provide the alternative solution for creating a personalized list of property recommendations.

Properties detail screen

This screen was configured alongside the list screen to display all the details of the selected property. However, we updated the source table on the list screen, and therefore, we will need to adjust the source here too. We also need to add the capability for the client to act on the recommendations. So, let's do that:

1. Select the **InputRow** data cell and change its format to **Auto**. This will clear the **Rowlink and picklist** format as well as the configured source table.
2. Select the block with objects to display the property details. Update its source from `=$[InputRow]` to `=$[InputRow][Property]`.
3. Lastly, add a new block. Add a **Picklist** object to it, with the source set as the `Buyer_Actions` table and the value of **Set shared source** as `=$[InputRow][Buyer action]`.

The completed **Properties detail** screen is shown in *Figure 14.17*:

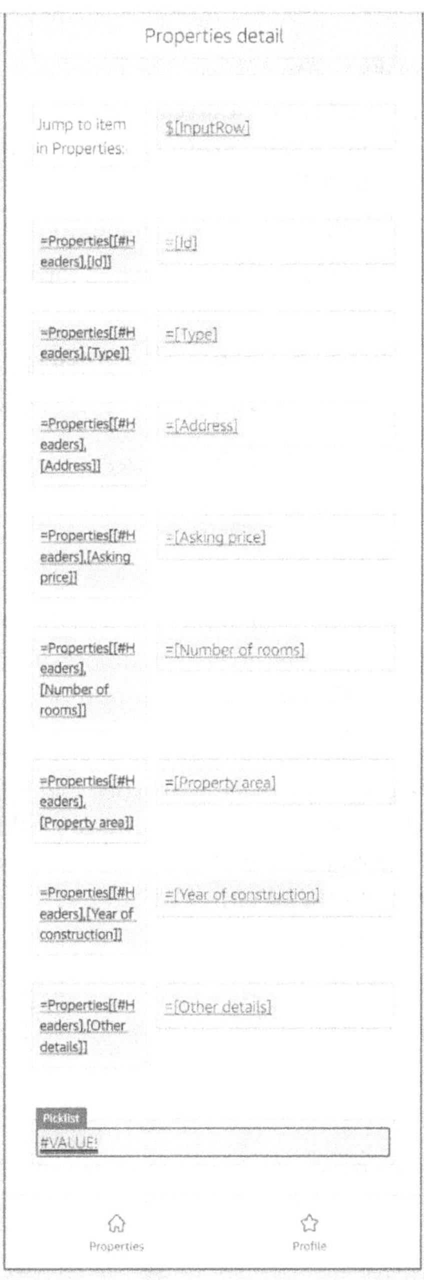

Figure 14.17 – The completed Properties detail screen

And with that, our Buyer app is complete.

Creating the Seller app

Similar to our Buyer app, let's make use of app wizard and build the views we need:

1. Launch app wizard to create a new app.
2. First, use the `Properties` table as a source and create the list, details, and forms view. In the list view, keep only the **Id**, **Address**, and **Asking price** fields. In the details and form screen, remove the `Seller client` column.
3. Rename the app from **Properties** (the default name) to `Seller`.
4. Next, we click on **+Add screens** and create a list and form screen for the `Viewing_Slots` table.
5. Click **Done** to exit the wizard to load the newly created app in the builder view.
6. Delete the `Property_Types detail` screen as it is of no use in this app.

Now that we have our basic app, let's customize the screens to meet the requirements of the seller clients.

Properties screen

The screen currently lists the properties, but as discussed in the previous section, for a seller client, a more meaningful view also shows the number of offers associated with properties on sale, as well as only offers that are made for properties they are selling. So, let's update the screen to enable this:

1. Let's first filter the list to only display the properties sold by the user. As with the Buyer app, we can do this either by personalization (which requires another column, as we did when personalizing the view in the Buyer app) or by updating the source to `=Filter(Properties, "Properties[Seller client][Client] = %", $[SYS_USER])`.

2. Add a content box in the block with column headers and set its content to `Offers`.
3. Add a corresponding data cell in the list and set its source as `=FILTER(Offers, "Offers[Property] = %", THISROW())`.

Now, let's look at the **Properties detail** screen.

Properties detail screen

For the seller, a detailed screen is a place to either update property details, review and act on offers, or add additional viewing slots for a property.

We have already set up the screen to display the property details with fields marked as editable when we created the app using the wizard. If you did not do it then, you can go to each relevant data cell and make it editable.

Here are the missing functions:

- Adding a list of offers with actions
- A list of viewing slots for a property with the ability to add new ones

We already know how to build them, as we added these functions in the **Property detail** and **Client details** screens of the Realtor app, and therefore, it is left to you to add them and complete the screen.

> Tip
>
> The simplest solution here is to copy and paste the entire **Property detail** screen from the Realtor app, remove or update the objects needed in the context of the Seller app, and update the navigation target of the list on the **Property** screen.

One last thing to do is to rename the screen `Property details` if you had not done so at the time of configuring it in the wizard.

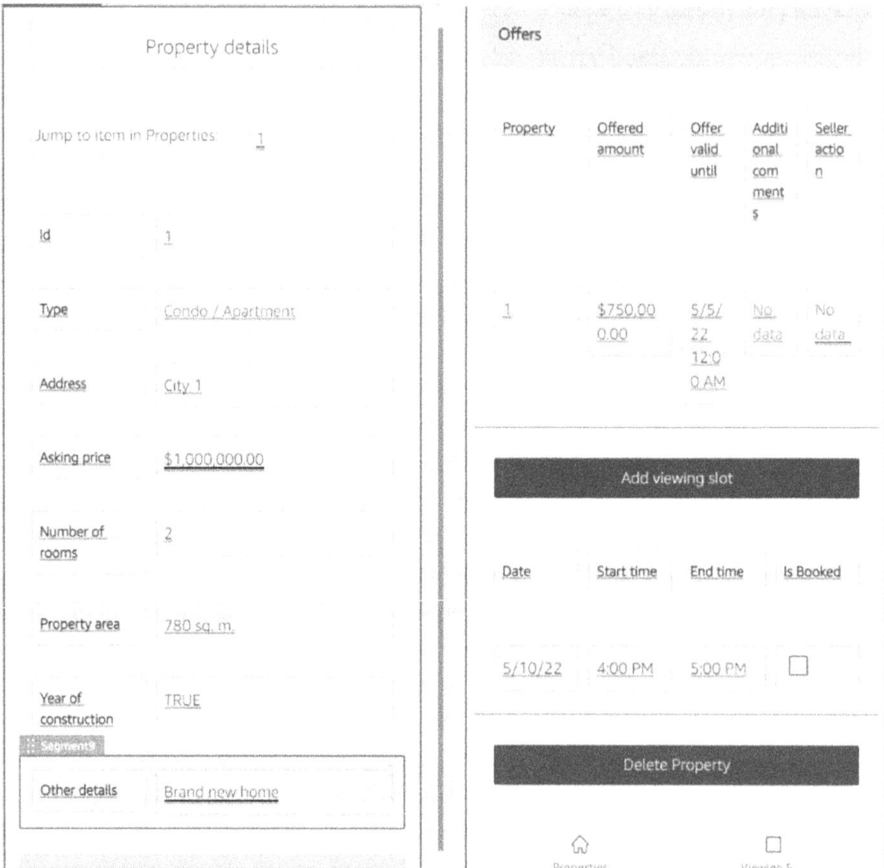

Figure 14.18 – The completed Property details screen in the builder view

The completed screen's builder view is shown in *Figure 14.18*:

The Properties form screen

When configuring this form screen in the wizard, we removed the field for **Seller client**, as we already know who the client is, but we need to add that information when creating the new entry. So, let's update the screen to accommodate that:

1. Update the screen name to `Add Property for Sale`.
2. Next, go to the **Actions** tab of the **Done** button and click **edit automation**. This will open the automation, displaying the **Add row** section with seven fields configured to write to respective columns.

3. Scroll to the bottom of the **Add row** section and click **Add another**, as shown in *Figure 14.19*:

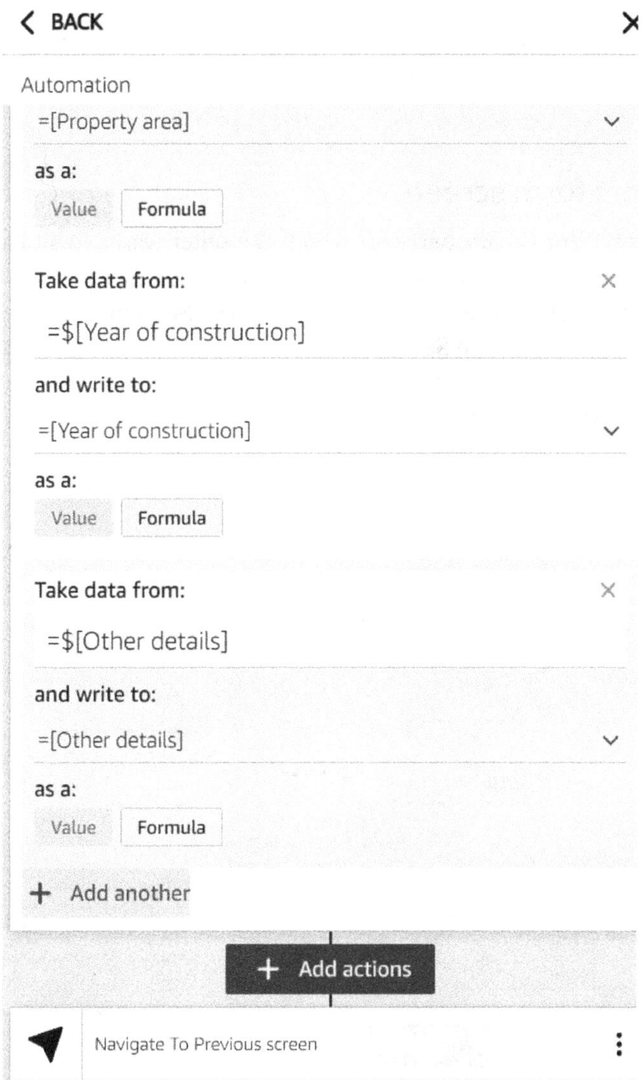

Figure 14.19 – Editing the automation on the Done button

4. Set the fields as follows:

   - **Take data from**: `=$[SYS_USER]`
   - **and write to**: `=[Seller client]`

That is all we need to change in this screen to complete the functionality again.

The Viewing Slots screen

The screen currently lists the viewing slots for all properties, but we should only show those for the user's property and the slots that are in the future. Again, here you can provide the filter using personalization or update the source to `=FILTER(Viewing_Slots, "Viewing_Slots[Property][Seller client][Client] = % AND Viewing_Slots[Date]>=%", $[SYS_USER], Today())`.

The Viewing Slots form screen

The form requires input for the property for which the seller wants to add a viewing slot, and we need to ensure that the only options available for the seller are those of the properties sold by them. Therefore, we need to update the **Set source type** option of the **Property** data cell to **Filter** and the **Set filter source** option to `=FILTER(Viewing_Slots, "Viewing_Slots[Property][Seller client][Client] = %", $[SYS_USER])`, as shown in *Figure 14.20*:

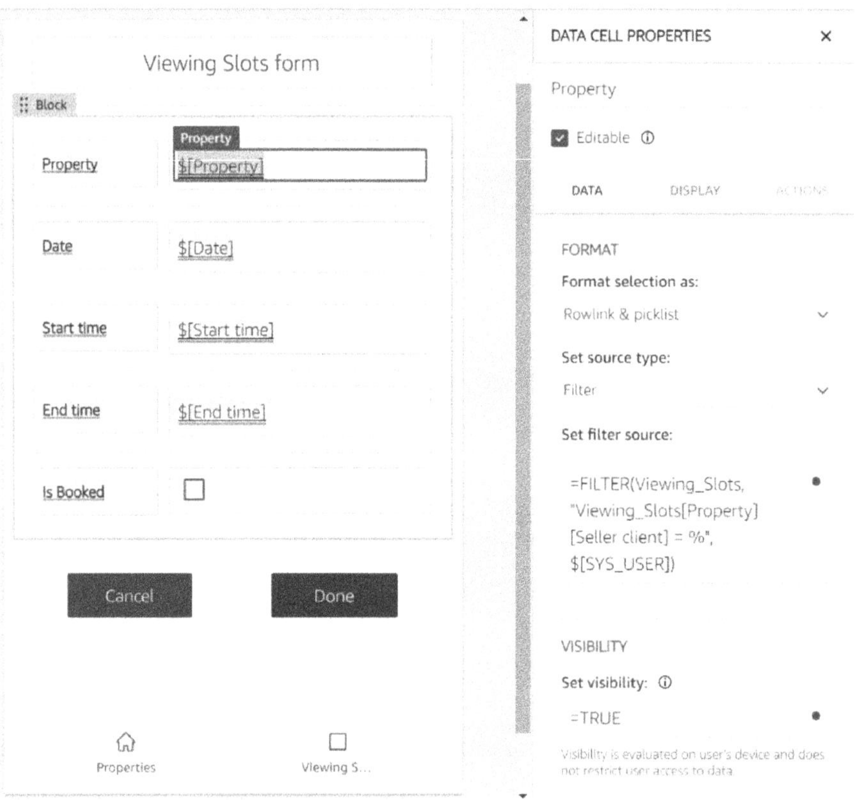

Figure 14.20 – Updating the source type and the filter source for the Property data cell

With this, we have now completed our Seller app.

Discussion

You may be thinking that this cannot be it, the apps do not feel complete, and there are so many things missing that I would like to see in a realtor app, no matter the role. And I agree with you on each of those points. However, the purpose of this chapter was not to build a complete app for each role but instead to demonstrate how we can solve problems with multiple apps, and to showcase how we can we use multiple apps on the same data source with differing views for different users. And we have been able to do this.

Summary

In this chapter, we solved the use case of a realtor trying to manage their clientele. We analyzed the problem, broke down the user interactions, and decided that the best solution would be to make use of three separate apps.

The key learning of the chapter was to be able to build multiple apps, over the same underlying data model, while catering to the needs of different user personas and their requirements as we went about solving the problem at hand. The other concept we covered was the use of personalization while also learning about the alternative solution to achieve the same outcome.

In addition, we again relied on app wizard to create the core of the app and then customize it. Through this journey of customization and exercises, we used or referenced various concepts we learned in the previous chapters, bringing together our learnings from the book.

Assessments

This section contains answers to questions from all the chapters.

Chapter 4 – Advanced Builder Tools in Honeycode

Exercise 1

This exercise can be completed by following these steps:

1. On the **My Tasks** screen, select the segment inside the list and open the **Conditional Styling** property for it.
2. Set the **WHEN** condition using the following formula: `=AND (Today() - Tasks[Due date] >= 0, Today() - Tasks[Due date] <= 1)`.

3. In the **THEN STYLE AS** block, we'll use the **Custom** style and set the fill color to a lighter shade of yellow, with dark yellow borders and large rounded corners to match the default style:

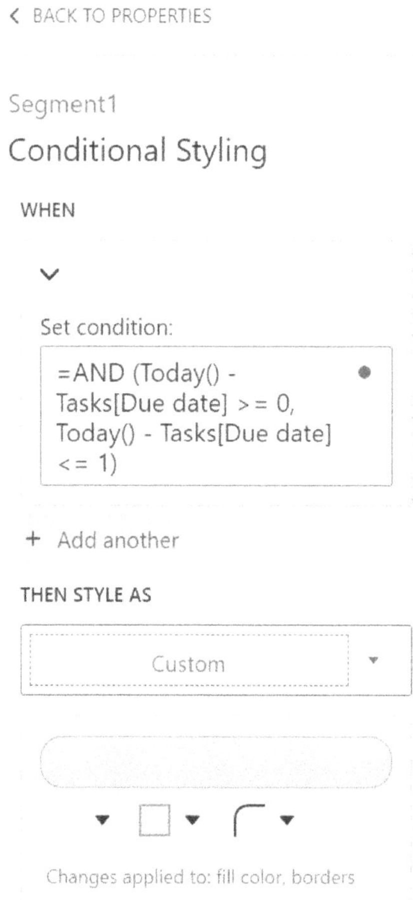

Figure 4.1 – Adding conditional style to a segment inside the lists on the My Tasks screen

4. Click on the **BACK TO PROPERTIES** button. The segment will now have two conditional styles defined, as shown in *Figure 4.2*:

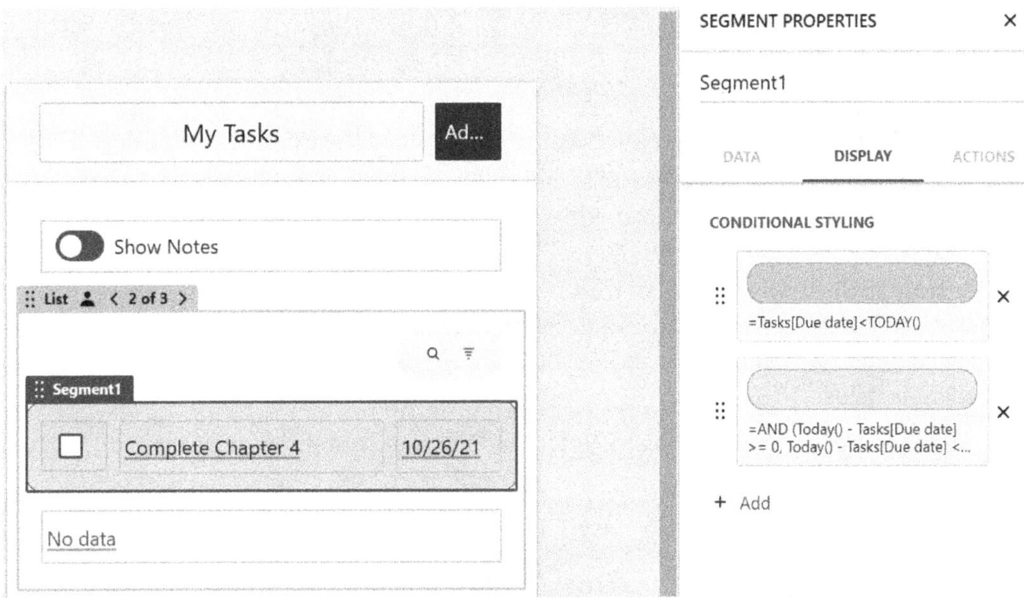

Figure 4.2 – A segment with two conditional styles defined

5. The result will look like *Figure 4.3*:

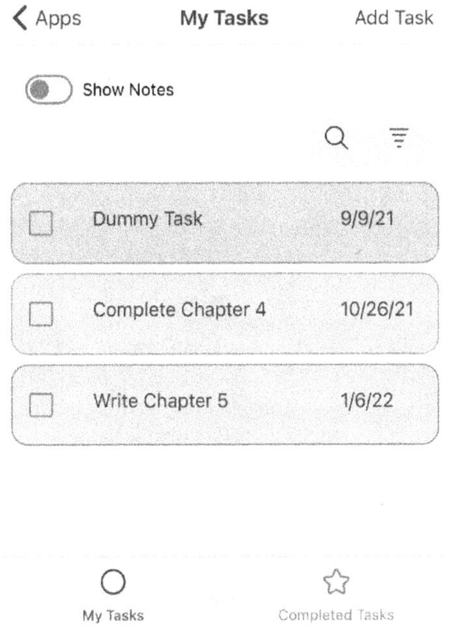

Figure 4.3 – A My Tasks list with tasks due the next day highlighted in yellow

Exercise 2

This exercise can be completed by following these steps:

1. Go to the **Edit Task** screen. As we did when we added the existing fields of **Due date** and **Reminder preference** in *Chapter 3, Building your first Honeycode Application* select the block containing these editable fields and, using the **Add Objects** control, add a content box and a data cell to the screen.
2. Rearrange the newly added controls by moving and resizing the content box to the left of the screen and the data cell right next to it, so that the content box aligns under the field names and the data cell aligns under the corresponding editable fields.
3. Set the text of the content box to **Notes:** and the source of the data cell as `=[Notes]`.
4. The result will look like *Figure 4.4*:

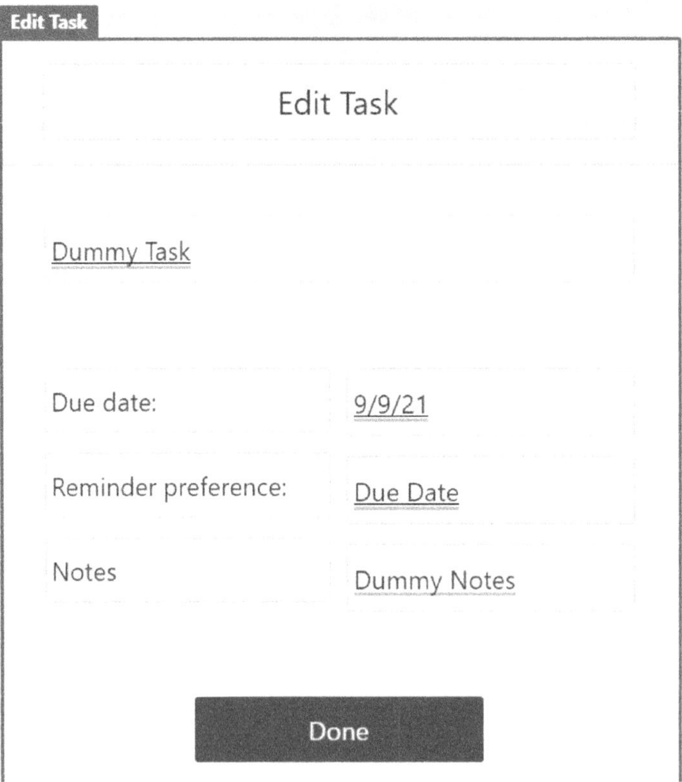

Figure 4.4 – Adding an editable field for Notes

Exercise 3.1

This exercise can be completed by following these steps:

1. Go to the **Edit Task** screen. Similar to how we added the existing fields of **Due date** and **Reminder preference** in *Chapter 3, Building your first Honeycode Application*, select the block containing these editable fields and, using the **Add Objects** control, add a content box and a data cell to the screen.
2. Rearrange the newly added controls by moving and resizing the content box to the left of the screen and the data cell right next to it, so that the content box aligns under the field names and the data cell aligns under the corresponding editable fields.
3. Set the text of the content box as `Assignee:` and the source of the data cell as `=[Assignee]`.
4. The result will look like *Figure 4.5*:

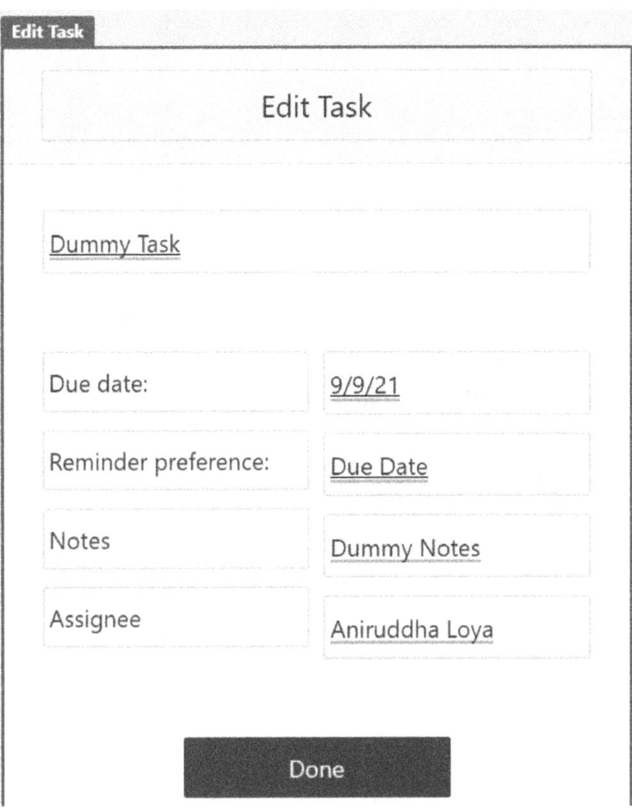

Figure 4.5 – Adding an editable field for Assignee

Exercise 3.2

This exercise can be completed by following these steps:

1. Go to the **Completed Tasks** screen. Select the list and, using the **Add Objects** control, add a data cell to the screen.
2. Set the source of the data cell as `= [Assignee].`
3. The result will look like *Figure 4.6*:

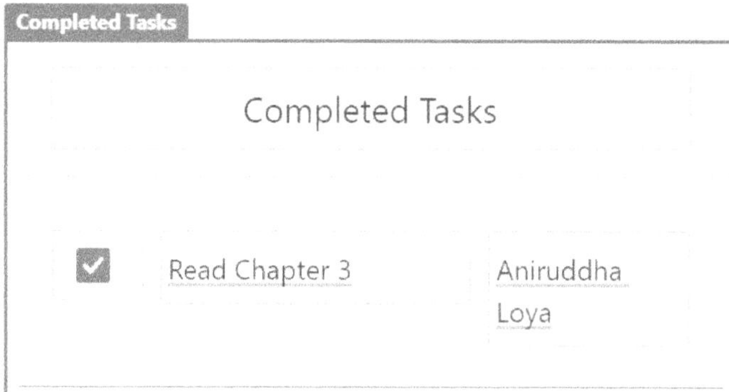

Figure 4.6 – Adding the Assignee field to the Completed Tasks screen

Exercise 4

This exercise can be completed by following these steps:

1. Go to the **Completed Tasks** screen. Select the list and enable the fields for **Search**, **Filter**, and **Sort** under the **DISPLAY** tab.
2. Customize the fields for **Filter** and **Sort** as needed.

3. The result will look like *Figure 4.7*:

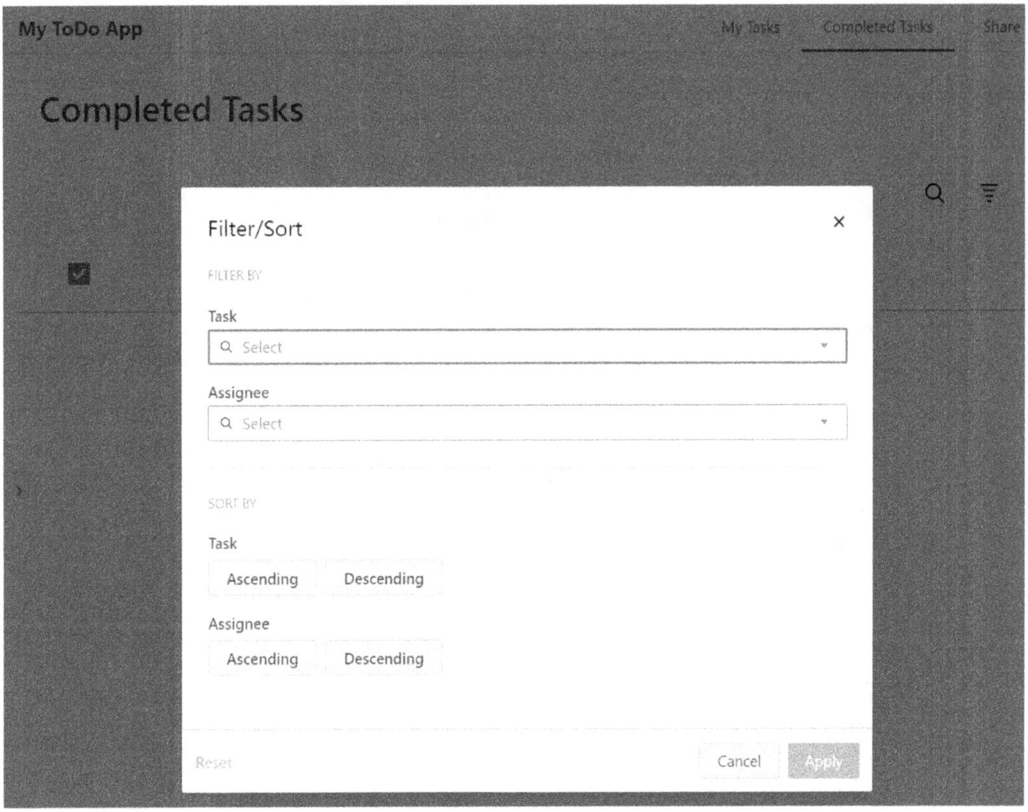

Figure 4.7 – Adding a search, filter, and sort capability to the Completed Tasks screen

Chapter 5 – Powering the Honeycode Apps with Automations

Exercise 1

This exercise can be completed by following these steps:

1. In the **Tasks** table, add a new column and rename it `Created by`. Set the column format to **Contact**.
2. In the **Add new tasks** screen, edit the automation on the **Done** button by adding another section in the first block of automation. Set **Take data from** as `=$ [SYS_USER]` and select the **Created by** column in the **and write to** field (see *Figure 5.1*):

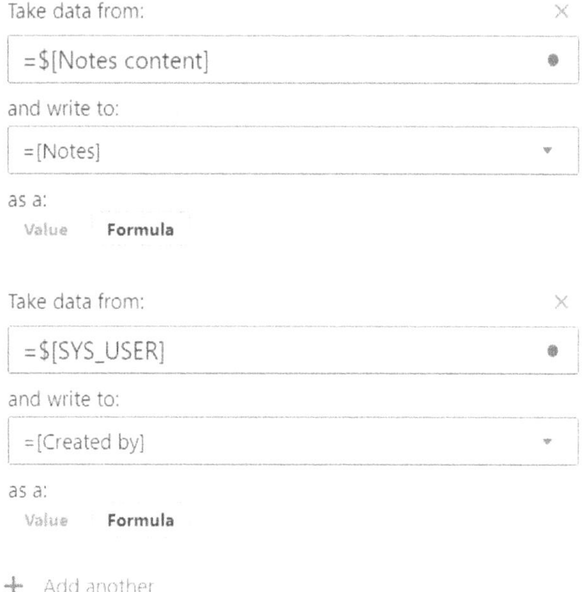

Figure 5.1 – Editing the Done automation to capture the information of the user that created the task

3. A sample task created after this will have the **Created by** field populated as shown in *Figure 5.2*.

> **Note**
>
> For existing tasks, the column will have to be manually updated.

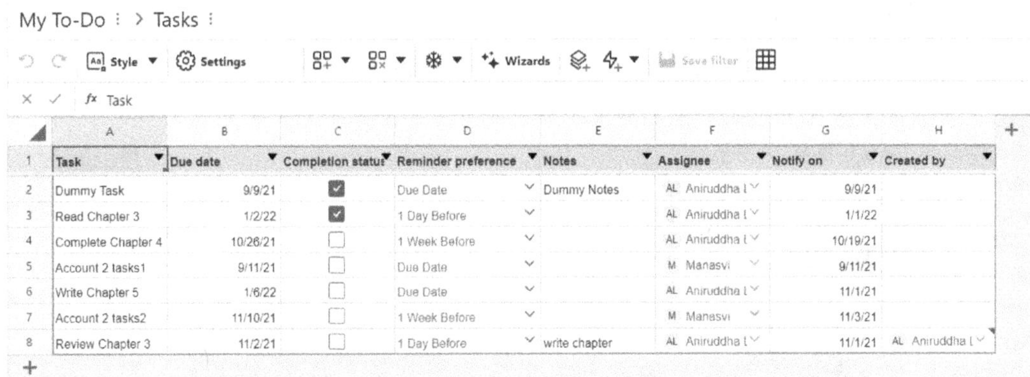

My To-Do > Tasks

	Task	Due date	Completion status	Reminder preference	Notes	Assignee	Notify on	Created by
2	Dummy Task	9/9/21	☑	Due Date	Dummy Notes	AL Aniruddha L	9/9/21	
3	Read Chapter 3	1/2/22	☑	1 Day Before		AL Aniruddha L	1/1/22	
4	Complete Chapter 4	10/26/21	☐	1 Week Before		AL Aniruddha L	10/19/21	
5	Account 2 tasks1	9/11/21	☐	Due Date		M Manasvi	9/11/21	
6	Write Chapter 5	1/6/22	☐	Due Date		AL Aniruddha L	11/1/21	
7	Account 2 tasks2	11/10/21	☐	1 Week Before		M Manasvi	11/3/21	
8	Review Chapter 3	11/2/21	☐	1 Day Before	write chapter	AL Aniruddha L	11/1/21	AL Aniruddha L

Figure 5.2 – The Tasks table with the Created by field updated

Exercise 2

The exercise can be completed by following these steps:

1. Create a new automation and rename it `Reassignment notification`.
2. In the **Start** block, select the **Column Changes** option under **Automation Trigger**. Choose the **Tasks** table and the **Assignee** column.
3. Now, using **Add action**, add a **Notify** block.
4. Fill this automation block as follows:

    I. Set the **To** field as `=[Assignee]`.

    II. Set **Subject** as `=[Task] assigned`.

    III. Set **Message** as `=[Task] is now assigned to you`.

5. Add another **Notify** block.
6. Fill this automation block as follows. Refer to *Figure 5.22* for the completed block:

    I. Set the **To** field as `=$[Previous]`.

    II. Set **Subject** as `=[Task] unassigned`.

    III. Set **Message** as `One of your task, =[Task], is now assigned to =[Assignee]`.

7. Publish the automation.

   Sample notifications created on reassignment are shown in *Figure 5.3*:

Review Chapter 3 assigned Inbox ×

notifications@honeycode.aws

12:17 AM (24 minutes ago)

amazon Honeycode

Review Chapter 3 is now assigned to you.

This email notification was generated by a Honeycode automation. Learn more

Need help? Visit Help & Community

Review Chapter 3 unassigned

N notifications@honeycode.aws

Mon 11/1/2021 12:17 AM

To:

One of your task, Review Chapter 3, is now assigned to Manasvi .

This email notification was generated by a Honeycode automation. Learn more

Need help? Visit Help & Community

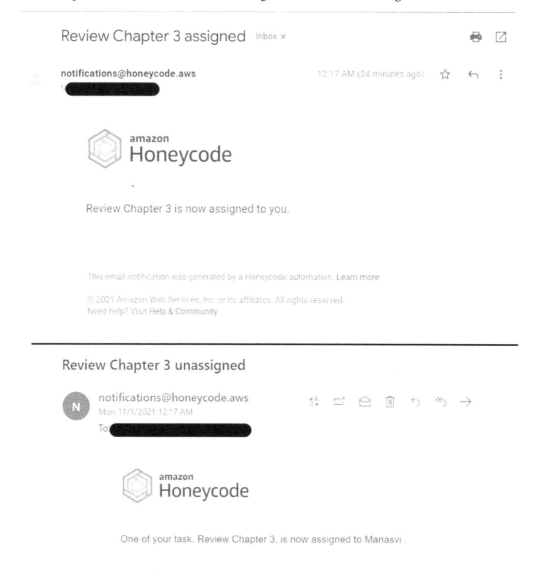

Figure 5.3 – Tasks reassignment notifications

Chapter 7 – Simple Survey Template

Exercise 1

> **Note**
>
> There are no values set for the responses to the questions.

1. Go back to the home screen and start the survey again.

> **Note**
>
> Your responses from the previous submission are shown.

2. Now, update any value. Do not submit the survey and return to the home screen.
3. Start the survey again.

> **Note**
>
> The value of the response that you had changed in *step 3*, not the value submitted in *step 1*, even though the change in *step 3* was never submitted.

Chapter 8 – Instant Polls Template

Exercise 1. Can you think of another way of achieving the same result of separate text color based on a specified condition?

We can apply conditional styling on the content box instead of the visibility condition by following these steps:

1. Select the bottom content box and copy its visibility condition.
2. Select the top content box, clear its visibility condition, and then click on **Conditional Styling**.

3. Set the `when` condition by pasting the formula from the visibility condition and then style as **Custom** to set the font color to be the lighter shade of green used in the other content box, as shown in *Figure 8.1*:

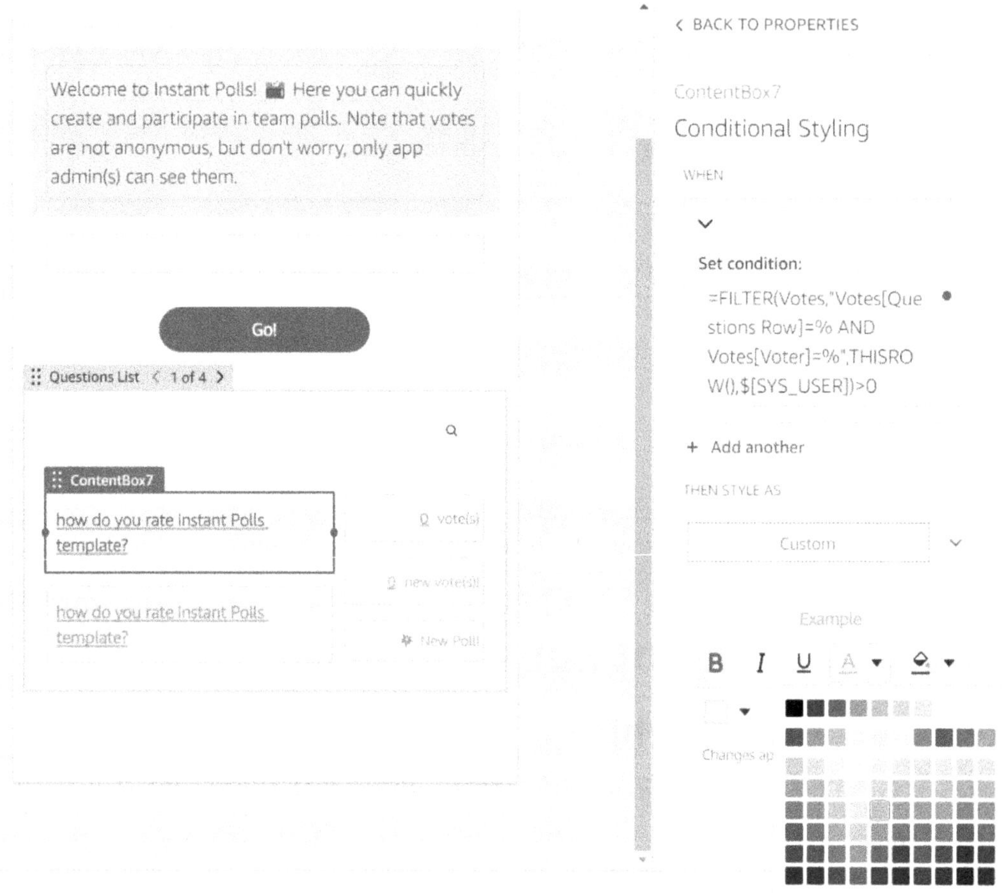

Figure 8.1 – Set the conditional styling on the top content box

4. Delete the second content box.

Chapter 9 – Event Management Template

Exercise 1

Currently, in the **A_Sessions** table, data from the **Location** and **Capacity** columns is added manually for each session that is added to the table. To keep the mapping consistent, we need to have a single place where the data for the location and its capacity can be added. This exercise can be done in two ways.

Solution 1

Perform the following steps:

1. Create a new table and name it `M_Locations`.
2. Rename the first column of the table `Location` and the second column `Capacity`.
3. Fill this table with unique **Location** values from the **A_Sessions** table along with the correct **Capacity** value in the respective columns:

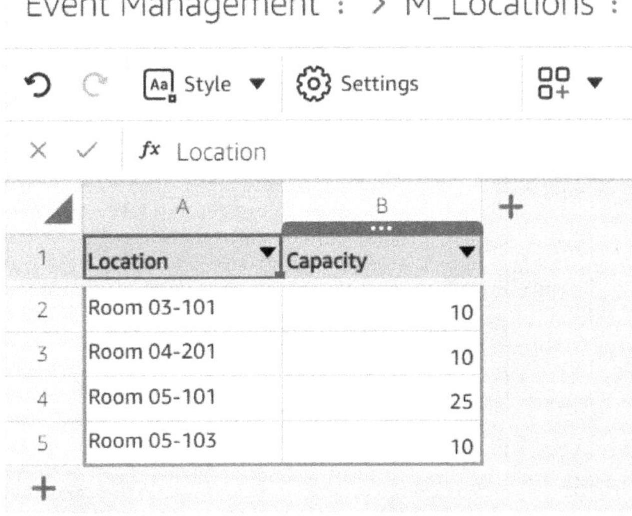

Figure 9.1 – The M_Locations table

4. Select the **Location** column in the **A_Sessions** table, set its format to **Rowlink and picklist**, and select the **M_Locations** table as the source. This will change the format of the entire column and automatically convert the cells into **Rowlinks** as long as the text values are an exact match with the **M_Locations** table's **Location** column.
5. Select the **Capacity** column, and set its formula as `= [Location] [Capacity]`.

Solution 2

Perform the following steps:

1. Click on **WIZARDS** in the toolbar and select the **Create Picklist** card from the panel on the right-hand side, as shown in *Figure 9.Ex2*:

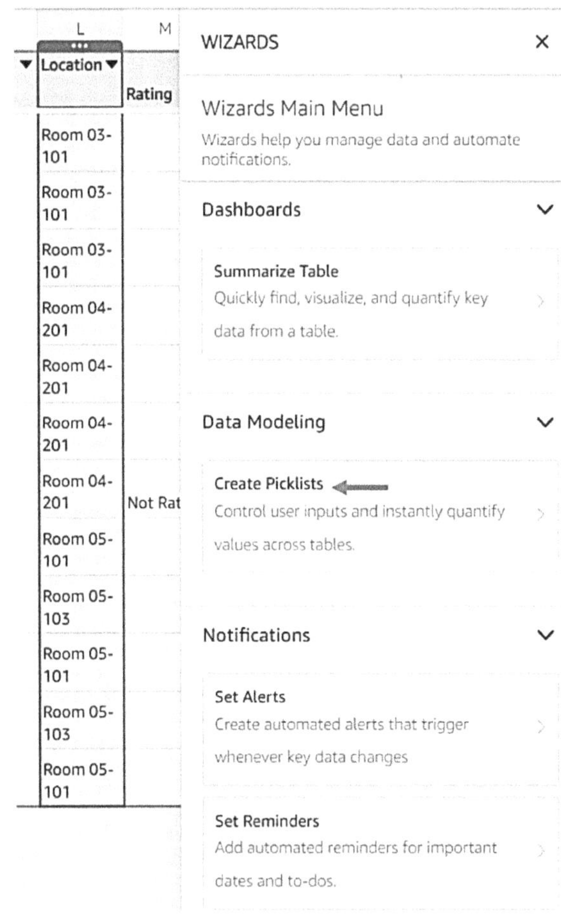

Figure 9.2 – Creating the Picklists card from WIZARDS

2. Clear the default suggestion by using the cross. Then, click on the + **Add new** button:

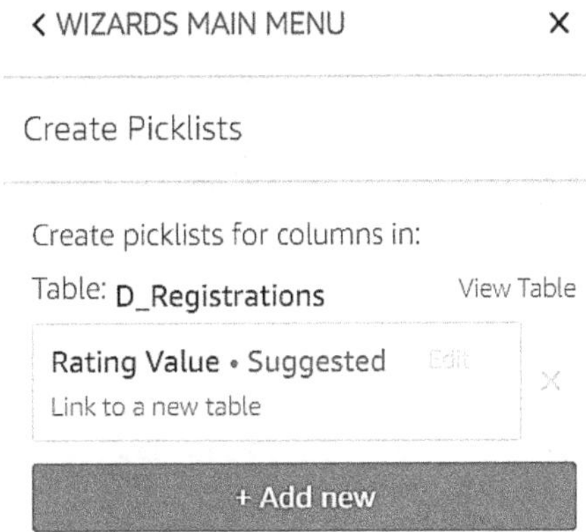

Figure 9.3 – The picklists wizard

3. On the next screen, select the **A_Sessions** table and the **Location** column, as shown in *Figure 9.4*. Leave the default value for **Link to data in field**, and click on **Apply**:

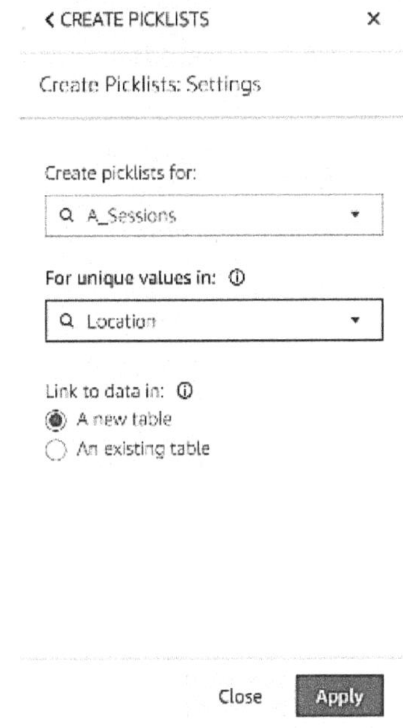

Figure 9.4 – Selecting the table and column for creating the picklist

4. On the next screen, click on **Go**.

5. Upon successful completion of the wizard, there will be a new **A_Sessions_ Location** table created, and the values of the **Location** column will be automatically converted into rowlinks.
6. Open the **A_Sessions_Location** table. It will contain two columns – **Location** and **Related A_Sessions**.
7. Add a new column to this table and name it `Capacity`. Fill in the `capacity` values against each `location` value.
8. Navigate back to the **A_Sessions** table. Select the **Capacity** column, and set its formula to `= [Location] [Capacity]`.

Chapter 10 – Inventory Management Template

Exercise 1

While reviewing the **Inventory** table, we observed that the **Assigned To** and **Requested On** columns are being used to capture values that are inconsistent with the column names. The **Assigned To** column is updated to capture both the requester name and the name of the person to whom the inventory was assigned. Similarly, the **Requested On** column is used to capture the date of request and the date of assignment.

Now, let's consider the case where the device is a test device or a loaner device and has been assigned to someone. To record that, we need to update the assignment information in the previously mentioned columns. Now, another person in the organization would like to make use of the device whenever it is available and submits a request for the same. Where do you capture this? What if you want to make use of the number of people waiting for a device as an input to decide the quantity to purchase for that device?

It is important to note that there is nothing wrong with not allowing a user to request unavailable devices, and it could be exactly how you want your app to work. The purpose of this exercise was to identify the limitations and help you make an informed choice.

Exercise 2

Try the following steps with a colleague of yours:

1. Both of you open the **My Devices** app.
2. Select the first device on the list. Wait until both of you are on the **Confirm** screen.
3. Now, confirm your request.

What do you expect will happen? Both of your requests should be registered, correct?

Now, check what has happened by viewing the inventory table or by opening the selected device in the **Devices Detail** screen of the **My Devices – Manager** app. Do you see both requests?

No; you will see that there is only one request, and it will be against the name whose request was registered last. Let's see why that is.

On the **Devices** screen of the **My Devices** app, we noted how the following `FindRow` function is used to fetch a specific row from the **Inventory** table to pass, as context, for the **Confirm** screen:

```
=Findrow(Inventory,"Inventory[Device]=Thisrow() AND
ISBLANK(Inventory[Assigned To])")
```

This formula will return the first row containing the selected device that has an empty **Assigned To** column, where the item is neither assigned nor requested. However, we also learned that the **Assigned To** column is only updated after submitting the request on the **Confirm** screen using the **Confirm** button or by changing the **Status** setting of the **Devices Detail** screen of the **My Devices – Manager** app.

Now, when you both selected the same device in *step 2*, the same row was set as input to the **Confirm** screen. Then, when you both confirmed your request, the automation updated the same row. Therefore, whoever makes the last request will overwrite the previous value. In the programming paradigm, this phenomenon is referred to as the **Race Condition**.

Exercise 3

On the **Inventory** screen of the **My Devices –Manager** app, we noted the visibility condition was set on the **AddButton** option, using the following formula:

```
=AND((FILTER(Inventory,"Inventory[Asset ID]=$[ID] AND
Inventory[Device]=$[DeviceRow]")=0),$[ID]<>"")
```

The preceding formula is validating two conditions:

1. There is no existing item in the **Inventory** table of the device being added that has the same **Asset Id** value that you have provided in the field.
2. The **Asset Id** field is not empty.

This is because we do not want to add an inventory item without an **Asset Id** field, and we want to have unique asset IDs.

The shortcoming of this formula is that instead of enforcing the uniqueness of **Asset Id**, it is making a check on the uniqueness of the **Asset Id** and **Device** pair. Therefore, it allows you to use the same **Asset Id** value for different device types, as shown in *Figure 10.Ex1*:

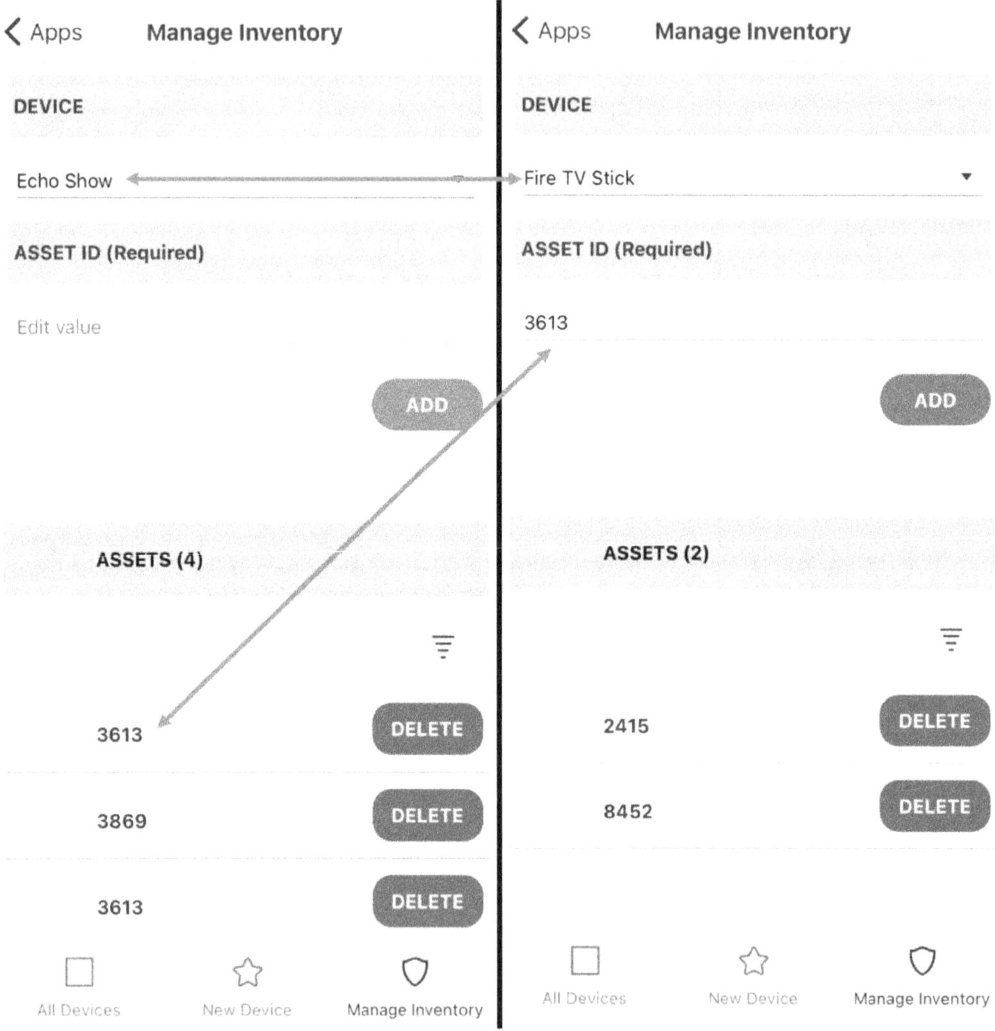

Figure 10.1 – App allowing you to add asset ID 3613 for a Fire TV stick when an Echo Show device exists with the same asset ID

Similar to *Exercise 1*, there is nothing wrong with this approach in the context of the use cases the template is currently built for. However, it can add to the level of complexity and possible errors if you decide to extend the apps to add more features that might require this value to be unique.

Chapter 11 – Building a Shopping List App in Honeycode

Exercise 1

```
=Filter(Items, "Items[Store] = THISROW() AND Items[Bought] =
FALSE")
```

We set this formula in the **ItemsToBuy** column of the **Stores** table to return the set of rows from the **Items** table with items to be bought from this store. Having this column simplified the creation of the screens based on the **Stores** table, as we already had the formula set and it can simply be assigned to a source. Furthermore, the formula this is stored in the table is pre-computed, whereas that set gets computed every time the app loads the screen, thereby reducing on-the-fly computations, which is extremely useful when working with large datasets.

Exercise 2

To set the default value of the store on the **Items** form, we need to pass the reference of the store we are currently viewing. Recall that we are on the **Store detail** screen and have navigated to it using the list view. The list view itself passes the store reference to the **detail** screen by setting the value to the `InputRow` variable and, therefore, we can simply pass this variable to the next screen with the following steps:

1. Click on the **Add Item** button and go to the **Actions** tab in the **Properties** panel.

2. Under the **Set variables** field, click on **Set variable (2 available)** and then choose **$[Store content]**, as shown in *Figure 11.Ex2.1*:

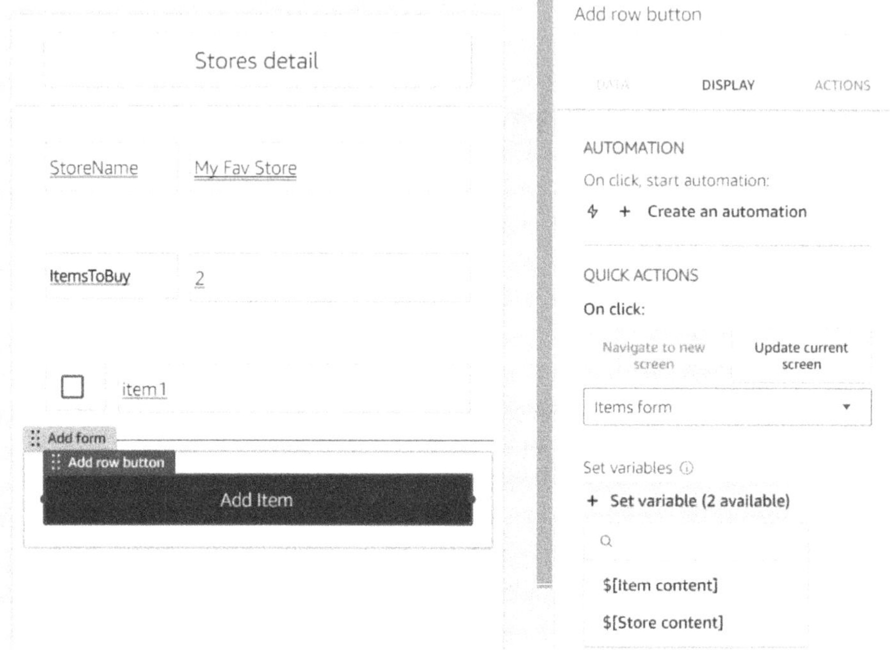

Figure 11.1 – Select the Store content variable to be set

3. Set `=$[InputRow]` as the value for the **Take data from** field:

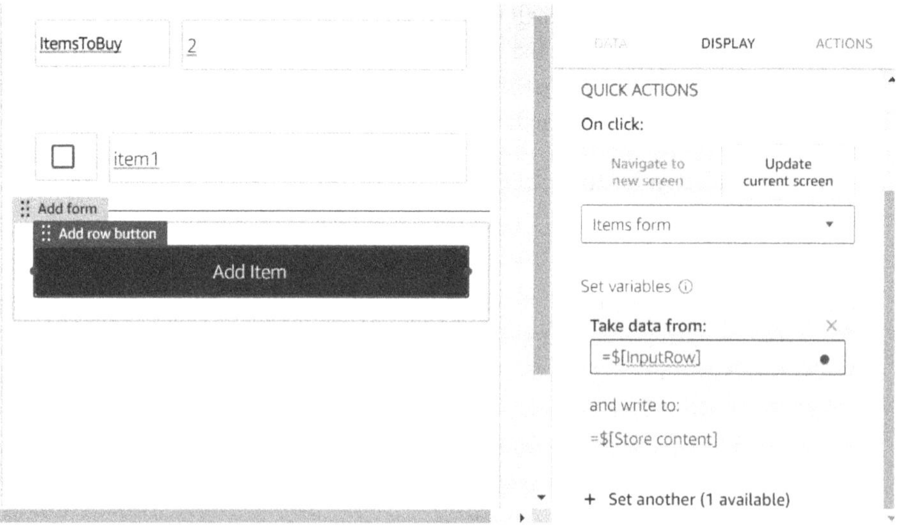

Figure 11.2 – Set the value to be passed for the Store content field

Exercise 3

The list of bought items can be added to the screen by following these steps:

1. Copy **Items List** and paste it on the screen to create a duplicate of it.
2. Set the source of the newly added screen as `=FILTER(Items,"Items[Bought] =TRUE")`:

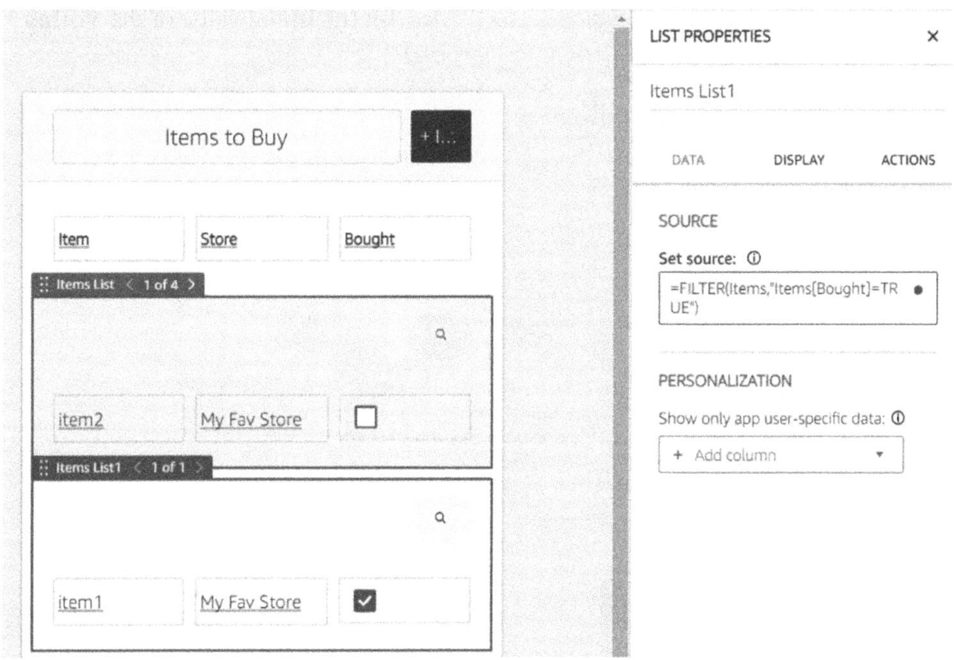

Figure 11.3 – Items to Buy screen with list to show bought items

Chapter 12 - Building a Nominate and Vote App in Honeycode

Exercise 1

The process of enabling this feature is already outlined in the chapter. Here are the detailed steps that you need to follow:

1. Create a table named `Voting_Types`, with a single column named `Voting type`, containing the following values:

    I. `Public`

    II. `Judging Panel`

2. Similar to the **Organizers** table, create a table named `Judges` to list the panel of judges.
3. Add a column to the **Contests** table, name it `Voting type`, and format it to **Rowlink & Picklist**, with the source set as the **Voting_Types** table.
4. In the middle section of the **Contests** screen, we added a field to configure **Max Votes Allowed**. Below that, add another set of fields to set the type of voting allowed for the contest, as shown in *Figure 12.Ex1*. Also, set the initial value of the **Voting type** data cell as `=FINDROW(Voting_Types)`.

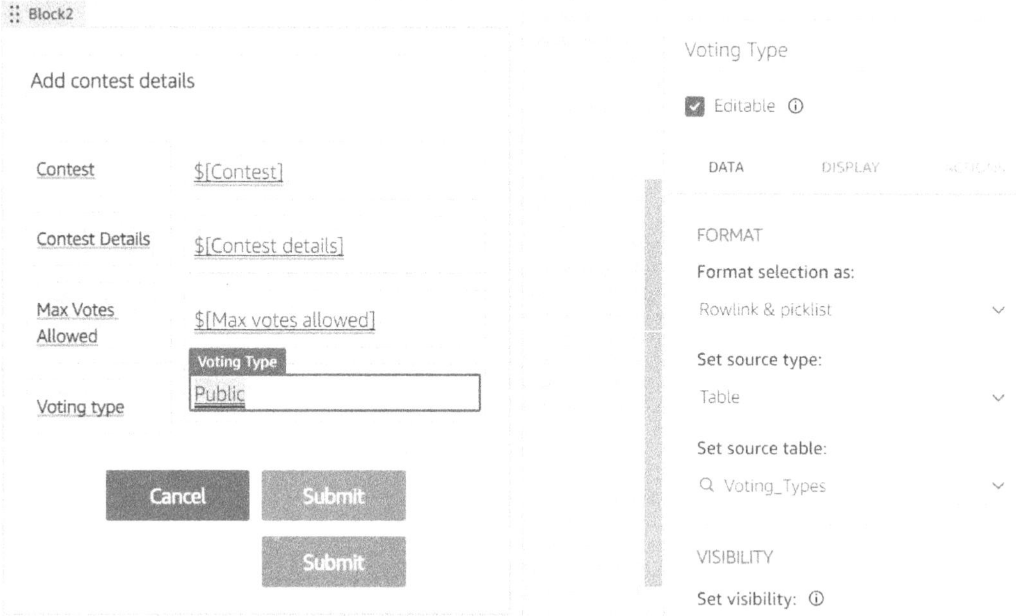

Figure 12.1 – Adding fields for configuring the voting type for the contest

5. Update the visibility conditions on the two **Submit** buttons to ensure that the value of the voting type for the contest is also set.

   Update the top (green-colored) button's visibility condition to the following:
   ```
   =AND($[Contest]<>"",NOT(ISBLANK($[Contest])),$[Contest
   details]<>"",NOT(ISBLANK($[Contest
   details])),$[ContestExists]=FALSE, $[Voting
   Type]<>"",NOT(ISBLANK($[Voting Type]))).
   ```

And, set that of the bottom (gray-colored) button to the following: `=OR($[Contest]="",ISBLANK($[Contest]),$[Contest details]="",ISBLANK($[Contest details]),$[ContestExists]=TRUE, $[Voting Type]="",ISBLANK($[Voting Type]))`.

6. Update the automation on the top **Submit** button to update the value of the **Voting type** column of the **Contests** table with the value set to the `$[Voting Type]` variable.
7. On the **Contest details** screen, update the **AddVoteButton** button's visibility condition to the following: `=AND(FILTER(Upvotes,"Upvotes[Voter]=% AND Upvotes[Nomination]=%",$[SYS_USER],[Nomination])=0,$[VotesCasted]<>$[InputRow][Max Votes Allowed], IF($[InputRow][Voting Type] = FINDROW(Voting_Types), TRUE, $[SYS_USER] IN Judges[Judge]))`.

Chapter 13 – Conducting Periodic Business Reviews Using Honeycode

Exercise 1

This automation is the same as the **Epic Updates** automation we wrote earlier in the chapter, and can be created by following these steps:

1. Click on the + icon to add new automation and name it `Feature updates`.
2. Select **Column changes** under **Automation trigger**, select the `Features` table and the `Updates` column.
3. Set `=NOT($[PREVIOUS] = [Updates])` as the condition to run the automation.
4. Next, add the overwrite action and set the value of the **Last updated** column to `=Today()`, and then publish the automation.

Chapter 14 – Solving Complex Problems through Multiple Apps Within a Workbook

Exercise 1

When adding a new property, we want the ID to be unique, and in typical databases, it is achieved by setting the column as `primary key` and making it auto-incrementing. The formula we added is meant to achieve the same.

The `GetRow` function is provided with the table as a source of rows, the current row as the context row, and `-1` as the offset to get the row above the current row. The returned row is then dereferenced to get the value of the `Id` column and we add `1` to that to set the value of the cell. The value of the IDs is instantiated in row `2` by setting the value `1` in the cell.

Another way of achieving the same result is to use cell references by setting the `=A2+1` formula in cell `A3` and then copying and pasting the formula to all cells in the `Id` column.

Exercise 2

As discussed in the **Client details** screen section, the functionality to add is similar to the changes made on the **Property details** screen. So, let's use that as a starting point and customize it by following these steps:

1. Update the visibility condition of the content boxes with client details that are not relevant to a seller client such as **Area of Search**, **Maximum property price**, and such, to `=$[InputRow][Type] = FINDROW(Client_Types, "Client_Types[Type] = %", "Buyer")`.
2. Copy the controls starting from the **Recommend to client** button until the **Delete Property** button and paste them into the **Client details** screen, as shown in *Figure 14.Ex1*:

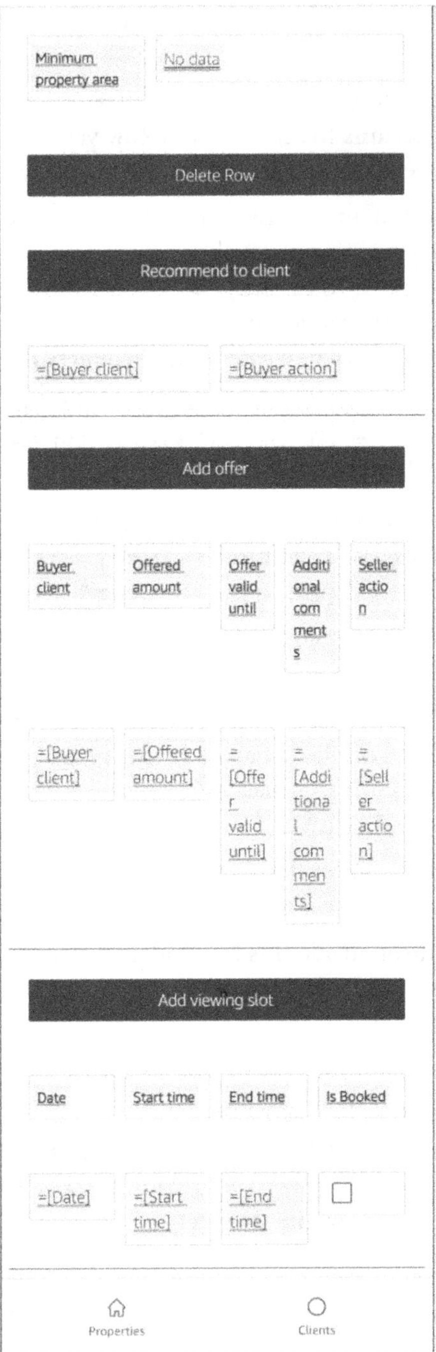

Figure 14.1 – Client details screen with copied over controls

3. Move the **Delete Row** button to the bottom and update the text to `Delete client`.
4. Update the **Recommend to client** button text to `Recommend property` and update its **Quick Actions** to set the **InputRow** variable value to a variable for the buyer client instead of the variable for the property.
5. In the list below the button, update the source to filter on the client instead of property: `=Filter(Property_Recommendation, "Property_Recommendation[Buyer client] = %", $[InputRow])`. Then, change the first data cell to show the value from the **Property** column instead of **Buyer client** and set **Quick Action** to navigate to the **Property details** screen.
6. Also, update the visibility condition for the button and list control to `=$[InputRow][Type] = FINDROW(Client_Types, "Client_Types[Type] = %", "Buyer")`.
7. Add a new **Content** box and set its text as `Properties on sale`. Set the visibility condition as `=$[InputRow][Type]=FINDROW(Client_Types,"Client_Types[Type]=%","Seller")`.
8. Add a column list and set the source as `=Filter(Properties, "Properties[Seller client] = %", $[InputRow])`; choose only the `Id` column and do not create the detail screen.
9. Set the visibility condition on the list as `=$[InputRow][Type]=FINDROW(Client_Types,"Client_Types[Type]=%","Seller")`.
10. Add a data cell to the list and set the source as `=FILTER(Offers, "Offers[Property] = %", THISROW())`.
11. Move the **Add offer** button inside this list, update its visibility condition to `=True`, and its **Quick Actions** to use `=ThisRow()` as source instead of `=$[InputRow]` for setting the variable.
12. Add a new **Content** box and set its text as `Offers`.
13. Update the source of the list below that to `=Filter(Offers,"Offers[Property][Seller client]=% OR Offers[Buyer client] = %", $[InputRow], $[InputRow])`, and update the first column to show the **Property** column instead of **Buyer client**. Also, update the corresponding column header configured in the block above the list.

14. Update the **Add viewing slot** button's **Quick Actions** to remove setting the variable.
15. Update the visibility condition on the list and block containing the column header for viewing slots to `=True`, add a data cell to the list to display the property, and add a corresponding data cell to display the column header.
16. Finally, update the list source to `=Filter(Viewing_Slots,"Viewing_Slots[Property][Seller client]=% AND Viewing_Slots[Date]>=%",$[InputRow],TODAY())`.

Exercise 3

Do you recall the automation we created in *Chapter 13* for the same purpose? We had created two automations for capturing when **Epic** or **Feature** was last updated so that we know the freshness of the information. We can apply the same logic for updates to the **Client** profile. We can add a similar automation on columns associated with the profile fields that, when updated, should be indicated to the realtor.

Exercise 4

The app view got fixed with the simple addition of `[Property]` as the first **dereference** because the existing set of dereferences was for a row of the **Properties** table. When we dereferenced the row of the **Property_Recommendation** table to its **Property** column, we essentially returned a row of the **Properties** table and we then dereferenced that row to retrieve the column values.

Exercise 5

Instead of setting the source as the **Property_Recommendation** table, we could set the source as `=Filter(Property_Recommendation, "Property_Recommendation[Buyer client][Client] = %", $[SYS_USER])` and then we do not need the **Client** column and the **Personalization** configuration.

Index

A

Action Items (AIs) 157
actions
  existing Actions, extending to persist values for fields 114-120
  functionality, for deleting task 120-123
  using, for data input through forms 109
Add New Task screen
  fields, adding 110-113
Amazon Connect 142
Amazon Honeycode
  account, creating 4
  reference link 4
app
  Add New Task screen, creating 69-71
  building 348
  completed Tasks screen, creating 68
  creating, from scratch 61, 62
  creating, with App Wizard 76-82
  Edit Task screen, creating 72-76
  My Tasks screen, creating 62-67
  tables, creating 349
  workbook, creating 348
app components
  styles, applying to 87
app data model
  creating 46
  data formats and relations 54-57
  tables, creating 48
  workbook, creating 47, 48
app interactions
  requirements, translating 346
app interface
  app screen, adding 59, 60
  app screen, renaming 57, 58
  building 57
Applicant Tracker template 138, 139
app objects
  data & display objects 29
  layout 31
app requirements
  defining 344, 345
app screen
  about 26, 28
  body 27
  global navigation 27
  header 26
automations
  about 31
  app automations, accessing 32
  debugging 131, 132
  workbook automations 31

B

Boolean field 254
brainstorm hub, Nominate and Vote app
  bottom section 293-295
  middle section 290-292
  sections 289
  top section 290
Builder interface
  about 24
  app objects and controls 28
  app screen 26, 28
  components 25
Buyer app
  Clients detail screen, customizing 364-367
  creating 362, 363
  Properties details screen, customizing 370, 371
  Properties screen, customizing 368-370

C

Client app
  about 346
  primary actions 347
Collaborative Brainstorming template 140, 141
component visibility
  controlling, with conditions 95-97
conditional style
  setting, for overdue tasks 93, 94
conditional-styling app components 88, 89
Connect Manager template 142-145
Content Tracker template 146
Contest details screen, Nominate and Vote app
  bottom section 299-302
  second section 297, 298
  sections 295
  third section 298
  top section 296, 297
controls, Honeycode
  checkbox/switch 30
  contact 30
  picklist 30
  radio 30
Customer Tracker template 146, 147

D

data
  importing, to existing table 22-24
  importing, to new table 21
  importing, to tables 20
  processing, with automation based on triggers 123
data access per user
  restricting, with personalized views 98-101
data and display objects
  block 30
  button 29
  column list 30
  content box 29
  data cell 29
  form 30
  list 30
  stacked list 30

data cell
  about 107
  shared data cell 108
  variable data cell 108
data formats, app data model
  about 54
  setting 54-57
data model
  defining 348
data model, Instant Polls app
  Last_Visits table 187, 188
  Options table 189
  Questions table 190
  reviewing 186, 187
  Votes table 191
  z_Readme table 192
data model, Inventory Management app
  reviewing 234
  tables 235
data model, Nominate and Vote app
  defining 282
data model, periodic review meetings
  defining 311, 312
data model, Survey app
  _README table 175
  Results table 171
  reviewing 170
  Scale table 172, 173
  Survey table 173, 174
data views, on fly
  filtering 101-103
  searching 101-103
  sorting 101-103
debugging 131
dereferencing 175
DRY principle 201

E

Employee Onboarding template 147-149
Epics table 314, 315
Event Management app
  block 219
  By Category screen 226
  creating 206, 207
  data model, reviewing 208
  Detail screen 223
  FAQ 226, 227
  list 220-222
  My Agenda screen 224
  requirements 204, 205
  reviewing 216
  Sessions screen 217
  solutions, to exercises 391-394
  Speakers screen 225
  tables 208
Event Management template 150, 151

F

feature 319
Features table 319, 320
fields
  adding, to Add New Task
    screen 110-113
Field Service Agent template 152, 153
FILTER function 188
FindLastRow function 178
FindRow function 178
foreign keys (FKs) 202

G

GetRow function 351

H

Honeycode
- about 136
- app objects 28, 29
- controls 29, 30
- pricing tiers 41
- style controls 87, 88

Honeycode Dashboard
- Dashboard section 16
- exploring 14, 15
- Learning & Resources section 17
- Teams section 16

Honeycode Teams
- about 34
- team, managing 40
- team members, adding 34

Honeycode templates
- about 137
- Applicant Tracker 138, 139
- Collaborative Brainstorming 140, 141
- Connect Manager 142-145
- Content Tracker 146
- Customer Tracker 146, 147
- Employee Onboarding 147-149
- Event Management 150, 151
- Field Service Agent 152, 153
- Instant Polls 153
- Inventory Management 154, 155
- Launch Manager 156
- Meeting Runner 157
- PO Approvals 158, 159
- reference link 137
- Simple Survey 160
- Simple To-Do 161
- Team Task Tracker 161
- Timeoff Reporting 162
- Timesheet Manager 163, 164
- Weekly Demo Schedule 165

Honeycode workbook
- exploring 17
- Left Navigation Bar 17, 18
- tables 19

I

IFERROR function 175

Instant Polls app
- Block section 194
- creating 185, 186
- data model, reviewing 186, 187
- home screen 193
- poll, creating 196-199
- polls 193
- Questions List section 194, 195
- requirements 184
- reviewing 192
- vote, casting 200, 201

Instant Polls template 153

Inventory Management app
- creating 231-233
- data model, reviewing 234
- My Devices app 238
- My Devices - Manager app 243
- requirements 230, 231
- reviewing 238
- solutions, to exercises 394-397

Inventory Management template 154, 155

L

Launch Manager template 156

layout, app objects
- screen 31
- segment 31

Left Navigation Bar 17, 18

M

Meeting Runner template 157
mobile device
  To-Do application, running 10, 11
My Devices app, Inventory Management app
  about 238, 239
  Confirm 240, 241
  Devices 239, 240
  My Devices 242
My Devices - Manager app, Inventory Management app
  about 243
  All Devices 244
  Devices Detail 244
  Inventory 247, 248
  New Device 246

N

Nominate and Vote app
  building 282, 283
  data model, defining 282
  functionalities, for organizers 304, 305
  Organizers table 305
  requirements, defining 280, 281
  requirements, translating to app interactions 281, 282
  tables 284
  template app, creating 283
  template app, editing 288
  updating 306
  voting, restricting to panel of judges 307

O

OKRs_Epics table 320, 321
OKRs table 313

P

Past_Updates table 321, 322
peeking sheet 265
periodic review meetings
  app, building 312
  app, creating 322-325
  app requirements, defining 310, 311
  data model, defining 311, 312
  Epics detail screen 327-331
  Features detail screen 331, 332
  Features screen 334-336
  OKRs detail screen 334
  OKRs screen 333
  reminders, sending for providing updates 341
  requirements, translating to app interactions 311
  Review screen 325-327
  review updates, sending 336, 337
  tables, creating 312
  updates, archiving 337-340
  workbook, creating 312
personalized views
  used, for restricting data access per user 98-101
picklist 254
PO Approvals template 158, 159
polls 183

R

Realtor app
  about 346
  Client details screen,
    customizing 360, 361
  Clients screen, customizing 358, 359
  creating 353, 354
  primary actions 346
  Properties screen, customizing 355, 356
  Property detail screen,
    customizing 356, 357
reminders
  sending, based on set
    preference 124-127
resolved questions, Nominate
    and Vote app 303, 304
rowlink 175

S

segment 263
Seller app
  creating 372
  Properties detail screen,
    customizing 373, 374
  Properties form screen,
    customizing 374, 375
  Properties screen, customizing 372, 373
  Viewing Slots form screen,
    customizing 376
  Viewing Slots screen, customizing 376
shared data cell 108
Shopping List app
  bought items list, clearing 273-275
  building 254
  building, wizard used 276-278
  creating 260
  data model, defining 254
  Item form screen 268, 269
  Items detail screen 272
  Items table, setting up 257-259
  Items to Buy screen 270, 271
  requirements 252, 253
  requirements, translating,
    to interactions 253
  Shopping List by Store screen 261-263
  Stores detail screen 263-267
  stores table, setting up 255, 256
  tables, creating 254
  workbook, creating 254
Simple Survey template 160
Simple To-Do template 161
single source of truth (SSOT) 191
States table 317
Status table 316
style controls, Honeycode
  object styles 87
  text alignment 88
  text configuration 87
  text formatting 87
styles
  applying, to app components 87
Survey app
  creating 169, 170
  data model, reviewing 170
  home screen 176, 177
  requirements 168, 169
  reviewing 175
  Survey screen 178-180
  Thanks screen 178
system variables
  about 107, 108
  PREVIOUS 109
  SYS_USER 108
  SYS_USER_GROUPS 108

T

tables
  about 19
  Buyer_Actions table, creating 349
  Clients table, creating 350
  Client_Types table, creating 349
  data, importing to 20
  data, importing to existing table 22, 24
  data, importing to new table 21
  Offers table, creating 352
  Properties table, creating 350, 351
  Property_Recommendation
    table, creating 352
  Property_Types table, creating 349
  Seller_Actions table, creating 350
  Viewing_Slots table, creating 352
tables, app data model
  creating 48
  reminder options table, creating 54
  renaming 49
  table column, adding 50, 51
  table column, deleting 52, 53
  table column, renaming 49
  tasks table, creating 53
tables, Event Management app
  A_Sessions 209, 210
  B_FAQ 211
  D_Registrations 212
  M_Category 212, 213
  M_Dates 213
  M_Ratings 214
  M_Speakers 214
  R_Dashboard 215
  _Readme 215
Tables interface
  components 19
tables, Inventory Management app
  Categories 235
  Devices 235
  Inventory 236
  Manufacturers 237
  _Readme 238
  Status 237
tables, Nominate and Vote app
  Contests table 285, 286
  Ideas table 286
  Nominations table 286
  Questions table 285
  setting up 284
  Upvotes table 287
tables, periodic review meetings
  Epics table 314, 315
  Features table 319, 320
  OKRs_Epics table 320, 321
  OKRs table 313
  Past_Updates table 321, 322
  States table 317
  Status table 316
  Teams table 318
task completion notification
  sending, to task creator 128-130
team 318
team members
  adding 34
  adding, by approving access
    requests 38, 39
  inviting 35
  inviting, by sharing app or
    workbook 36, 37
Teams table 318
Team Task Tracker template 161
template 135

template app, Nominate and Vote app
brainstorm hub 289
Contest details screen 295
creating 283
editing 288, 289
My questions 304
resolved questions 303, 304
ThisRow function 191
Timeoff Reporting template 162
Timesheet Manager template 163, 164
To-Do application
conditional style, setting for overdue tasks 93, 94
creating 6-8
default style, setting for each task 90-93
mobile device, running 10, 11
requirements, defining 86
running, on web browser 8, 9
styling 90

U

user-defined variables 107

V

variable data cell 108
variables
about 107
system variables 108
user-defined variables 107
visibility property 96

W

web browser
To-Do application, running 8, 9
Weekly Demo Schedule template 165
wizards 276
workbook automations
about 31
Builder components 32

Other Books You May Enjoy

If you enjoyed this book, you may be interested in these other books by Packt:

Building Low-Code Applications with Mendix

Bryan Kenneweg, Imran Kasam, Micah McMullen

ISBN: 9781800201422

- Gain a clear understanding of what low-code development is and the factors driving its adoption.
- Become familiar with the various features of Mendix for rapid application development.
- Discover concrete use cases of Studio Pro .
- Build a fully functioning web application that meets your business requirements

- Get to grips with Mendix fundamentals to prepare for the Mendix certification exam
- Understand the key concepts of app development such as data management, APIs, troubleshooting, and debugging

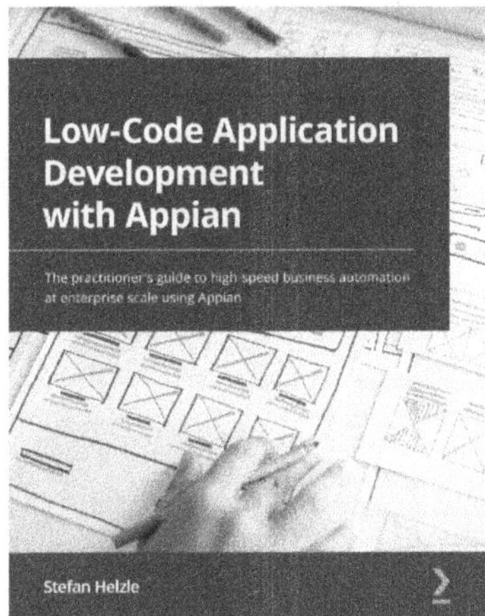

Low-Code Application Development with Appian

Stefan Helzle

ISBN: 9781800205628

- Use Appian Quick Apps to solve the most urgent business challenges.
- Leverage Appian's low-code functionalities to enable faster digital innovation in your organization
- Model business data, Appian records, and processes
- Perform UX discovery and UI building in Appian
- Connect to other systems with Appian Integrations and Web APIs
- Work with Appian expressions, data querying, and constants

Packt is searching for authors like you

If you're interested in becoming an author for Packt, please visit `authors.packtpub.com` and apply today. We have worked with thousands of developers and tech professionals, just like you, to help them share their insight with the global tech community. You can make a general application, apply for a specific hot topic that we are recruiting an author for, or submit your own idea.

Share Your Thoughts

Now you've finished *Build Customized Apps with Amazon Honeycode*, we'd love to hear your thoughts! Scan the QR code below to go straight to the Amazon review page for this book and share your feedback or leave a review on the site that you purchased it from.

`https://packt.link/r/1-800-56369-8`

Your review is important to us and the tech community and will help us make sure we're delivering excellent quality content.

www.ingramcontent.com/pod-product-compliance
Lightning Source LLC
Chambersburg PA
CBHW081500050326
40690CB00015B/2867

* 9 7 8 1 8 0 0 5 6 3 6 9 8 *